BHAGAVAN

with 24 tables and 25
figures

Review copy ship laid in

60.-

# NEW GENERIC TECHNOLOGIES IN DEVELOPING COUNTRIES

*Also by M. R. Bhagavan*

TECHNOLOGICAL TRANSFORMATION OF THE THIRD WORLD

AFRICAN INDUSTRIALISATION (*with C. E. Barker, P. von Mitschke-Collande and D. V. Wield*)

ENERGY MANAGEMENT IN AFRICA (*co-edited with S. Karekezi*)

ENERGY FOR RURAL DEVELOPMENT (*co-edited with S. Karekezi*)

# St. Martin's Press

Scholarly and Reference Division
175 Fifth Avenue
New York, NY 10010

We are pleased to submit the following ST. MARTIN'S PRESS book for review:

TITLE: NEW GENERIC TECHNOLOGIES IN DEVELOPING COUNTRIES

Editor Bhagavan, M.R., ed.

**Publication Date**     Monday, November 17, 1997

**Cloth**     $79.95     0-312-17643-0

**Paper**

For further information, please contact:

Meredith Howard, Scholarly and Reference Publicist (212) 982-3900

We would appreciate receiving tearsheets of any reviews that appear

# New Generic Technologies in Developing Countries

Edited by

M. R. Bhagavan
*Senior Research Adviser*
*Department for Research Cooperation*
*Swedish International Development Cooperation Agency (Sida)*

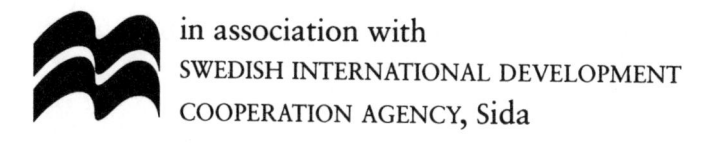

in association with
SWEDISH INTERNATIONAL DEVELOPMENT
COOPERATION AGENCY, Sida

First published in Great Britain 1997 by
**MACMILLAN PRESS LTD**
Houndmills, Basingstoke, Hampshire RG21 6XS and London
Companies and representatives throughout the world

A catalogue record for this book is available from the British Library.

ISBN 0–333–65048–4 hardcover
ISBN 0–333–65049–2 paperback

First published in the United States of America 1997 by
**ST. MARTIN'S PRESS, INC.,**
Scholarly and Reference Division,
175 Fifth Avenue, New York, N.Y. 10010

ISBN 0–312–17643–0

Library of Congress Cataloging-in-Publication Data
New generic technologies in developing countries / edited by M.R. Bhagavan.
p. cm.
"In association with SIDA, the Swedish International development Cooperation Agency."
"Present volume is the outcome of an international seminar held in Stockholm in 1994"—P. xii.
Includes bibliographical references and index.
ISBN 0–312–17643–0 (cloth)
1. Technology—Developing countries—Congresses.   I. Bhagavan, M. R., 1934–   .   II. Sweden. Beredningen för u-landsforskning.
T49.5.N483 1997
338.9'27'091724—dc21                                                97–11909
                                                                                        CIP

© Swedish International Development Cooperation Agency, Sida 1997

Edited and typeset by Robin Gable and Lucy Morton, London SE12

All rights reserved. No reproduction, copy or transmission of this publication may be made without written permission.

No paragraph of this publication may be reproduced, copied or transmitted save with written permission or in accordance with the provisions of the Copyright, Designs and Patents Act 1988, or under the terms of any licence permitting limited copying issued by the Copyright Licensing Agency, 90 Tottenham Court Road, London W1P 9HE.

Any person who does any unauthorised act in relation to this publication may be liable to criminal prosecution and civil claims for damages.

The authors have asserted their rights to be identified as the authors of this work in accordance with the Copyright, Designs and Patents Act 1988.

This book is printed on paper suitable for recycling and made from fully managed and sustained forest sources.

10  9  8  7  6  5  4  3  2  1
06  05  04  03  02  01  00  99  98  97

Printed and bound in Great Britain by
Antony Rowe Ltd, Chippenham, Wiltshire

# Contents

List of Tables and Figures     vii

Foreword
*Anders Wijkman*     ix

Introduction
*M.R. Bhagavan*     1

## PART I   The Technological Scene: Main Features, Capabilities and Trends

1 Contemporary Technological Revolutions: Characteristics and Dynamics
*Ian Miles*     25

2 Information and Communications Technologies in Developing Countries
*José E. Cassiolato*     43

3 The Advanced Materials Revolution: Effects on Third World Development
*Helena M.M. Lastres*     68

## PART II   Case Studies of Developing Countries

4 High Technology Programmes in China
*Jian Song*     95

5 Development of New Generic Technologies in India
*P.K.B. Menon*     99

6 Public Policy and New Generic Technologies: The Case of Biotechnology in Sub-Saharan Africa
*Calestous Juma and John Mugabe*     115

7  New Materials Technology in Developing Countries
   *Hugo F. Lopez and P.K. Rohatgi*                                        140

8  Generic Skills of Management and Organization:
   The Energy Sector in Africa
   *Stephen Karekezi*                                                      174

## PART III  Views from the Industrialized Countries

9  Functional Markets and Indigenous Capacity for
   Sustainable Development: What Can Transnational
   Corporations Do through Technology Transfer?
   *Richard Adams*                                                         195

10 Recycling Technologies and Engineering Challenges
   *Donald V. Roberts*                                                     214

11 Underutilized Capabilities in the Transfer of New Generic
   Technologies from Sweden to the Developing Countries
   *Carl-Göran Hedén*                                                      221

## PART IV  Donors' Experiences and Policy Approaches

12 Information Technology Support to Developing Countries:
   The Canadian Experience
   *Keith A. Bezanson*                                                     233

13 Support for Biotechnology in Developing Countries:
   The Dutch Experience
   *T.J. Wessels*                                                          251

14 Generic Technologies: New Factors for Swedish Aid for
   Technology Development
   *Jon Sigurdson*                                                         264

15 Development Aid Policy Options for the North
   *Charles Cooper*                                                        285

## PART V  Conclusions

16 The Major Issues under Debate
   *M.R. Bhagavan*                                                         297

Notes on the Contributors                                                  320

Index                                                                      326

# List of Tables and Figures

| | | |
|---|---|---|
| Table 1.1 | Three transformative technologies compared | 30 |
| Table 1.2 | Key features of new information technology | 32 |
| Table 1.3 | The new biotechnology | 35 |
| Table 1.4 | Characteristics of new materials technology | 37 |
| Table 2.1 | World R&D expenditures by region ($), 1970–90 | 55 |
| Table 2.2 | World R&D expenditures by region (%), 1970–90 | 56 |
| Figure 3.1 | Declining trend in the intensity of use of major metals in market economies | 69 |
| Figure 3.2 | World consumption of engineering plastics | 71 |
| Figure 3.3 | Advanced ceramics production in Japan | 72 |
| Figure 3.4 | Evolution of transition temperature of the best superconducting material | 75 |
| Figure 3.5 | Progress in materials' strength-to-density ratio as a function of time | 76 |
| Figure 3.6 | Declining trend in world consumption of major metals | 79 |
| Figure 3.7 | Weighted index of metal and mineral prices | 80 |
| Table 3.1 | New materials in Brazil: number of institutions, researchers, research projects and patents granted, 1987 | 85 |
| Table 3.2 | Number and type of institutions performing activities in new materials in Brazil, 1987 | 86 |
| Table 3.3 | Distribution of researchers in new materials in Brazil by degree, 1987 | 87 |
| Table 5.1 | Selected research projects in biotechnology commercialized in India | 101 |
| Table 5.2 | Selected biotech products and processes in India | 103 |
| Table 5.3 | Demand for biotech products by 2000 | 104 |
| Table 5.4 | Biosensor technologies transferred to industry | 105 |
| Table 5.5 | Development projects in 'home-grown' technologies | 110 |
| Table 5.6 | Trends and implications in information technology | 111 |

## List of Tables and Figures

| | | |
|---|---|---|
| Figure 5.1 | Integration of communication, transmission and networking | 113 |
| Table 6.1 | Research focus of US biotechnology firms | 117 |
| Figure 7.1 | Location and production of refractory, steel, cement and glass products in Mexico | 141 |
| Figure 7.2 | Evolution of federal expenditure in science and technology in Mexico | 142 |
| Figure 7.3 | Published works in science and technology by Mexican scientists, 1980–93 | 143 |
| Figure 7.4 | Members of the National System of Mexican Scientists by area, 1990–93 | 144 |
| Figure 7.5 | National expenditure on R&D activities by country as % of GDP | 145 |
| Figure 7.6 | Main sources of support for R&D activities by country | 146 |
| Figure 7.7 | Impact of electricity on villages of a developing nation where the majority of the population live in rural areas | 151 |
| Figure 7.8 | US net import reliance on selected minerals and metals | 162 |
| Table 7.1 | Per-capita consumption of metals in selected countries, 1985 | 165 |
| Table 7.2 | Per-capita consumption of metals by region, 2000 | 166 |
| Table 7.3 | Some important targets for materials technology for development | 167 |
| Table 7.4 | Examples of new materials that will be important in the future | 168 |
| Table 7.5 | Some new materials-processing technologies | 168 |
| Table 7.6 | Examples of materials applications/markets | 169 |
| Table 7.7 | Examples of material problem areas | 170 |
| Table 8.1 | Per-capita commercial energy consumption for selected developing countries, 1992 | 175 |
| Figure 10.1 | Humans have used a linear approach to date | 215 |
| Figure 10.2 | Elements of a sustainable system for humans | 216 |
| Figure 10.3 | Environmental studies – now | 217 |
| Figure 10.4 | Environmental studies – the future | 218 |
| Figure 11.1 | An NGO's likely path in aid-management | 223 |
| Figure 11.2 | A decision-making ladder for biotechnology-based entrepreneurship | 225 |
| Figure 14.1 | Mode 2 networks | 268 |
| Figure 14.2 | Conditions affecting the support of generic technologies in selected countries | 269 |
| Figure 14.3 | Areas of operation and potentially overlapping functions of Swedish overseas development agencies | 274 |

# Foreword
## Anders Wijkman

A new society is emerging. To describe it many use the notion of a neo-industrial society, others the notion of a knowledge society – a generic description for many of the new trends currently shaping our lives, ranging from a more liberalized world where borders disappear and economies globalize, to higher levels of mechanization and computerization of tasks earlier carried out manually, to information technologies opening up new channels of communications, all revolutionizing the way we do business, interact with people, are entertained, express views, learn, and so on.

We have always lived in a knowledge society, ever since we learned how to make fire and where to find food. The difference today, of course, is the pace of change in terms of knowledge build-up and technology development. It has been estimated that the stock of knowledge in the world doubled during the first half of this century. Now it doubles every fourth to fifth year. New findings, discoveries and speculations are challenging centuries' old ideas about the nature of human beings and the place of humanity in the order of things. The striking characteristic of this emerging new order is that it is global but not integrated; that is, an order that puts all of us in contact with one another but simultaneously maintains deep – and deepening – fissures between different groups of nations and also between different groups of people within nations. It is an order that today benefits a small percentage of humanity and segregates a large percentage of the world's population.

Opinions vary greatly as regards the long-term consequences of the many new forces at play. As always, there are optimists as well as pessimists. There are enormous possibilities out there, not the least in terms of technologies. Whether we have the capacity to turn these possibilities into positive developments for society is an open question. One thing seems clear: we will not succeed in building a better world unless we are guided by a strong vision, a vision that to a large extent is missing today.

Developments in science and technology have been phenomenal in recent decades. The resulting increase in the material standard of living in parts of the world has been truly impressive. However, these developments have not managed to prevent increasing poverty, inequality, environmental degradation and, most recently, increasing unemployment. We have the knowledge today to give everyone a decent standard of living, to manage natural resources sustainably, and to offer everyone something meaningful to do. But we seem to lack both the imagination and the will to do so.

Developments in the South have been uneven. Some countries – in particular in Southeast Asia – have been very successful. But most developing countries lag desperately behind. It is not that development processes in general do not work. On the contrary, considerable achievements have been made also in the so-called least developed countries (LDCs). But the combination of growing populations, misdirected policies, too much spending on defence, an unrealistic belief in the theory of 'trickle-down economics', corruption, ethnic conflicts, and so on, has left Southern countries as a group with one-third of its population deprived of the most basic human necessities.

What we are confronted with in vast regions is an appalling situation of poverty and underdevelopment. More than one billion people live in absolute poverty. The great challenge is how to make these people part of the overall development process. It is a formidable challenge not least because of continued rapid population growth in regions where poverty is widespread and because of lack of financial resources. Private investments are increasing in the South but concentrated in some ten better-off developing countries. The LDCs as a group gets less than 2 per cent of foreign direct investment. Their share of world trade is minimal at 0.2 per cent. What is more, foreign aid is shrinking.

Unfortunately, far too little of the power of modern science and technology has been directed at development for the poor. Mobilization of scientists in the rich countries to deal with the problems of the poor has not been successful, and the technological capacity of the poor countries is at present far too limited to bring about adequate responses. Only 4 per cent of all research in the world is carried out in the South. It may be an oversimplification, but it seems that many of the technology developments of the rich countries are characterized by 'solutions in search of problems' – that is, innovations in search of applications – while, simultaneously, there are a great number of problems in the North, as well as in the South, 'in search of solutions'. Much would be gained if only a small part of global research funding were diverted to tackling these problems.

The 'knowledge revolution' will accentuate the importance of science and technology even further. New and radically different ways of communication and manufacturing, as well as delivery of services, have greatly improved productivity, but also led to profound changes in the way we do things: hence the far-reaching implications for work and employment, trade and environment, as well as for power relations in society.

So far, the most debated issue in relation to the 'knowledge revolution' is employment. Two schools of thought have emerged. On the one hand, there is the position that technological change is not to blame for rising unemployment. The USA and Japan, two countries that have shifted towards high-tech, knowledge-based societies, are among the advanced countries with the lowest unemployment rates. According to this view, we do not live in some kind of a 'zero-sum game' with a given amount of output and jobs. Technology change will create more output and hence more demand, leading to more jobs. Similarly, improved productivity will raise company profits and give more purchasing as well as investment power to society at large.

The other view stresses the risk that automation will result in fewer jobs and more unemployment. Some even predict that the world's rich economies will have virtually no need for workers in the future. They point to the fact that automation has already made millions of people dispensable and vulnerable to decreasing wages and widespread unemployment, particularly among the less skilled. Entire occupations and professions can change radically. Information technology further tends to favour thought and language skills, rewarding a few highly knowledgeable individuals – the so-called 'knowledge workers'. If this is true, we have managed to develop systems where only parts of the labour force are required to produce what is needed. How then do we organize our daily lives? One thing is obvious: if we want to avoid large parts of our populations being left behind in this new era of knowledge and information, we have to make tremendous efforts in terms of education and training. Only in that way can we avoid widening gaps between individuals and groups of individuals.

Much of what can be said about the effects of the technology revolution on certain groups in the rich countries – in terms of risks of unemployment and marginalization – is true for entire countries or regions of the South. Without access to information and knowledge, poor nations can be forced into exclusion or marginalization. We would be perpetuating an already segregated world. Southern countries used to have a few distinct comparative advantages like good access to raw materials, cheap labour and lower standards for the environment and working conditions. Such factors will be of less importance in the years to come as knowledge, skills and information come to predominate.

Obviously, one cannot discuss developing countries across the board. Some countries like China, India, Brazil, Mexico and Korea have developed a strong research and technology capacity. For most of the LDCs, however, the picture is much bleaker. The problem is not only one of lack of capacity and competence. It also includes corrupt and disintegrating political regimes. Who wants to invest under such conditions?

Some commentators have argued that the poor countries may have an advantage where the new technologies are concerned. Unlike the industrialized nations, they have not yet invested large amounts of capital in the infrastructural and educational systems of an earlier vintage. For this very reason – in theory, at least – they are at greater liberty to commit themselves more fully to the technologies of tomorrow. This is a telling argument, but it presupposes a certain threshold in technological capacity and competence to begin with, a capability which most LDCs do not yet have.

The present volume is the outcome of an international seminar held in Stockholm in 1994. It was organized by the then Swedish Agency for Research Cooperation with Developing Countries (SAREC), which is now integrated into the new Swedish International Development Cooperation Agency (Sida). The objective was to focus attention on the 'knowledge revolution' and its implications for the South. It was one of the first seminars of its kind and of great importance as an eye-opener. For many of the participants, like myself, the seminar was the beginning of a process of developing a set of proposals aimed at capacity-building in the South to prevent large parts of the LDCs from being totally excluded from the new technology developments. We still have a long way to go, whether we belong to bilateral or to multilateral institutions. But the awareness is there, which is one important requisite.

# Introduction
## M.R. Bhagavan

Governments and policy-makers in developing countries have over the last four decades invoked the injunction 'science and technology for development'. The conviction that technology is one of the keys that unlocks the development door is based on the perception that the dramatic rise in the material prosperity, economic strength and political power of Western Europe, North America and Japan over the last one hundred years is due in no small measure to the conscious and systematic application of modern science and modern technology to the social processes of production, distribution, communication, organization and control. The translation of this perception into action has varied greatly from region to region in the Third World, depending on the varied outcomes of the historical impact of the West on the economies and societies of Africa, Asia and Latin America. More specifically, it has depended on the magnitude and character of a base in modern industry and infrastructure that these regions inherited at decolonization.

The West has undergone three modern industrial-technological revolutions. The first, stretching from the 1760s to the 1850s, was based on steam power, with textiles and steel leading the transformation. During the second revolution, which lasted the next hundred years, electricity and petroleum demoted steam, with railways, automobiles, chemicals and electricals in the vanguard of radical change.

The onset of the third era, which in reality comprises not one but several technological revolutions, can be dated roughly to the 1960s. Over the last three decades, advances in three broad technological areas have been radically altering the economic systems in the North and contributing significantly to the ongoing process of globalization of national economies. They have been instrumental in causing across-the-board structural changes of a generic nature in the leading sectors of the economy in the OECD region. These *new generic technologies* are in the realms of *information and communication*, *biotechnology* and *new materials*.

Although generated almost entirely in the North, these technologies are profoundly affecting the South. But, by and large, the South finds itself unprepared to derive benefits from these momentous changes.

In order to address the question of how the international donor community could cooperate with the South in responding to this challenge, an international seminar was held in Stockholm in April 1994, organized by the Swedish Agency for Research Cooperation with Developing Countries – SAREC (integrated since July 1995 into the reorganized Swedish International Development Cooperation Agency – Sida). Over 30 eminent scholars, policy-makers, industrialists, and development officials from several developing and industrialized countries were invited in their personal capacity to participate in the seminar. The 15 invited papers published in this volume were presented at that gathering and generated a highly informed discussion.

As a prelude to the volume, we will begin this introductory section with some concepts and definitions of technology, a differentiated technological classification of developing countries, and a summary of the technological characteristics of new generic technologies. This will be followed by brief presentations of the principal themes dealt with in each of the following 15 chapters, grouped into four sections. The concluding section gathers together the major issues raised for debate in this book.

## Some Concepts and Definitions of Modern Technology

The concept of modern technology we adopt is a broad one. Its 'embodied' or 'hardware' form consists of tools, machinery, equipment and vehicles, which together make up the category of capital goods. Its 'disembodied' or 'software' form encompasses the knowledge and skills required for the use, maintenance, repair, production, adaptation and innovation of capital goods, often also labelled in the literature as the know-how and the know-why of processes and products. Knowledge and skills refer not only to scientific, engineering and technical abilities, but also to the skills associated with organization, management and information. Hereafter the terms 'knowledge' and 'skills' are used interchangeably, and the term 'technology' connotes an integrated combination of capital goods and skills.

We find it analytically useful to classify modern technology into two broad types, standard-modern and highly-modern, on the basis of five indicators: automation, science-relatedness, research-intensity, dominant innovative skills and leading sectors.[1] These two broad types, which we hereafter denote by 'standard-tech' and 'high-tech', are characterized respectively by 'medium' and 'high' values in their degrees of automation, science-relatedness and research intensity.[2]

Roughly speaking, the transition from the standard-modern to the highly modern era began in Western Europe and North America in the 1960s. Some examples of the leading sectors that exemplify standard-tech capital goods and skills are steel, railways, electricals, automobiles, plastics, synthetic textiles and synthetic dyes. However, all these 'old' products have made the transition and are now available in their high-tech avatars. Examples of entirely new leading sectors that have arisen with high-tech are microelectronics, computers, digital and satellite-based telecommunications, robotics, informatics, biotechnology (based on genetic engineering) and new materials.

## Differences in the Technological Levels of Developing Countries: Not One South, but Several

There are many ways of conceptualizing and measuring the differences in the technological levels of countries. Technology can be measured, for instance, in terms of the existing stock of (i) capital goods and its most skill-intensive part, viz. the machinery subsector, (ii) professionally skilled people, in particular scientists, engineers and technicians, and (iii) R&D personnel and R&D investment. It can be computed in terms of the magnitudes and growth rates of the domestic production of capital goods, skills and R&D inputs. And so forth. Thus there are static (technological snapshot) and dynamic (technological change) indicators, both kinds being essential for an understanding of the relative positions of countries. A number of authors have conducted quantitative and qualitative analyses based on such static and dynamic indicators.[3] The broad picture that emerges has the following contour.

There is, as would be expected, a great diversity among developing countries. A few are relatively strong in science and technology, while the great majority are very weak, with some taking an in-between position. We can denote them respectively as the 'strong', the 'medium' and the 'weak' South. The worldwide distribution of highly modern technology is at present grossly unbalanced, with the 'weak' South more or less excluded. In order to focus on this issue, we will often refer below to the 'weak' South as the 'excluded' South.

The 'strong' South has made the transition to the standard-tech era, but has just embarked on the road to the high-tech destination. It is nearly self-reliant in meeting its domestic demand for standard-tech through domestic production, and also exports some of that technology to the world market. However, its technological indices are between one-and-a-half to three times lower than that of the OECD, and it has considerable technological distance to cover in order to draw level with the OECD. Prominent examples of the 'strong' South are Brazil, China, India and Mexico.

The 'medium' South has not yet the same degree of self-reliance in standard-tech that the 'strong' South has achieved, but is firmly on the road to it. At present, it imports not only all of its high-tech from the OECD, but a good deal of its standard-tech as well. Its technological indices are lower than that of the 'strong' South by a factor of two to three. Examples are Indonesia, Malaysia, Pakistan, Thailand, Turkey, South Africa, Argentina and Chile.

The 'weak' South imports almost all of its technology from the OECD region, with perhaps a smattering from the 'strong' South. Its technological dependence on the North is as heavy today as it was before decolonization. Its technological indices are roughly one-half of the 'medium' South. This vast group of countries is exemplified by sub-Saharan Africa (with the exception of South Africa), parts of South and West Asia, the Caribbean, Central America and the northern areas of the Andes.

In the South, in general, less than 10 per cent of scientists and engineers are directly involved in activities related to technical change. The great majority work with existing technology in various sectors of the economy and society. Issues related to the systematic upgrading of the skills of this huge corps of S&T workers have been neglected by Southern governments as well as the donor community.

## R&D Systems in the 'Strong' South

Among several models currently in existence in the 'strong' South for structuring and organizing R&D, a three-tiered system is finding strong advocates: (1) government-supported basic research; (2) joint public/private-sector development of advanced technologies; and (3) entirely market-driven private sector support to the development of other (less advanced) technologies.

It is argued that the substantial saving of public funds generated through the second and third tiers could be invested in the first one –

that is, basic research. What is more, this would provide scientists and engineers with the choice to work inside or outside the government system. The competition would promote efficiency and better performance in the R&D laboratories. The recent moves to reform the R&D systems in Brazil, India and Mexico point in this direction.

Realistically speaking, the full three-tier model can perhaps be made operational only in countries with largish domestic markets and substantial actual resources. In some respects, the corresponding S&T policies would be similar to those of the North. However, even in these countries, given their wide internal disparities, the technological spectrum will continue to span a very broad range. In response to this, implementable policy regimes would have to be heterogeneous mixes, catering to widely differing constituencies.

## Existing Technology in the 'Excluded' South

In the 'excluded' South, governments have been either unwilling or unable to invest adequate resources (domestic or foreign-supplied) in higher education and research in the natural and engineering sciences, which are the foundations for technological advance. The creation of 'threshold' levels in S&T infrastructure on which to base the move towards highly modern technology therefore remains a distant prospect. Since indigenous technological development seems to be a long way off in most of the 'excluded' South, some recent thinking favours a strategy that would concentrate on improving the use of existing imported technology, with primarily management and organizational intervention, in particular in repair, maintenance and training. Further, the governments of the 'weak' South are enjoined to rely more on the advice of independent technology analysts and the indigenous private sector while negotiating terms of technology transfer.

## New Generic Technologies: A Class of their Own

In the category of highly modern technologies (high-tech) that we considered above, there are three which stand out as being in a class of their own: information and communication technology (ICT), biotechnology (BT) and new materials technology (NMT). They are termed 'generic' because they are radically altering techniques and systems in wide areas of production and distribution, bringing about fundamental changes in economic sectors. These generic changes are at present largely confined to the OECD region, but are bound to spread to the developing world, where they are already causing profound indirect impact through world trade.

The new generic technologies involve manipulation of processes at the level of the atom, the molecule and the gene. By intervening at such deep levels of the physical and biological world, one brings about a diversity of profound changes in the behaviour of matter, which translate into a diversity of techniques. As can be inferred, the generic technologies are extremely science- and research-intensive. In what follows, we will give a very brief sketch of the main characteristics and areas of impact of the three technologies.

## *Information and Communication Technology (ICT)*

The physics of microelectronics and the semiconductor, as applied to integrated circuits printed on thin silicon wafers, called microprocessors, is the science behind ICT. The transformation of this science into the hardware and software of ICT involves a whole range of new engineering and mathematical skills. The microprocessor ('the chip') is not only the basic device in all ICT equipment but is also now a near universal element in all high-technology processes and products. Hence its generic impact.

The hardware is getting progressively smaller and less expensive, according to Moore's Law that the information-processing capacity of the chip has roughly doubled and its cost has fallen by one-third every year. The reliability of ICT hardware is considerably higher than in other technologies because of the relative absence of mechanical components. The decreasing size and cost, and high reliability, increasingly facilitate the insertion of ICT hardware into a great variety of equipment across a wide range of industry and infrastructure, promote decentralized and networking operations, and are transforming work organization.

The development in the software is even more dramatic than in the hardware. It is now possible to write a program to perform any logically sequenceable and networkable operation, in most sectors of society, including education and the media.

The programmability of the ICT system makes it into a powerful 'virtual machine' that can decide which performance to call forth from a real machine or vehicle, and regulate that performance. Its flexibility reduces the time and cost of retooling in manufacturing industry.

As with the hardware, the costs of software are rapidly decreasing, making it more accessible to wider sections in the economy and society. Steadily on the increase are the ICT systems' performance, utility and user-friendliness. This has, in turn, led to dramatic improvements in the quantity, speed and quality of information-processing, and reductions in costs.

One among many interrelated breakthroughs is the instantaneous transfer of bulk information and knowledge from one medium to another, from one part of the world to another. This has been made possible by digitalization of information and communication by satellites and optic fibres (the so-called 'information highways').

Finally, the information and knowledge-processing capabilities of ICT systems, their character as 'virtual machines', their seemingly unlimited potential for decentralization and networking, and their increasing user-friendliness, are radically altering *management and organization* practices. As these new practices cascade through to other levels in society, they too will have across the board generic impact.

## *Biotechnology (BT)*

One must distinguish between traditional and modern biotechnology. The former is as old as settled agriculture, involving the use of bacterial cultures to ferment and preserve food and to make alcoholic beverages, and in our own time to produce antibiotics. Modern biotechnology has its origins in the new biology inaugurated by the discovery of the structure of the DNA molecule in the early 1950s and the way DNA operates in genetic material to determine the character of organisms. Since then, a range of biological and biochemical tools have been developed to alter the structure of genetic material, which go under the collective term 'genetic engineering'. The scientific disciplines behind modern biotechnology are molecular, micro and cell biologies, in combination with biochemistry.

The current and future generic impact of BT can be gauged from the fact that over 60 per cent of the current manufactured output of the OECD countries is of biological origin and that no other technology is at present in sight to ensure the production of food, energy and industrial raw materials for a world population approaching ten billion in the next four or five decades, let alone waste management.

The impact of modern biotechnology is already evident in some areas of agriculture, forestry and bio-energy, industry, medicine and environmental protection. Here are some major examples:

*In agriculture*: plant and animal genetics for building-in resistance to diseases and stresses, and thus attaining higher yields.

*In forestry and bio-energy*: cell and tissue culture for the rapid cultivation and propagation of trees and biomass.

*In industry*: bio-catalytic methods for rapid, high-volume and high-efficiency production, extraction and purification of materials for use in food processing and pharmaceutical industries.

*In medicine*: inexpensive diagnostic kits of potential relevance to the health care of the poor; affordable manufacture by developing countries of essential drugs for mass use.

*In environmental protection*: biological methods for cleaning up polluted soils, water bodies and industrial effluent.

## *New materials technology (NMT)*

New materials technology differs significantly from what one may call standard-modern material technology in two ways: First, it is much more science-intensive and less engineering-intensive than standard-tech; and second, the science intervention takes place at the levels of atomic and molecular physics and chemistry rather than at levels of bulk physics and chemistry. That said, NMT is more the outcome of a whole collection of new insights and processes in both advanced science and advanced engineering, involving many sub-disciplines, rather than of breakthroughs in any one key area like microelectronics in ICT and molecular biology in BT.

In comparison with standard-modern materials, new materials display higher values in the following indicators: costs of R&D; design and processing which reflect the higher knowledge-content, with attendant lower costs in energy and raw materials; microstructure complexity, and functional integration through higher density of performance characteristics per unit area and volume; value added, variety, and customization to suit user requirements.

The great future importance of NMT consists of three processes: first, bulk production of high performance materials (in terms of toughness, strength, malleability, heat and corrosion resistance) from elements which are in abundance in nature like aluminium, carbon, oxygen, nitrogen and silicon, rather than as at present from the scarce so-called strategic metals like chromium, cobalt, nickel and tungsten; second, increasing the performance properties of biologically derivable, renewable and environmentally benign material, in conjunction with biotechnology; and third, efforts to build in higher degrees of energy-efficiency, durability and recycling.

New materials technologies are now in commercial use in the fields of metals, ceramics, polymers and composites, semiconductors, optical fibres, prostheses, organic-inorganic bone replacements, and so on.

## The Structure and Themes of the Present Volume

The main body of the book is divided into four parts. The first part analyses the main features, capabilities and trends in the present scene in new generic technologies, in the global as well as the developing-world context. This is followed in the second part by case studies of developing countries, describing the achievements to date in the 'strong' South and obstacles facing the 'weak and excluded' South. A discussion of the actual and potential roles of corporate and individual actors in promoting new generic technologies in developing countries is taken up in the third part. It looks at transnational corporations, the scientific and engineering professions, the entrepreneurial classes and non-governmental organizations. The fourth part opens with a presentation of the experiences of the donor agencies that have been in the lead in making generic technologies available to developing countries, and follows this up with analyses of, and recommendations on, policy approaches and options for the donor community in general, and Sweden in particular. The concluding chapter forms the fifth part, which brings together the major issues under debate in the realm of new generic technologies, as they have emerged from the contributions to the volume.

We will now briefly summarize the themes that are dealt with by the contributors under each section.

In the first chapter of Part I, Ian Miles begins his article by discussing several central concepts that have shaped our understanding of, and thinking about, the radical technological changes that have occurred in recent times. Among these are 'heartland technologies', 'paradigm shift in technology', 'technological trajectory' and 'technological revolution'. 'Heartland technologies' are innovations that fundamentally change the techniques used for carrying out a wide range of activities. At the basis of heartland technologies are new key discoveries in science. Defined this way, 'heartland technology' and 'generic technology' have much in common, but the latter connotes a much greater across-the-board impact on the economy and society than the former. A 'paradigm shift in technology' is said to occur when heartland technologies become integrated into the economic, social and knowledge sectors of society, resulting in entirely new ways of perceiving and using technology. The 'quantum jump' in productivity that a paradigm shift brings about leads in turn to a shift of investment onto new 'technological trajectories' that production, distribution and marketing must henceforth tread. A 'technological revolution' unfolds as all these phenomena and their causations reinforce each other, sweeping aside earlier heartland technologies.

Miles makes a comparative analysis of the characteristics and dynamics of IT, BT and NMT within the above conceptual framework, bringing out their common features as much as their specificities. He shows how IT is the most generic of the three, itself becoming an indispensable building block for both BT and NMT. BT in turn has become a principal tool for creating a range of new material inputs in industry and health care. He underlines the revolutionary changes in management and organizational practices brought about by IT. Because of their mass replicability and their ability to penetrate to all levels, these changes have themselves become generic. It is no longer possible to understand and influence economic, social and political change in the contemporary world, without grappling with management as a generic technology.

In Chapter 2, José Cassiolato reads a pattern of 'integration–exclusion' in the way in which trade, investment and new technologies have impinged on the world over the last few decades. On the one hand, the OECD countries, the transnational corporations and the 'triad' comprising the USA, Japan and Western Europe have become more integrated with each other on a more equitable basis. On the other, the developing countries have been excluded from this 'integration on an equitable basis', but at the same time constrained to incorporate themselves into the 'new order' on deleterious terms. This is most evident in the area of foreign trade, where new technologies constitute a significant and growing share. The North has largely succeeded in enforcing a regime of economic and import 'liberalization' on the South, while effectively protecting its domestic industries and markets through subsidies and non-tariff barriers.

Cassiolato points to the radical changes being wrought by new generic technology in the structures of production and trade of the advanced industrialized countries, especially the rapid rise in the share of the services sector in the overall economy. There is an attendant decline in the attraction offered by the so-called traditional comparative advantages of developing countries in labour and raw material costs. He considers the argument that a better diffusion of new technologies from the North to the South may be a way of reducing the problems brought on by the 'new order'. His analysis reveals several major shortcomings inherent in the 'transfer of technology' approach, in particular the dramatic decline it can bring in its wake to the indigenous development of science and technology in developing countries, unless strong corrective action is taken. This is especially true of investment, skills and expertise.

Helena Lastres (Chapter 3) has investigated the patterns in the global demand for materials over the last two decades and discovered two trends: a break in the historical trend of rising demand for and production of conventional bulk materials, accompanied by a trend of increasing demand for new materials. The two trends are interlinked not only by the fact that new materials are being used as substitutes for conventional materials but also by the sole use of new materials in conventional applications in both industrial and service sectors.

New materials require less input of minerals, metals and energy than conventional materials. Being very knowledge-intensive, NMT fully benefits from the powerful tools developed by IT. New materials can be created to customized specifications of lightness, hardness, tensile strength, temperature and corrosion resistance, and so on, for use in particular types of applications. As such, their R&D, and production, require close interaction between the customer, the developer and the manufacturer. This in turn calls for novel and flexible forms of industrial organization. One consequence is that producers and sellers of traditional materials, in both developing and industrialized countries, are being bypassed. This, in combination with the rapid growth in the demand for new materials, has led to NMT's rising influence on the world materials' markets.

Lastres finds that developing countries have been badly hit by the falling rates of growth in the production and export of conventional materials (in particular, the major metals), in addition to the well-established secular trend of falling real export prices. Further, she raises the question of how well prepared these countries are for meeting this situation and what strategies are open to them. In this, one has to take into account that the processing plants for new materials have to be located geographically near the potential customers to respond rapidly to changing user-specifications.

She points out that a situation of 'over-choice' has arisen, with several materials competing to satisfy any given function – yet another example of a 'solution in search of a problem'. She ventures the idea that the profitability of NMT is contingent on how well IT is deployed in the promotion of new materials both locally and globally.

Over the last fifty years, some of the large nations of Asia and Latin America, like China, India, Brazil and Mexico, have systematically implemented a set of industrial and technology policies whose goal was the creation of a diversified and fairly strong base in industry, infrastructure, science and technology. The state has played the dominant role in this successful effort. One consequence of this achievement is that these nations, which together with a few others define the 'strong South', have the capability and the capacity to absorb and further develop generic technologies.

In Part II, Jian Song, P.K.B. Menon, and Hugo F. Lopez and P.K. Rohatgi, describe the achievements to date in China, India and Mexico in new generic and other high technologies. (The earlier chapters by Cassiolato and Lastres present the Brazilian experience as part of a general global analysis.)

According to Jian Song (Chapter 4), the Chinese approach to entering the age of high-tech was to launch a three-tier science and technology (S&T) strategy in the mid-1980s. The first tier concentrated on applying the existing S&T capabilities to the production sector in order to upgrade its technology and raise its productivity. The second tier tackled the task of bringing on board the new generic and other advanced technologies through transfer of technology from the leading economies of the world, combined with indigenous research. The technologies so absorbed were disseminated widely within the country. To implement the second-tier programmes, China created over twenty R&D centres staffed by 15 000 professionals and established over fifty science parks in provincial capitals. Song points out that so far some five hundred new technologies have been made available to the agricultural and industrial sectors. The third tier is dedicated to fundamental and long-term research at the frontiers of science and technology, unconstrained by any short-term need to arrive at applicable results.

The large scale and the wide scope of the effort being made in India in new generic technologies becomes apparent in Chapter 5, by P.K.B. Menon. The advances made in the absorption of imported technologies and the development of new ones rest on an infrastructure of scores of state-funded national laboratories dedicated to BT, NMT and IT.

Until the early 1990s, however, there was little pressure on these laboratories to venture beyond research into the arena of turning research results into innovations that would attract the interest of industry. That attitude underwent considerable change as the Indian government, as part of its structural adjustment and economic reform programme, began to reduce its level of funding and insisted that the laboratories take on industry-sponsored projects. Such a reorientation was long overdue. It has produced a number of commercially applicable innovations, as Menon's account shows.

In Chapter 7, Hugo Lopez and P.K. Rohatgi take us through the various stages of the incorporation of NMT into the R&D establishments and the industrial sectors in Mexico and India. In the case of Mexico, the proximity of the USA has acted both as incentive and as obstacle, depending on the specific material being considered and on the historical context, in particular the emergence of the North American Free Trade Area. Lopez lays particular stress on the question of quality and pricing, the role of technology transfer, the linkages between R&D institutions and industry, and appropriate government policies. Both authors deal with the issue of institutional structures and specialist training that NMT calls for. They see international institutional collaboration, not only between the South and the North, but among the group of countries that make up the strong South, as one of the most effective ways of developing NMT. They go into considerable detail on the process, shape and content of policy and strategy for promoting international collaboration.

In large parts of the developing world, new generic technologies are virtually non-existent or have only recently begun to make their presence felt. Sub-Saharan Africa falls into this category. It is also a region which exemplifies the formidable challenges facing the 'weak and excluded' South. From Bezanson's chapter in Part IV one can note that, thanks to the effort of Canadian and other donor agencies, information technology has been introduced into a number of institutions and countries in Africa south of the Sahara, though not so far into the production sectors, except perhaps in a marginal way.

It transpires from Wessels' chapter in Part IV that agricultural biotechnology is emerging as a potentially important generic technology in a few sub-Saharan African (SSA) countries. There is interest in exploring and promoting this possibility on the part of SSA governments and scientists, matched by an equal interest on the part of agri-business corporations and scientists in North America and Western Europe. But the transfer of agricultural biotechnology across national boundaries is intimately bound up with very sensitive issues such as biosafety and intellectual property rights as enshrined in the global Convention on Biodiversity. These and other related issues lead to the question of the capabilities and capacities that SSA must needs acquire not only in the 'hard' S&T of biotechnology, but also in the 'soft science' of policy analysis, research and implementation.

It is the latter theme that is the focus of Chapter 6, by Calestous Juma and John Mugabe. They point out that, with a few exceptions, SSA countries have yet to work out national policies and priorities specific to biotechnology. What there is tends to be a part of the overall S&T policy. Research is fragmented and conducted on the basis of individual initiatives, relying on support from foreign donor agencies. Juma and Mugabe describe in detail the stages through which the process of policy formulation and institutional development has progressed, and the current situation at which the process finds itself, in the few countries that form part of the exception to the general lack of initiative mentioned above. The countries concerned are Ethiopia, Kenya and Zimbabwe. In Cameroon, Tanzania and Uganda one can detect active signs of interest and movement. The authors pin their faith on international cooperation as a prime vehicle for helping to create S&T capabilities in SSA, for facilitating technology transfer, and for integrating biotechnology into the production sectors. They regard policy-oriented research as of critical importance in the tasks of capacity building, monitoring global technology trends, participating in global consultations and acquiring information, and put forward proposals on how SSA countries ought to go about this.

Information and communication technologies are radically altering management and organization practices in the industrialized world. These changes are bound to penetrate developing countries as well. As argued earlier, with the integration of IT into their systems, management and organization (M&O) have themselves acquired a generic character. Generic M&O have the potential to jack up rapidly the performance of individual companies to substantially high levels through short-cut innovations that would otherwise have been impossible. In addition, being knowledge-intensive rather than equipment-intensive, generic M&O entail significantly less financial costs than conventional M&O in the medium to long term. In Chapter 8, Stephen Karekezi examines how generic M&O can be geared to the task of raising the performance and efficiency levels in the energy sector in sub-Saharan Africa. Within this framework, Karekezi analyses what is achievable through demand management measures, change in the patterns of energy use, repair and maintenance schedules, energy-saving routines, and the decentralization of electric power generation and distribution. This leads him to identify some key policy options.

Transnational corporations (TNCs) see themselves as being in the forefront of making new technologies functional in developing countries. Although small- and medium-sized firms do play a role, TNCs dominate the process of innovating, transferring and using new technologies, which require large and long-term investments in R&D. In Chapter 9, which opens Part III, Richard Adams argues that there is more than just technology involved in this process. It entails complex arrangements requiring state-of-the-art management, marketing and training, which only the TNCs can offer at present. As Adams implies, governments in the South that are keen on bringing new technologies on board in their countries have to woo the TNCs. The pressure of this imperative is all the greater under the new GATT regime, which provides for the protection of intellectual property rights, characterized by Adams as the 'corporatization of knowledge'.

However, it is not an entirely one-way process. To stay competitive in the global market against rivals, TNCs need access to the markets and the dynamic comparative advantages that some parts of the developing world can offer. This provides governments, the public and the private sectors, and the NGOs in some developing countries with some leverage, which they can use to create a partnership with TNCs that promotes rather than hinders sustainable development. Adams reflects on the forms this partnership could take and the benefits that it might be made to yield.

Any trajectory of sustainable technology must incorporate the principle of recycling 'waste' products and 'waste' energy. In Chapter 10, Donald Roberts suggests that the way ahead is for industry and infrastructure to model themselves on the nutrient and energy recycling that occurs ceaselessly in nature. He proposes that the present unsustainable systems be replaced by industrial ecologies or artificial ecosystems, and outlines what this would imply for future methods of extraction and use of materials and energy. In this context, Roberts is particularly concerned about the challenges that face the engineering profession. On another front, he makes a case for replacing the by-now conventional methodology of environmental impact studies, which he regards as being wasteful and ineffective, by a longer-term feedback approach that begins at the pre-planning stage of a project and ends much later than its completion.

The themes raised, on the one hand by Adams on how companies based in the industrialized world can be motivated to think in terms of environmental and social sustainability in their operations in developing countries, and on the other by Roberts on how to integrate the principle of recycling into industry, are considered by Carl-Göran Hedén in the Swedish context. In Chapter 11, Hedén foresees great promise in combining new generic technologies of the industrialized world with traditional recycling technologies of the developing world. He espouses a model in which these 'blended' technologies are promoted by a coalition of Swedish industry, entrepreneurs, science and engineering talent, and official development assistance, in close collaboration with developing-country entrepreneurs and non-governmental organizations (NGOs). He argues for a switch from the emphasis on bilateral assistance through governments to direct assistance through entrepreneurs and NGOs.

Hedén sees the link between entrepreneurs, as mediated by scientists, engineers and NGOs, on both sides of the equation, as the key to the future success of the development process, not least in 'blended' technologies. To be meaningful, the transfer of technology (which includes know-how and know-why) presumes the availability of local expertise to absorb and use the technology. The creation of such expertise, among both entrepreneurs and technical professionals, through training based in Sweden and the developing countries, emerges as a decisive factor.

Part IV explores the approaches of donor countries. As far back as 25 years ago, Canada was unique in grasping the importance to developing countries of the information technology (IT) revolution, which was then in its early stages. As Keith Bezanson underlines in Chapter 12, one of the principal reasons prompting Canadian support, through the Ottawa-based International Development Research Centre (IDRC), was the recognition that if developing countries did not participate in the IT revolution, they would become ever more uncompetitive and disenfranchised in the globalizing economy.

Bezanson identifies partnership and capacity building as the two watchwords of the support begun by IDRC. These were translated into operational guidelines encompassing the transfer of the latest technology, locally conducted result-oriented research, the building up and strengthening of indigenous IT skills, networking and international collaboration involving both developing and industrialized countries, and support to IT policy research as a backup for decision-making and allocation of resources.

The projects and programmes that IDRC has launched and supported over the last two decades can be grouped under three generic kinds of IT applications: telematics, informatics and geomatics. The telematics programme has contributed significantly to the creation of computer- and telecommunication-based networks connecting research and other development-oriented institutions within and across countries. Several nets are now in operation in Africa and Latin America, for exchanging information and knowledge in agriculture and health.

The informatics programme concentrates on building up and strengthening indigenous capacity in software for the management and use of computer-based information systems. The areas of applications are very diverse, ranging from museums to genetic research institutes, from urbanization to environmental problems. It includes the ability to access and use international databases, meteorological information, demographic information, geological survey inventories, and so on. The various software developed under the programme have been installed in a very large number of developing-country organizations, and followed up by training on the use of the software.

The geomatics programme deals with spatial information, in particular with remote sensing, geographical information systems and global positioning systems. These also happen to be areas where Canada has been and continues to be part of the technological vanguard. Geomatics is of great value in the analysis and management of natural resources. Geomatic projects have been funded in over 25 countries across all the three developing continents. The approach used is three-pronged: the validation of the technology and its adaptation to the local environment; assessment of its potential for the developing country concerned; and technology research and development.

Bezanson points to the need to consider not only the promises but also the pitfalls of the spread of IT in developing countries, not the least of which is the risk of widening the IT gap, and with it the opportunity gap, between different social groups within developing societies.

The Netherlands is one of the few countries in the North to have systematically developed its aid policies to promote biotechnology in selected developing countries, and to have committed substantial resources to that effort. In 1992, it set up a five-year Special Programme in Biotechnology to assist four developing countries selected according to a set of criteria: India, Colombia, Kenya and Zimbabwe. Agriculture, human health care and environmental management were picked as the three thematic areas for provision of support. As T.J. Wessels explains in Chapter 13, the Dutch initiative is characterized by a participatory 'bottom-up' approach, in which needs and priorities are identified, and activities decided upon, with the involvement of local policy-makers, researchers, non-governmental organizations and end-users.

One highly unusual feature of the Dutch approach was to try and incorporate the concerns of developing countries into the process of formulating the internal biotechnology policy for the Netherlands. This was to prevent the development of biotechnology within the Netherlands that would be detrimental to the developing countries. Wessels indicates that the Netherlands has attempted to promote the same stance in international fora where crucial issues are discussed and decided on, such as intellectual property rights, biodiversity and biosafety.

The Special Programme comprises three components: (1) the integration of developing country dimensions into the Dutch internal biotechnology policy; (2) problem-solving and user-oriented technical cooperation between the Netherlands and the four selected developing countries, based on the demands expressed by the latter; and (3) collaboration and coordination with other donors and international institutions.

The technical cooperation component incorporates not only technological projects but also socioeconomic, legal, cultural and ecological aspects, to make sure that it addresses the problem in an integrated way, in order to be able to make a contribution to poverty alleviation. The focus is on small-scale farmers and women. As means to this end, the Special Programme promotes research capacity building, research projects, technology assessment, socioeconomic studies, risk analyses and policy capacity building in the areas of biotechnology and genetic resources.

The international component took a leading part in having biotechnology integrated into the research, service and policy activities conducted by the institutions that operate under the umbrella of the Consultative Group on International Agricultural Research. Examples of this are the Cassava Biotechnology Network and the Intermediary Biotechnology Service.

Central to Jon Sigurdson's understanding (Chapter 14) is the perception that systems of innovation in industrialized countries are no longer predominantly national in character. Precisely like production and marketing, R&D is also being internationalized, with industries relying on networks of researchers and laboratories. One can see a marked difference in the approaches adopted by national governments and transnational corporations (TNCs). In collaboration with the corporate sector, the governments of the leading economies of the OECD region have tried to foster breakthroughs in high-tech R&D either within the national context or in regional groupings, such as the European Union. These attempts have been only modestly successful. TNCs have taken a different route. On the one hand, they are translocating parts of their R&D effort from their home base to other countries, including a few of the technologically more advanced developing countries. On the other, they are creating internationally linked project teams, involving universities and industries, to tackle specific problems. In high-tech, the TNC approach has delivered more impressive results than the national one.

Sigurdson questions the continuing relevance and dynamism of national systems of innovation in the era of new generic technologies. He contrasts the roles, approaches and performances of national governments and transnational corporations in their efforts to achieve R&D breakthroughs in high-tech. Linked to these concerns is the question of the relevance and usefulness of the 'traditional' methods of knowledge production in academia within the confines of narrowly defined disciplines, whereas the dispensers of public and private corporate funds for research are increasingly calling for 'problem-solving and applicable' research, which presumes transdisciplinarity.

Examining the contours of Swedish R&D and Swedish aid in this unfolding global context, Sigurdson proposes three future scenarios for Swedish aid for promoting generic technologies in developing countries: they are, respectively, about utilizing (i) the technological strength of Sweden, (ii) the management capability of Sweden, and (iii) portfolio investments in the economically dynamic developing countries. It transpires that several conditions need to be met, both in developing countries and in Sweden, for these scenarios to be effective.

Reflecting on why technology issues have become so important to developing countries in recent years, Charles Cooper in Chapter 15 identifies two interlinked forces at work: the dramatic acceleration in technological change in conjunction with the advent of generic technologies, and the move from protected to open economies. As he points out, generic technologies have brought about significant technological change in the traditional sectors of manufacturing, such as food processing, textiles, garments and leather. And in a regime of open economies and relatively low transport costs, the entry of foreign companies into a country erodes the difference between domestic and world markets, exposing local industry to the gales of global competition.

To understand what happens to the comparative advantages that developing countries have hitherto enjoyed, Cooper examines the differences that obtain across the spectrum from price competition and innovative competition. Starting out from this vantage point, his analysis identifies a number of changes that technological factors have brought about in the open economies of both the industrialized and the developing world. Among these are privatization of technology, temporary monopolies, the struggle for dominance between technology user and technology supplier, the erosion of previous assumptions and rules governing international trade, and the conditions that govern the entry of firms into manufacturing. These have implications for the policies of the World Bank and the IMF, as well as for the trade and aid policies of individual donor countries in the North. The former concentrate on resource allocation under conditions of static efficiency, while the latter try out a mix of policies ranging from protection of domestic markets, exit from 'old' industries, and shifting to new technologies. Cooper examines how these impinge on the developing countries and what aid policies need to be put in place to assist the South in benefiting from global technological change.

In the final part of the volume, M.R. Bhagavan presents the major issues under debate on the importance of new generic technologies to developing countries. This chapter takes up for discussion not only the views expressed by the other contributors to this volume but also those being debated elsewhere by leading scholars and decision-makers involved in examining the role of technology in development. Among the principal themes considered are the capacity and capability of developing countries to grapple with the new technologies; the predicament of the technologically 'excluded' South; the impact on the economy and foreign trade; the role of the public and private sectors and of the transnational corporations; investment patterns, competition and efficiency; globalization of R&D; access to genetic resources; free trade in technology and intellectual property rights; biosafety and bioethics; environmental and social sustainability of the new technologies; institutional reform and innovation; and policy approaches to development assistance by the international donor community.

## Notes

1. C. Freeman, *The Economics of Industrial Innovation* (Harmondsworth: Penguin, 1974).
2. For a detailed exposition of this approach to technology analysis, see M.R. Bhagavan, *The Technological Transformation of The Third World: Strategies and Prospects* (London: Zed Books, 1990).
3. For overviews, see for example Bhagavan, *The Technological Transformation of The Third World*; 'Technological Change and Restructuring in Asian Industries: Implications for Human Resources Development', in M. Muqtada and A. Hildeman, eds, *Labour Markets and Human Resource Planning in Asia: Perspectives and Evidence* (Geneva and New Delhi: UNDP and ARTEP/ILO, 1993); J.L. Enos, *The Creation of Technological Capability in Developing Countries* (London: Pinter Publishers, 1991); M. Fransman and K. King, eds, *Technological Capability in the Third World*, (London: Macmillan, 1984); OECD, *Managing Technological Change in Less-Advanced Developing Countries* (Paris: OECD, 1991); UNIDO, *Industry and Development: Global Report 1990/91* (Vienna: UNIDO, 1990); *The Changing Technological Scene: Trends in Selected Developing Countries*, mimeo (Vienna: UNIDO, 1991); S. Wyatt, *The Changing Technological Scene: The OECD Countries*, mimeo (Vienna: UNIDO, 1987).

## References

Bhagavan, M.R., *The Technological Transformation of the Third World: Strategies and Prospects* (London: Zed Books, 1990).
Bhagavan, M.R., 'Technological Change and Restructuring in Asian Industries: Implications for Human Resources Development', in M. Muqtada and A. Hildeman, eds, *Labour Markets and Human Resource Planning in Asia: Perspectives and Evidence* (Geneva and New Delhi: UNDP and ARTEP/ILO, 1993).
Enos, J.L., *The Creation of Technological Capability in Developing Countries* (London: Pinter Publishers, 1991).
Fransman, M. and K. King, eds, *Technological Capability in the Third World* (London: Macmillan, 1984).
Freeman, C., *The Economics of Industrial Innovation* (Harmondsworth: Penguin, 1974).
OECD, *Managing Technological Change in Less-Advanced Developing Countries* (Paris: OECD, 1991).
UNIDO, *Industry and Development: Global Report 1990/91* (Vienna: UNIDO, 1990).
UNIDO, *The Changing Technological Scene: Trends in Selected Developing Countries*, mimeo (Vienna: UNIDO, 1991).
Wyatt, S., *The Changing Technological Scene: The OECD Countries*, mimeo (Vienna: UNIDO, 1987).

*Part I*

The Technological Scene:
Main Features, Capabilities and Trends

# 1

# Contemporary Technological Revolutions: Characteristics and Dynamics

*Ian Miles*

## Introduction: Evolution and Revolution in Technological Change

It is rare today to find a country which has not formulated some sort of technology strategy. With almost monotonous regularity, three areas of advanced technologies are listed as priorities for national development: Information Technology (IT), Biotechnology (BT), and Advanced Materials (AM). These are hailed as revolutionary new technologies whose potential for changing the fundamentals of economic life are such that no country can afford to ignore them.

Talk of technological revolutions is so common and casual that we may be inclined to dismiss it as mere hyperbole. Suppliers of new products ranging from aircraft engines to zirconium alloys are liable to claim that their products are revolutionary. But there are more systematic ways of thinking about technological revolutions.

These new approaches are part of important shifts that have been underway for some time now in the analysis of technological change and innovation, both for scholarly purposes and to inform policy. There has been a shift of attention away from individual innovations and their diffusion, to more systemic and more evolutionary views. New products are seen as competing with other solutions to the same problems, as evolving through the product cycle, and as being influenced by feedback and more active inputs from users (rather than being seen as isolated innovations diffusing to a market of passive adopters). Suppliers and users are seen as being located together in regional and national systems of innovation, and the course of technological development as being determined by the sociotechnical constituencies established within such networks. And innovations are not all of equal weight, since the knowledge bases on which they draw are related systemically.

Thus Freeman distinguishes technological revolutions from incremental and radical innovations.[1] *Incremental innovations* occur more or less continuously, involving small modifications in products or processes, with at most minor changes in training and work organization. They often derive from improvements and suggestions supplied by engineers and production workers on the job, or from users themselves. *Radical innovations*, in contrast, involve more substantial change. In the twentieth century, these have often originated from formal research and development (R&D) activities, which are usually carried out in the suppliers' laboratories. They may involve substantial changes in production processes and organizational arrangements, and/or new products which establish new markets or displace familiar products from established markets.

A *technological revolution*, though including many examples of incremental and radical innovation, is more profound still.[2] Typically it will be based on scientific discoveries, possibly carried out by pure scientists in non-industrial settings, for example 'basic research' in universities and specialized laboratories, which yield new basic knowledge about fundamental chemical, physical, biological or other processes. This new knowledge, and the techniques associated with it, allows for changes to be made – not just improvements limited to a particular class of product (as with an incremental or perhaps a radical innovation), or even to a whole industrial sector (as with many radical innovations). The knowledge is employed to develop fundamentally new and far-reaching inventions – allowing for the introduction of what Freeman terms new '*heartland technologies*'. These are revolutionary technologies which can be used to carry out operations common to a wide spectrum of economic activities.

For example, the application of motor power is required by many activities in manufacturing, transport, agriculture and construction, and substantial changes in these areas of economic activity have been achieved by applying such technologies as water and wind power, steam power, electric power, and petroleum engines (involving distinct heartland technologies). Steam power, in particular, by substantially reducing motor power's dependence on the weather and geography, allowed for the substitution of machinery for human and animal effort in many factory and transport applications: it was a key component of the great industrial revolution of the early nineteenth century.

A technological revolution, then, involves the application of the new heartland technologies across broad swathes of the economy, and the associated changes in products and processes, in working practices and interfirm relationships, and in the centres of economic power that develop as new opportunities are recognized and seized. The steam engine is frequently depicted as an exemplary new revolutionary technology. It became the focus of a great deal of innovative effort as smaller and lighter, more efficient and more robust engines were developed. Opportunities to create new products were seized: the railway engine, the traction engine, and a great many industrial devices that could use the unprecedentedly powerful and reliable source of motor energy. Applications proliferated across a vast number of industrial and commercial processes. Innovation in processes allowed, in turn, for further innovation in products: a whole wave of new materials, components, and industrial and consumer goods emerged as the engineering industries developed. These developments unfolded over a long period of diffusion and experiment, during which period many social, institutional and organizational changes were also tried out.

Perez further specifies the conditions for innovations being revolutionary technologies, arguing that a new 'key factor' has to have:

- clearly perceived low – and descending – relative cost;
- unlimited supply for all practical purposes;
- potential all-pervasiveness;[3]
- a capacity to reduce the costs of capital, labour and products as well as to change them qualitatively.'[4]

Furthermore, these characteristics may well only be realized in a technological revolution if the prevous 'key factor' and technologies based upon it are facing limits themselves. Once such a conjunction occurs, then efforts will increasingly be focused on developing and applying the revolutionary technology. In the words of René Kemp, a successful new paradigm involves 'a combination of interrelated product and process, technical, organizational and managerial innovations, embodying a quantum jump in potential productivity for all or most of the economy, and opening up an unusually wide range of investment and profit opportunities'.[5]

The process whereby a breakthrough becomes a heartland technology is liable to be a protracted one. Recognition has to grow that substantial opportunities can be created and seized in many areas of production and new products. There is then a 'swarming of innovations' (and of innovators), and a concentration of (often mutually reinforcing) innovative practices. Suppliers see opportunities for products that will reap profits in new or old markets; users see opportunities to achieve their objectives – for example, higher productivity in the case of industrial users, more convenient or more pleasant leisure pursuits on the part of consumers. Such opportunities are demonstrated through social networks – in trade shows, journals, the mass media, conferences, wholesale and retail outlets, in the course of actual use of the products, through government awareness and promotion campaigns, and so on.

In a technological revolution many innovations are liable to be developed, with many unsuccessful products as well as many successful ones. As attention is paid to the characteristics of and opportunities offered by a new technology – and of the nature of the market for these opportunities – a framework of understandings and expectations will be developed by innovators. This framework (a 'paradigm') leads to a concentration of efforts along lines determined by perceptions about what is technologically feasible to achieve, and what users will pay for.

The concentration of effort means that technologies tend to develop in particular ways: there is a *technological trajectory*, involving broad trends in technological performance, as numerous firms and research establishments compete to provide successful improvements. Thus the trajectory is not inherent in the technology itself, but the outcome of search behaviour and technological efforts of firms acting within technological paradigms. Freeman and Perez talk about technological revolutions as involving change in 'techno-economic paradigms'. These are 'commonsense' notions about the feasibility and value of particular types of activity, how they constrain the ways we behave, and the things that we attempt to do; and these derive from the frameworks of ideas associated with the use of heartland technologies. Heartland technologies allow for substantially new common sense to be developed about where production and consumption can be located in space and time, how costly energy will be, and so on. Kemp notes that,

> The concept of a technological paradigm ... involves two elements: an exemplar ... and a set of heuristics that guide research and development activities ... a basic design and a set of approaches for solving technological problems. An important characteristic ... is that there exists a core technological framework which is shared by the entire community of technological and economic actors as the basis upon which one looks for improvements in process efficiency and product performances.[6]

Such shifts in common sense have substantial implications for the organization of work and social activities. The skills required, the communication and coordination links between different agents, the costs and benefits routinely expected to be associated with different activities, may all be rethought.

But this can take an extremely long time to accomplish. A classic case is the organization of factories around power sources: electric motors allowed for each machine tool to possess its own power source, and thus to be located wherever there was an electrical outlet, but for long after their introduction the organization of factories continued to reflect the contingencies of steam power, with machine tools to be grouped together around a central power source.[7] The notion of 'cultural lags' has effectively been resurrected in some of the innovation research literature to describe – or account for – this process. Thus the problems in adapting social norms to the potentials of new technologies is held to be responsible for the apparent paradox that, despite high levels of investment in new IT over the past decade, our economies have relatively little to show for this in terms of an increased rate of productivity growth.[8]

The concept of transition across paradigms (some authors prefer the less theoretically loaded notion of technological regimes) has proved to be immensely fruitful in the study of technological revolutions. More precisely, the contrast between one paradigm and another has been a compelling one, while the study of transition paths is relatively underdeveloped.) One consequence is that there is a great deal of interesting speculation about the nature of the managerial, organizational and other changes appropriate to facilitating the exploitation of the new technologies, but less attention to the blockages, bottlenecks and obstacles to immediate realization of the opportunities that are emerging.

A technological revolution is a time of great excitement and uncertainty. There is much hyperbole – some justified, other not. There are liable to be a great many new products emerging onto the market, with various design configurations, aimed at applying the revolutionary new technology to industrial and consumer activities. Innovations will be competing with each other, in many cases, as well as with traditional ways of doing things. Numerous products are liable to offer overlapping functionality, and it is by no means clear which of the 'solutions in search of a problem' will come to dominate. The process of change is liable to be long-drawn-out, with considerable uncertainty as to how the constellation of successful new products will be used. Uncertainty is accentuated by the rate of change that is continuing in the heartland technology itself.

## The Triple Revolution: Information, Biology, Materials

### *The long view*

IT, BT and AM are all candidates for being new heartland technologies, and the basis for technological revolutions. The three technologies are very different from each other, but each has characteristics that fit the Perez definition of a revolutionary technology.

**TABLE 1.1** Three transformative technologies compared

|  | Information technology | Biological technology | Materials technology |
|---|---|---|---|
| Traditional technologies (based on empirical knowledge and crafts) | Data stored and transmitted in form of signs embodied in materials, and crafted there by human skills. | Harnessing of natural processes of cultivation and fermentation (agriculture, brewing, baking, etc.) | Shaping and casting of part-processed raw materials: early ceramics, metals, alloys. |
| Industrial era technologies (products of applied sciences and application of industrial processes) | Data stored and transmitted in analogue form using electricity and electronics. (Key science: electrophysics) | 'Industrial' fermentation using enzymes and microorganisms. (Key science: biology) | Industrial transformation of materials (esp. steel); later, new materials (e.g. plastics). (Key science: chemistry) |
| Emerging technologies (products of strategic science and application of new industrial processes) | Data manipulated in digital form using microelectronics, optronics, and associated software. (Key sciences: physics, computer science) | Genetic and microbiological techniques applied to microscopic engineering of living material. (Key science: molecular biology) | Microscopic control of structure of materials. (Key science: physics) |

Of course, we have had technologies that deal with information, biological processes and materials for a very long time. Information technologies have a long history (smoke signals, books); even the application of electricity and electronics to information goes back a long way (telegraphs, telephones, radios). Biological processes have been submitted to technological intervention since the agricultural revolution in prehistory, too, and the tools used for digging, cutting and treating crops used the materials of the day such as flints, bronze and the like. But there have been successive transformations of our abilities to transform things. Table 1.1 identifies certain very broad elements of these transformations, noting in particular the developments associated with the industrial revolutions of the nineteenth and twentieth centuries, and those that are characteristic of the current period. The picture presented here is very broad and schematic, but has the virtue of drawing attention to the key features distinguishing different epochs. In particular, we see that the important development in modern technologies is the ability to control phenomena at incredibly detailed levels – the bit of information, the gene, the atom in a new material.

The three types of technology have undergone transformation as growing knowledge of underlying processes has been developed and applied. Increasingly knowledge-intensive methods of instrumentation and manipulation have been available, allowing for increasingly fine control of data, organisms and materials. These micro-level achievements have huge consequences at the macro-level of the products that result, as the microscopic properties mean qualitatively different features of the systems to which they contribute. These new technological opportunities have macro-consequences for economies and societies, too, as they are taken up and used by various social actors. In common among the opportunities now being offered by the new IT, BT and AM are the possibilities for increased variety of output based upon the capability of controlling the fine detail of products. For some commentators, this would seem to fit the 'post-modern' nature of the 'information age', which is often characterized as featuring more diversity and differentiation.

## New information technology

However, it is important to be sensitive to the specificities of the three revolutionary technologies. Consider first IT. What is new about new IT is the application of microelectronics, based upon semiconductor technology. Microelectronics has meant that it is possible to apply technology to producing, storing, retrieving, communicating, manipulating and displaying information in ways that are considerably cheaper and more powerful and convenient than was previously possible. The new heartland technology addresses a process that is implicit in *all* human activities, and thus in all economic production and distribution processes – information processing. In this respect, which means IT is applicable to services as well as to goods production, this may be a uniquely pervasive technology. Since IT can be applied itself to the further design and production of IT, among other reasons, the rate of progress continues to be dramatically fast – as is indicated by 'Moore's Law' (effectively that the power you can expect from a chip will double

**TABLE 1.2** Key features of new information technology

*Hardware*
- Decreasing *size* of equipment (so that new technology can be incorporated unobtrusively into existing devices, or installed in locations where large artefacts would have rendered the application unviable).
- Increased *reliability* of equipment (fewer mechanical components and non-integrated components to go wrong).
- Reduced *cost* of data processing means that distributed rather than centralized processing facilities become feasible. This may change the 'common-sense' methods of handling problems (as, for instance, the processing power of 'terminals' or network nodes may outstrip that previously possessed by the hosts). The shifting costs of information processing as compared to data inputs, and of distributed as compared to centralized processing, makes network systems more attractive: data can be rapidly transferred through space rather than reinput, efforts in data processing can be shared in new ways.

*Programmability*
- IT systems are essentially *programmable* (though this capacity may be reduced by design of dedicated systems). Any computer can simulate any other computer (though this may be very poor and slow if the architectures are very different). IT devices are 'virtual machines', where the programming determines what type of machine performance is forthcoming.
- Programmability means the growing importance of the role of *'software'* of all kinds, both in terms of contribution to systems performance and utility, and in terms of production costs, though reproduction costs typically decrease very substantially. (The argument about 'software' here includes texts and broadcast programmes as well as systems and applications programs for computers and microprocessor controlled devices i.e. media-type programmes and instructional programs).
- Scope for increased systems *flexibility*, as devices can be reprogrammed to carry out distinct tasks; reducing time taken for retooling.

*Information processing activities*
- Decreasing *cost* of information processing (which means that bulk or very intensive operations can now be carried out that would heretofore have been unthinkable). All forms of information processing are effectively being cheapened and rendered more effective: data capture, communication, storage, manipulation and presentation are all subject to dramatic quantitative and qualitative improvement.
- *Speedier* information processing (largely in consequence of overcoming the batch processing system, but also reflecting more rapid data transmission and more integrated systems).
- *Digitalization* of information of all kinds (speech, video, text, graphics, etc.) makes it possible to transfer readily data generated via one medium to

different media (thus the same material may be embodied in a printed paper text, as 'hypertext' on a CD-ROM, as interactive source material on an online database, and so on). This facilitates the shift to networking.

*Human–machine interaction*

- Increased quality and *user-friendliness* of systems (as 'spare' processing power is applied to error checking) to provide more choices to users, for 'intelligent' or 'user-friendly' front ends, etc.

- Behaviour of information systems in an *interactive* fashion: IT has the capacity to engage in 'conversations' with users, using everyday language or 'common sense' icons, requesting 'dialogue', being interruptible, and producing unique output customized to user inputs.

- The *tasks* subject to transformation through IT enter realms relatively unaffected by earlier technological revolutions: the sphere of management, of design and coordination, etc. This changes diffusion dynamics (since more powerful users are involved). It allows for the application of IT to make IT (e.g. software engineering, CAD of chips). And it might be appropriate to think that we are moving on a trajectory from data processing (computers), through information processing (early IT), to knowledge processing (the full IT paradigm).

---

SOURCE: Ken Ducatel and Ian Miles, 'New Information Technologies and Working Conditions in the European Community', PREST, Manchester: report to the EC, presented at EEC Conference 17–18 October 1991.

every year – while the costs will fall by just under a third). Table 1.2 sets out key features of New IT.

This would seem to fulfil the requirements for a heartland technology. The explosion of IT applications across our economies and societies seems strong evidence that a new revolutionary technology has indeed emerged, and that we are already well into a technological revolution. Microelectronics has diffused with remarkable, perhaps unprecedented, rapidity across most sectors of industry and services in advanced industrial economies.[9] These applications span both generic information-processing systems (computers and telecommunications), and more specialized applications (robots, energy control systems, videogames, and so on).[10] A number of associated and complementary innovations are also rapidly developing and diffusing, including hardware – such as optical fibres and associated switches – and, equally fundamentally, software. Software is in many respects the main bottleneck to many IT applications, since its production is still labour-intensive and in some respects more like a craft than an engineering discipline, though much effort has gone into changing this. Despite this, it would be foolish to fail to recognize the tremendous advances that have been made in software, which have helped render modern IT much more 'user-friendly' and thus capable of dissemination to non-technical users.

## New biotechnology

BT is very different in origins and nature from IT. (Though at least one author has sought to portray it as a variety of IT, on the grounds that genes can be seen as information-handling devices).[11] As with IT, it is important to distinguish the new biotechnology from traditional and earlier industrial technologies. Earlier technologies have provided ways in which living organisms could be applied to the production of products such as food and drugs (for example, brewing alcohol, fermentation of dairy products, the manufacture of antibiotics). The critical development here was not the discovery of the properties of a new set of materials (as was the case with semiconductors in the development of microelectronics). Rather, it involved the growing scientific understanding of how DNA operates in genetic material to determine the characteristics of organisms. This has led to the creation of a series of powerful new techniques for, in particular, genetic engineering (as with IT, other related, and often complementary, innovations are emerging here). Table 1.3 portrays key elements of this evolution. It is borrowed from a FAST report which also stressed the potential pervasiveness of BT, suggesting that over 60 per cent of the manufactured output of industrial countries was of biological origin.

In addition to obvious applications to agriculture, food and medical, and pharmaceuticals industries, there are already commercial applications in areas as diverse as waste remediation, metal extraction, energy (biomass), and other sectors. This may not be as diverse a set of applications as those of IT, although we can expect BT applications to proliferate; yet it is unlikely that they will pervade every nook and cranny of social and economic activity in the way that IT seems set to do – unless BT and IT fuse, for example in 'biochips'. Even so, BT's applications are still sufficiently wide-ranging and sizeable for the claim of a technological revolution to seem eminently justified.[12]

An example of a medical application indicates the economic and social significance of new BT. Blood products was previously a relatively undynamic area of medical science. It is now being revitalized, since BT makes it possible to manufacture some of the components of blood in hitherto unthinkable volumes. Examples are those hormones which stimulate the production of red and white blood cells by the

**TABLE 1.3** The new biotechnology

- Fundamental discoveries in the life sciences, particularly of the role of DNA as the molecular carrier of the stored information in all genetic material.
- Techniques for the manipulation, alteration and synthesis of genetic material (either directly or via cell fusion) to create new life forms
- Techniques based on microbiology for cultivation, screening and selection of useful cells or microorganisms, and manipulation of their behaviour under controlled conditions
- Techniques for plant cell and tissue culture for accelerated propagation of useful plants
- Downstream processing techniques for extraction, treatment, purification and conversion of useful materials following the biomass production stage.

SOURCE: Forecasting and Assessment in Science and Technology (FAST), 1984, p. 12.

bone marrow. The stimulation of red cell production reduces the need for expensive blood transfusions for patients undergoing dialysis for kidney disease; that of white cells reduces the vulnerability to infection of patients undergoing chemotherapy for cancer. The drugs are expensive, but these new medical treatments offer the chance to save money – and lives – in established high-tech areas of medical treatment. Industry commentators forecast that these will be among the world's top three best-selling drugs by the end of the century, with combined sales equivalent to the health budget of many industrial countries.[13]

## New materials technology

Again in the case of AM it is necessary to distinguish the new materials technology from the ages-old methods of producing and applying materials which have been established in human societies. Many 'new' materials have been available in some form for decades – for example, plastics have a relatively long history. Completely novel substances have recently been developed – in the past few years whole new classes of materials, such as the fullerenes and metcars, have been 'discovered', for example. But new AM is not a matter of finding a particular class of new material, nor even of applying a particular technique or set of instruments to producing materials. As with BT, the key element appears to be less the development of one specific new material or technique, than the development of knowledge that allows for new processes to be applied to the production of materials, so that effectively the details of AM can be defined down to the atomic or molecular levels.[14] IT could also be seen as requiring a development in knowledge – of semiconductor electronics, in particular, as well as of relevant engineering and computational methodologies. But AM appears to be less triggered by a specific step forward than do IT and BT, and to depend more upon the concatenation of a series of cumulative leaps in knowledge. So is it a technological revolution, or generic technology?

Again, the technology is one with extremely wide applicability. Barker[15] indicates the potentially great pervasiveness of new materials technology by noting that the materials sectors of industrial economies may well account for 5–10 per cent of output, and that materials can account for as much as 60 per cent of the cost of manufacture. Though AM are so far only a few per cent of the total materials markets – and thus the AM revolution, if such a thing exists, has a long way to go – their role in many strategic applications is already substantial. Barker identifies a series of characteristics that distinguish new from traditional materials technology: these are summarized in Table 1.4. Again we have the makings of a technological revolution, although this may be rather more inchoate and less profound than those associated with IT and BT. If there is a new 'heartland technology', this is less clearly crystallized than in the latter two cases.

Nevertheless, the developments around AM are clearly of substantial importance to the modern world; if there were not other technological revolutions underway they would no doubt be hailed as the defining technologies of our time. AM will certainly require close attention by both producers and users of traditional materials (including agriculture-derived materials such as wood), and their new properties may be the basis for new products and processes of many kinds.

## Clean Technology: A New Regime?

As if all the excitement about IT, BT and AM were not enough, we have another set of highly significant developments to deal with. The last decade has seen a growing consensus that the world is facing severe environmental problems associated with human industrial activities. To 'regional' problems like acid rain are being added 'global' ones such as the greenhouse effect and the diminution of the ozone layer. These immense problems are intimately associated with the mode of industrial development that has been undertaken by industrial soci-

**TABLE 1.4**  Characteristics of new materials technology

- *Information content*  R&D, processing and design expertise are much higher proportion of total costs, energy and raw materials lower.
- *Complexity*  Greater control of materials microstructure, so AM often constituted by a series of phases yielding a desired microstructure with specific properties. Requires multidisciplinary knowledge inputs.
- *Integration of function*  Packing of more performance characteristics into smaller areas and volumes, reduced steps in manufacturing process.
- *Added value*  High unit prices related to information content and level of processing required.
- *Variety*  Broad and diverse range of materials, reflecting variety of manufacturing methods and raw material inputs, and amount of scientific and engineering knowledge now available; so scope for more customization to user requirements.
- *Market size*  Already having impact on nearly all sectors of manufacturing industry, especially high-tech sectors; likely to have multiplier effect across whole economy.
- *Market growth*  Whereas many traditional materials have mature or saturated markets, AM display rapid growth.
- *Life cycle*  Apparently short, reflecting increased competition among continually evolving materials, and shorter life cycles of products in which used.

SOURCE: Brendan Barker, 'Engineering Ceramics and High-temperature Superconductivity: Two Case Studies in the Innovation and Diffusion of New Materials', D.Phil. thesis, PREST, Manchester, 1990.

eties, and thus, in the absence of a groundswell of opinion in favour of restricting consumption in the richer countries, there has been much attention to whether new technologies and ways of using technologies can succeed in overcoming and averting the environmental problems. Indeed, whether or not standards of living need to be restrained, 'clean' technologies are bound to be an important element of a shift to more sustainable patterns of development. The challenge will be not just one of shifting to less polluting and more conservationist technologies, but of shifting to *new technological trajectories* which promote the continued development of clean and cleaner technologies.

How does this relate to the discussion of technological revolutions? In two ways. First, IT, BT and AM can all contribute significantly in positive or negative ways to environmental affairs. As 'solutions looking for problems' there is already much rhetoric – and much of it is soundly based – concerning the application of the new technologies to cleaner production and consumption patterns. IT can be used to monitor and manage energy use for example, resulting in significant savings in consumption; BT can produce organisms which can clean wastes with fewer by-products than conventional chemicals; AM can provide more durable structures with lower raw material inputs; and so on. On the other hand, the semiconductor industry was a notorious polluter of groundwater (though it has moved with exemplary speed in phasing out CFCs); there are fears of the consequences for ecosystems of the release of new organisms; new ceramic products pose disposal problems of their own; and so on. The technological trajectories of the three revolutions need to be reshaped themselves, by having a 'clean technology' trajectory imposed upon them.

But, second, what is this 'clean technology trajectory'? Are we talking about another technological revolution? Unlike IT, BT and AM, where a radical new heartland technology is being seized on by many key economic actors, spanning wide ranges of industrial operations, in response to their perceptions of its opportunities for improving productivity and performance characteristics, 'clean technology' is emerging under different imperatives. The three revolutionary technologies feature radical heartland innovations which have emerged from R&D labs and similar sites and are solutions looking for problems to which they can be applied (consider the almost frantic search for new product and process applications of microelectronics, genetically modified organisms, new ceramics). But with 'clean technologies' we have a complex of problems crying out for solution. The problems can be disaggregated into numerous categories based upon area of impact – for example, workplace problems; local soil, water, air pollution; the global atmosphere; biodiversity – or type of problem – for example, energy use, raw materials use, release of heavy metals, release of organics, release of halons.

The problems are diverse: does this mean that the solutions are equally diverse, or can there be some generic solutions? In terms of technology, it seems likely that the solutions may not be so diverse as the problems – in many cases similar technologies can be used for varied purposes. But they are very diverse nevertheless. There is not (currently) a core technology at the heart of the solutions, even though one can trace out some large clusters of solutions which do share a common core. (For instance, one sprawling and important class of innovations involves monitoring equipment, and here IT is proving extremely pervasive; as it is, too, in control systems.) But there are equally critical innovations dependent on chemical and physical transformations, rather than information-processing: catalytic converters, membranes, industrial scrubbers, non-energy-intensive building materials, CFC substitutes, biological methods of land reclamation, and many more. It may be that one of the clusters of innovation just mentioned would show explosive growth and widespread applicability, and thereby become the core of a new technological revolution. But even if we were to concentrate on industrial process technology and $CO_2$ emissions, which would be to narrow the scope of analysis dramatically, we would expect to find a range of 'solutions'. That is, we would find a range of quite dissimilar innovations (in terms of their technical basis) with high impact on emissions. On the supply side we would anticipate a shift to a mix of renewable energy sources and away from fossil fuels. On the demand side, energy-saving innovations will include redesigned products and processes, better controls on existing processes, increased attention to logistics, and so on. Many such innovations will be of sufficient value and applicability to be widely adopted, to establish industrial markets, to display economies of scale and accumulation of knowledge and performance, and so on.

If we were to see this as a change in 'techno-economic paradigm', the key factor would be the 'economic' rather than the 'techno-'. The driving force is *not* in the first instance cognitive change (and the institutional and behavioural forces promoting such change) concerning the trajectories of, and the opportunities of applying, a heartland technology. Rather, the critical change concerns a modification of conventional corporate practice: a change in the managerial 'common-sense' that then affects technology choices, rather than a change driven by awareness of technological opportunity (and action based on this awareness).[16]

In its starkest form, this argument is evidently too simplistic. Almost invariably there will be a dialectic between (a) perceived technological opportunities and (b) organization for effective performance. But though the two poles of the dialectic coevolve, one element may be dominant in certain instances. Indeed, most often we can expect one or other to dominate, though there are liable to be reversals of role over time as technologies mature and organizations restructure.

Can *management practice* acquire this role of reshaping technology choice – and thus establishing new technological trajectories? Historically, resource scarcities and regulatory pressures have had such a shaping force – consider synthetic rubber and smokeless domestic fuel, for instance. But these are instances of single technologies changing, not of technological revolutions or new paradigms. It seems that the shift toward more sustainable, clean technological trajectories will depend upon a revolution in managerial practices, nevertheless, rather than being triggered by revolutionary new heartland technologies. There are some grounds for optimism here. Perhaps the clearest evidence in the clean technology field involves the development of new management methodologies – organizational techniques that may be as important, or more so, as just-in-time and quality circles are to the IT paradigm in manufacturing. Probably the best established of these techniques is energy analysis, but considerable attention is currently being generated by efforts to establish standards for environmental management and environmental auditing. Just how much impact these systems will have on technological trajectories remains to be seen; to date, the literature on R&D and environmental management is very weak indeed.[17]

The confluence of these issues poses a major challenge as we approach the new millennium. How can technological development be put onto more sustainable paths, at a time where R&D activities are already reeling from the shock waves of the three technological revolutions that we have discussed? Clearly there are outstanding possibilities offered by IT, BT and AM for cleaner technologies. But it is vital not to be blinded to the opportunities that may be offered from other directions for cleaner production and consumption – including *social* innovation, as well as alternative new technologies and management techniques. A strong dynamic has been established by the three revolutionary technologies, as firms are increasingly aware of the associated opportunities for profit. Continual vigilance will be required to keep open alternative possibilities, so that efforts to generate cleaner technologies and living patterns are not subordinated to this dynamic.

## Notes

1. Chris Freeman, *The Economics of Industrial Innovation* (London: Frances Pinter, 1982).

2. Freeman also discusses a somewhat more constrained technological revolution – or somewhat more pervasive radical innovation – with his concept of a *new technology system*, a set of innovations that substantially affect a few branches of the economy.

3. 'All-pervasiveness' may be an overstatement – a very wide range of applications is more to the point. Note, too, an aspect implied by this definition: that

the new technology is socially acceptable, and that there are no major political or ethical barriers to its adoption. This is not the case for all innovations – for example, nuclear power has remained controversial. There is scope for controversy around some IT and BT applications.

4. Carlota Perez, 'Structural Change and Assimilation of New Technologies', *Futures*, vol. 15 (1983), no. 5, pp. 357–75, 361.
5. René Kemp, 'Environmental Policy and Technical Change', Ph.D. thesis, University of Limburg, Maastricht, 1995, p. 240.
6. Ibid., pp. 240–41.
7. And it took time for electricity to be seen as a source of power rather than as an energy supply for lighting – the first electricity suppliers were Electric Light Companies.
8. This highly contentious topic – some commentators argue that conventional estimates of productivity are not appropriate to a new paradigm, for example – is treated in OECD, *Technology and Productivity: The Challenge for Economic Policy* (Paris: OECD, 1991).
9. See, for example, D. Kimble and I. Miles, *Usage Indicators – A New Foundation for Information Technology Policies* (Paris: OECD, Information Computer Communications Policy ICCP 31, 1993).
10. I. Miles et al., *Mapping and Measuring the Information Economy* (Boston Spa, Wetherby: British Library [LIR Report 77], 1990).
11. Manuel Castells, *The Informational City* (Oxford: Basil Blackwell, 1989). BT may in the future contribute to IT, too, through the development of biochips and similar devices.
12. For an interesting analysis of the early development of genetic engineering from a perspective complementary to that set out here – but developed in much more detail and richness – see Maureen McKelvey, *Evolutionary Innovation: Early Industrial Uses of Genetic Engineering* (Linköping: Linköping University: Department of Technology and Social Change, 1994).
13. *Financial Times*, 27 May 1994, p. 14.
14. A particularly helpful discussion of this point is provided by P. Cohendet, M.J. Ledoux and E. Zuskovitch 'The Evolution of New Materials: A New Dynamic for Growth', in *Technology and Productivity: The Challenge for Economic Policy* (Paris: OECD, 1991). The approach of these authors has been drawn upon in our contrast of the three technological revolutions.
15. Brendan Barker, 'Engineering Ceramics and High-temperature Superconductivity: Two Case Studies in the Innovation and Diffusion of New Materials', D.Phil. thesis, PREST, Faculty of Economics and Social Sciences, University of Manchester, 1990.
16. 'Economic' is rather a restrictive term for describing this facet of social activity. Perhaps we should talk of 'techno-organizational' paradigms or 'sociotechnical ones' (see I. Miles, 'The New Post-Industrial State', *Futures*, vol. 17 [December 1995], no. 6, pp. 588–617).
17. The best review currently available, which reaches disappointing conclusions about the existing literature, is S.F. Winn and N.J. Roome, 'R&D Management Responses to the Environment', *R&D Management*, vol. 23 (1993), no. 2, pp. 147–60; see also N.J. Roome, 'Business Strategy, R&D Management

and Environmental Imperatives', *R&D Management*, vol. 24 (1994), no. 1, pp. 65–81; and Brian M. Rushton, 'How Protecting the Environment Impacts R&D in the United States', *Research-Technology Management*, May–June 1993, pp. 13–21.

## References

Barker, Brendan, 'Engineering Ceramics and High-temperature Superconductivity: Two Case Studies in the Innovation and Diffusion of New Materials', D.Phil. thesis, PREST, Faculty of Economics and Social Sciences, University of Manchester, 1990.
Castells, Manuel, *The Informational City* (Oxford: Basil Blackwell, 1989).
Cohendet, P., M.J. Ledoux and E. Zuskovitch 'The Evolution of New Materials: A New Dynamic for Growth', in *Technology and Productivity: The Challenge for Economic Policy* (Paris: OECD, 1991).
Freeman, Chris, *The Economics of Industrial Innovation* (London: Frances Pinter, 1982).
Kemp, René, 'Environmental Policy and Technical Change', Ph.D. thesis, University of Limburg, Maastricht, 1995.
Kimble, D. and I. Miles, *Usage Indicators – A New Foundation for Information Technology Policies* (Paris: OECD, Information Computer Communications Policy ICCP 31, 1993).
McKelvey, Maureen, *Evolutionary Innovation: Early Industrial Uses of Genetic Engineering* (Linköping: Linköping University, Department of Technology and Social Change, 1994).
Miles, I., 'The New Post-Industrial State', *Futures*, vol. 17 (December 1995), no. 6, pp. 588–617.
Miles, I., et al., *Mapping and Measuring the Information Economy* (Boston Spa, Wetherby: British Library [LIR Report 77], 1990).
OECD, *Technology and Productivity: The Challenge for Economic Policy* (Paris: OECD, 1991).
Perez, Carlota, 'Structural Change and Assimilation of New Technologies', *Futures*, vol. 15 (1983), no. 5, pp. 357–75.
Roome, N.J., 'Business Strategy, R&D Management and Environmental Imperatives', *R&D Management*, vol. 24 (1994), no. 1, pp. 65–81.
Rushton, Brian M., 'How Protecting the Environment Impacts R&D in the United States', *Research-Technology Management*, May–June 1993, pp. 13–21.
Winn, S.F. and N.J. Roome, 'R&D Management Responses to the Environment', *R&D Management*, vol. 23 (1993), no. 2, pp. 147–60.

# 2

# Information and Communications Technologies in Developing Countries

*José E. Cassiolato*

## 1. Introduction

It has long been recognized that radical technological change based on microelectronics technology is having and will continue to have far-reaching consequences for developing countries. Information technologies have provided the means for economic, institutional and technological changes that are altering world-wide patterns of production and distribution. Developing countries' insertion into the global economy is deeply bound up with the new technologies, and the efficient use of information technologies by these countries is on the policy agenda of all important international forums concerned with development. On the other hand, the 1980s and early 1990s – during which period information technologies have diffused rapidly in the developed world – have witnessed a dramatic increase in the gap between rich and poor countries. We live now in a more divided world, 'a world where OECD nations, Triadic-grouping, Multinationals or whatever, constitute 'a partially-integrated whole' and a greater majority who are without, or excluded'.[1]

In this 'Integrated–Excluded' world (I–E), one of the main features observed in the last few years – an acute globalization of economic activities – has been accompanied by a remarkable tendency towards the appearance of strong non-tariff protectionist measures by developed countries, particularly the USA and in Europe. The developing world has been affected by this protectionist wave in the more advanced countries, which was paradoxically accompanied by pressures for liberalization of their markets. The decade also witnessed falling prices of commodities and the explosion of the debt problem – both very damaging to developing countries.

Revolutionary scientific and technological developments based on information technologies present threats to and opportunities for developing countries. These are shaped by the fact that two of the main factors that have changed the environment for the development of new technologies – global interdependence in the economic and technological spheres, and growing involvement on the part of governments of industrialized and industrializing countries in policies to promote technical change and innovation – are blurring the boundaries between technology and trade policies of advanced countries.[2] Trade policies, particularly in advanced countries, have been increasingly dictated by technology issues and vice versa. As a consequence, new-technology factors are playing an important role in the trend towards the renewed concentration of trade within OECD.

In fact, significant changes are occurring in the composition of international trade. OECD countries are modifying their structure of production by making technological transformations that give services a greater share in the composition of both product and employment and that are in the main biased towards effecting substantial savings in the energy and materials components of products and processes. These structural changes are part and parcel of the above-mentioned 'technological revolution'. This process has already begun to alter the situation that existed with respect to 'traditional comparative advantages', not only between developing and developed countries but also among developed countries themselves. One of the consequences of these structural changes has been a significant reduction in the complementarity that existed between developed and developing countries, giving way to an international market characterized by intense competition and an upsurge of non-tariff trade barriers in advanced countries.

We can interpret the apparent paradox of growing international competition coexistent with increasingly protected markets in Western economies as simply a predilection to be in a better position in the new techno-economic paradigm. The potential benefits of the new technologies seem to be directed towards increasing the trade advantages of more advanced countries. The extent to which these trends move us closer to a global solution of world problems, or at least an attenuation of them, appears rather doubtful.

North–South trade has been increasingly influenced by new technologies. This influence has not been restricted to the new items of goods and services that are substantially altering the traditional division of labour between developed and developing countries. In the context of 'managed trade' such influence has also been felt at the level of the advanced countries' trade policies, which are hampering the diffusion of new technologies to the developing world. At the same time, information industries are going through a deep process of restructuring at the international level. This is still an ongoing process with as yet unclear results. However, despite the lack of consensus about the reasons for the process and its likely consequences on production, diffusion and competitiveness, an attempt can be made to understand its basic features.

Sales of the electronics industry in the USA (by far the world's largest market) had, during the 1980s, their highest rates of growth in the US manufacturing sector (an average of 11.2 per cent per year between 1977 and 1987).[3] The IT industry was the most dynamic sector of the US electronics industry. Nevertheless, the US IT industry (which is responsible for 45–50 per cent of world production) has recently experienced a very substantial slowing down in its growth path, which does not seem to be linked primarily to the economic conjuncture. This slowing down, which extended to European and Asian firms, brought falling profit rates.[4] Stagnation of sales and the fall in profits have been accompanied by comprehensive price reductions, particularly on 'commodities' and mainframes. As a consequence, the major firms have gone through an extensive process of restructuring, based on cuts (including of personnel) and the closing down of several plants.

The crisis period has been characterized by two converging trends.[5] From the suppliers' point of view, the increasing competition in the industry, with the growing importance of open systems – allowing for substitutability and connection of products with different technologies – has led to the design of strategies based on systems' customization and marketing. IT firms have ceased to be considered as 'vendors' of computers and emphasize as their main competitive weapon instead the aggressive supply of services and technical assistance to users. From the demand point of view, users have become increasingly influential in the development of products and systems since the so-called 'productivity paradox' of IT in the USA was noted.[6]

At a more general level, it is being recognized that the efficient diffusion of IT is not independent from the innovation process of IT systems. This is so since, although most computers can be characterized as 'commodities', the novel characteristic of the new technologies is that they process non-physical phenomena – information – which is intrinsically non-homogeneous. As a result,

> understanding the dynamics of achieving productivity gains from the mechanization of information-processing functions is not so much to do with the level and degree of performance of ITs, but more to do with the extent to

which the information requirements are defined, brought into light and highlighted by agents for each distinct area of application ... [with agents being] ... shapers of technologies rather than receivers of it.[7]

Consequently, from the user's point of view IT goods cannot be considered as 'commodities'. In this sense, their production and use are not dissociated and synergy between producer and user matters. As Freeman[8] pointed out, IT is a network technology 'par excellence': not only is the IT industry itself characterized by intensive technological networking for the development of its products but its diffusion throughout the economy to new sectors of application depends on the developments of its own networks in every sector.

It is in this dynamic setting that developing countries must pursue strategies to benefit from the new technologies. The basic aim of this chapter is to examine the problems that less developed countries are facing with IT and what they should base their strategies on for establishing technological capabilities.

Section 2 will present some characteristics of IT-centred technical change which pose novel problems for the development of capabilities in the Third World. Section 3 will discuss some of the obstacles to the diffusion of the new information technologies in the Third World. Particular emphasis will be put on the specificities of such technologies; namely, that its efficient use presupposes an upgrading of basic educational conditions in the Third World. Section 4 will discuss the human resources issue, and section 5 will analyse recent trends in science and technology expenditures in the developing world. The concluding section will address the policy issues.

## 2. The IT-Intensity of Technical Change

Within the overall complex of intensified technical change, the importance and pervasive impact of electronics and information technologies are well recognized and need no further emphasis here.[9] However, three characteristics of IT-centred technical change require some elaboration.[10]

First, to an extent that is perhaps greater than in other areas of technical change, the incorporation of electronics and IT elements into products, processes and organizational systems seems to require direct user-involvement in technology development and design. Compared with certain other areas of technology, the application of many areas of electronics/information technology requires, rather than standardized systems, those that are highly specific to the characteristics of individual firms, their products and processes, and their markets. These system specifications are not easily transferred in the form of 'ready-made' capital goods and blueprints, and their efficient introduction therefore requires much more localized technical change. Moreover, that localization must often go beyond the routine 'adaptation' of systems. It has to be deeply rooted in development and design of the hardware and, especially, the software, in the immediate context of use. Also, since that frequently involves relatively complex engineering and design, the importance of tacit knowledge is often particularly great.[11]

In particular, however, what is frequently involved is the integration of electronics/IT elements and systems within existing products, processes and organizational procedures, and large proportions of the tacit and other knowledge needed for localized development and design must therefore be drawn from the 'user' of those elements and systems. Thus, the technology users frequently need to play a particularly significant and direct role in the process of technology development and design. Then, of course, subsequent dynamic assimilation of the technology after its initial implementation requires, as with most other areas of technology, a yet greater direct involvement of the user in generating and managing technical change.

Second, most applications of electronics/information technologies involve systems and networks. This raises important issues about 'network externalities',[12] with progressive diffusion yielding falling transactions costs[13] and benefits to *all* users, not just the marginal adopters. At one level, this has important implications for change within individual firms. As Kaplinsky has emphasized,[14] the gains from using automation and information technologies rise disproportionately fast with increasing degrees of system integration. This does not mean that merely trivial gains can be captured from implementing only parts of the 'electronic jigsaw', but it does suggest that there are likely to be high returns to rapid intrafirm diffusion of the technology. Correspondingly, adopters and users of the technology are likely to gain high returns to investment not simply in 'the technology' itself, but in the bodies of knowledge and expertise that are needed to interact with users in developing and extending their IT systems.

The network characteristics of IT systems also have important implications at the overall interfirm level. Significant benefits accrue to individual firms (as 'externalities' from the actions of other firms) as the overall density of IT adopters and users increases within the total population of geographically related and market-linked firms. In particular, the efficiency of using IT systems grows with increasing local availability of (i) information about the technology from other users, (ii) a trained and experienced workforce, (iii) technical assistance and maintenance services, (iv) suppliers of equipment and software, and (v) complementary innovations – both supplier-developed and user-generated, and both technical and organizational.

Within such evolving structures and processes of collective learning, the diffusion of electronics/information technology is frequently accelerated by the presence of advanced user-firms that not only act as 'demonstrators' for others but also contribute to the development of innovations that improve the efficiency of the technologies in the specific local context of their use.[15] Given these patterns, it is not surprising that public policy in many of the advanced industrial countries has played a significant role in accelerating the diffusion of information technologies – in particular by stimulating the emergence of efficient technology users and the development of user-producer linkages. With respect to advanced automation technology in Sweden, for example, public policy and public institutions were crucial in setting up several 'demonstration plants' partly financed by the National Board for Industrial and Technical Development and the Board for Industrial Development.[16]

Third, information technology is not just an area of changing technology. It is frequently also a powerful instrument for generating innovation and technical change. This is most obvious in the case of computer-aided design systems, which not only permit more rapid and frequent changes in product and process design but also allow much more intensive and extensive exploration of design options. However, the same change-stimulating role of IT is evident in other ways that 'feed into' product and process design. In the various types of development and research, IT systems evidently play an enormously important role in accelerating the generation of new knowledge, in acquiring existing knowledge, and in developing new configurations of technology for incorporation into specific designs. Perhaps less evident is the change-stimulating role of IT when applied in production and management processes themselves. For example, the information that can be generated by various types of advanced process control technology, combined with the power of advanced computing, allows the acceleration of incremental process improvements. Similarly, the knowledge generated by IT applications for organization and administration permit more intensive analysis of changes in the 'organizational technology' of firms.

One of the key consequences of these characteristics of information technologies for developing countries is that traditional location-related factors (such as low labour costs and easy access to abundant natural resources) become increasingly ephemeral as a source of comparative advantage.[17] Another, more contentious, issue relates to how the above-mentioned features of the new technologies help or hinder their diffusion into the developing world and what their implications are for the accumulation of technological capabilities in those areas.

## 3. Problems of the Diffusion of New Technologies in the Developing World

It may be expected that an efficient diffusion to the Third World of the new technologies can help at least to alleviate some of these problems. Knowledge transfer from the North to the South should be a central concern of the present international economic system. There are, however, some important limits regarding what should be expected from this 'transfer' of technology. First, as the literature on the subject has uncovered, reliance on foreign technology should not substitute for an intense, costly and long *internal* commitment to science and technology. More specifically, the existence of local scientific and technological infrastructure, R&D efforts by local firms, and 'networks' of innovating firms are preconditions not only for a national system of innovation but also for an efficient use of the new technologies.

Second, one has to bear in mind the intrinsic limits of the 'transferability' of technology from the environment in which it was developed to other environments. Instead of the simple, neo-classical view that innovation and technological knowledge are 'commodities' that could be bought or sold, recent research suggests that firms produce things in ways that are differentiated technically from things in other firms and perform innovation largely on the basis of in-house technology, but with some contribution from other firms and from public knowledge. The outcome of such a complex process is a technology very specific to the firm.[18] The 'firm specificity' characteristic of the new technologies also suggests that even if firms in the North are prepared to 'transfer' its technology to the developing world, the recipient environment should be able to develop its own 'specific' technologies to assimilate fully the new technologies.

Multinational corporations (MNCs) have a very limited role to play in such a 'transfer' process. For instance, Dosi, Pavitt and Soete[19] have pointed out that, even if some positive contribution by MNCs to upgrade the technological capability of the recipient economy is likely to occur, firm-specific technologies are much less mobile and leaky than 'information' and than equipment-embodied techniques of production. There are important aspects of technology, related to manufacturing and innovative learning, which are not information and can hardly be traded since they relate to people-embodied and organization-embodied knowledge and capabilities.

As shown by recent research on the diffusion of new technologies, the 'absorption process' of the new technologies is itself a key determinant of the direction in which they evolve. In other words, the diffusion of new technologies and the development of these technologies are simultaneously determined.[20] This simply means that there are country/region specificities of technical change associated with the new technologies. Thus in a contemporary world setting, in order to exploit the potential benefits of new technologies developing countries should be able to develop capabilities to produce and use them.

The fact that the technology innovation and diffusion processes are dynamically intertwined has important implications for technological mastery in developing countries. At the core of such processes are the interactions between users and producers of a technology that is part of this new techno-economic paradigm. Each producer is likely to interact with a subset of all potential users. These connections, while being selective, will also, in important ways, be non-economic in character. Several case studies carried out in the Nordic countries show that closer interaction between producers and a competent and demanding domestic user sector are essential in generating technological capabilities in IT and in guaranteeing international competitiveness.[21]

In information technology, design, development and improvements of information systems are now the concern not just of suppliers, but also of specialized groups of 'software suppliers' within each of the user organizations. The maintenance of software systems is not only critical for improvements in technology but also accounts for an increasingly high proportion of total costs. If the South is not to be left behind it has to acquire capabilities to develop and use such systems. As an example, McKendrick[22] showed how the inefficient introduction of automation in the Indonesian banking system was caused by an excessive reliance on foreign vendors of hardware who supplied neither maintenance nor the software needed for the system to work properly (since the 'needs' of Indonesian banks were specific to their financial system). This is contrasted to the successful introduction of automation in Brazilian banks, which jointly developed their automation systems with local suppliers who were also owned by the main banks.[23]

Also, as the literature shows, the electronics industry is characterized by rapid growth in both labour and capital productivity,[24] which seems to be of particular relevance to industrializing countries 'where growth has been hampered by general capital shortage problems'.[25] As the labour displacement caused by such technologies is in fact a de-skilling at the level of the labour process, labour shortage acquires a completely different dimension in developing countries. In fact, during the mature phase of the technological paradigm of the postwar period, a variety of highly specialized technical (mechanical and electrical) skills were probably the major human bottleneck of developing countries.

The problem with the new technologies is that such de-skilling is accompanied by a growing need for very specialized skills of design, operation and maintenance of sophisticated systems and of basic skills. The question of efficient use of new technologies is also a question of meeting the requirements for structural changes which accompany the new technologies. For the effective exploitation of these technologies a much broader set of capabilities is required, and, as a consequence, much greater expenditures in financial and human resources are needed.[26] Unfortunately, what is happening in the majority of developing countries is precisely the opposite.

As we noted above, the efficient use of microelectronics technology reinforces the need to conceive new products and processes as 'systems'. It increases the importance of the project and other R&D activities. Even if microelectronics allows a greater production flexibility, this flexibility is defined *ex ante* – at the project design stage – leaving very little room for operators to introduce change in the concept of a particular product.[27]

Hoffman, in a survey about the diffusion of flexible manufacturing systems (FMSs),[28] provides evidence to show that most of the gains in flexible manufacturing arise from the preparation for, rather than the execution of, FMSs. Bessant and Haywood[29] suggest that the extent of these organizational benefits is around 75 per cent of the total acquired from flexible manufacturing.

These basic characteristics of the development of new information technologies imply that to be an effective user of them a country needs to acquire skills and capabilities that go beyond the ones received through the import of sophisticated machinery and products. Gaining access to available techniques may lead to static efficiencies; but to be an efficient user of the new technologies, a firm (or a country) also has to acquire some knowledge through a long interactive process of learning-by-innovating, where user–producer relations play a significant role.[30]

Empirical evidence suggests that the rate of diffusion of new technologies across countries is influenced by these systemic, strong cumulative effects of 'interrelatedness': late-adopting countries, with smaller amounts of the innovative capital goods that embody the new technologies, have lower rates of diffusion.[31] The reason is that it is difficult to obtain all the relevant information and skills for information-based technologies without having already reached a certain degree of interrelatedness with other complementary technologies.[32] In Less Developed Country (LDC) settings this implies that one can expect that an efficient use of the new information technologies will only happen if the recipient country is able to develop its own capacity to understand technologically all activities related to them. One can hardly expect that such countries could feel the potential benefits of advanced technologies simply by importing them.

As Kaplinsky points out:

> unless all, or most of the 'electronic jigsaw' is in place, the systemic advantages of automation are difficult to capture ... (and) ... in some parts of the third world only those technologies associated with product characteristics are diffusing. This obviously poses obstacles for longer-run systemic competitiveness.[33]

## The Human Resources Issue

The human resources issue is particularly important for new information technologies. In electronics this is perhaps more pertinent than in any other industrial sector. As O'Connor has argued, 'the computer industry is essentially a knowledge-intensive industry wherein skilled, highly trained scientific, engineering and technical labour power is probably the single most important asset.'[34] However, only when – particularly in Japan – new automation technologies started to be used in combination with radically distinct forms of organization, was their disrupting potential to the old paradigm fulfilled. Thus the advantages of the new technologies can only be exploited,

> when the process of production is understood and controlled at all levels of the work force. In the most successful firms today the role of production workers is shifting from one of passive performance of narrow, repetitive tasks to one of active collaboration in the organization and fine-tuning of production ... skill and flexibility of experiences in a variety of assignment and broadening responsibilities.[35]

Contrary to what happens with rigid automation, new occupational profiles and requirements are such that the importance of manual ability decreases and a broader knowledge about the production process, cooperative attitudes and intellectual abilities typical of good general education gain importance. These new expectations regarding the performance and participation of the workforce led to the valorization of certain basic skills which, independent of the 'specialized' knowledge of the direct worker, have to be part of occupational profiles. In short, these are:

- capacity to read and understand manuals, electronic panels, etc;
- capacity to write documents, reports, etc;
- capacity to talk and to communicate with superiors, colleagues and subordinates;
- capacity to work with computers, to interpret numbers, to measure time, distance, volume, etc;
- ability to understand, organize and analyse quantitative problems;
- capacity to identify and define problems, to formulate alternatives, compare different solutions and evaluate results;
- creativity, initiative, inventiveness, use of intuition and logical reasoning, to transform ideas into practical applications;
- self-esteem, motivation and capacity to take responsibilities;
- capacity to bargain and counter-argue and to collaborate.[36]

Thus the industry is characterized by an enormous demand not only for skilled labour but, most importantly, for labour with very good basic, general skills. The shortage of this type of labour has been well documented for the advanced industrial countries, particularly in software and hardware skills but also in areas of management.[37]

From the perspective of developing countries, the 1980s saw an important development. The old paradigm of labour demand can still be found in certain regions and sub-sectors of electronics in developing countries, but there has also been a decisive transition to what may be termed a new paradigm of demand for labour. This transition had been amply documented by Hewitt.[38] It may be characterized as using a much wider range of skills and, in particular, a high proportion of skilled engineering and technical labour. The origin of this increased demand for skills is the attempt by developing countries to establish a local high-tech sector and to stimulate a local capability in new technologies. Such analysis reinforces the argument that labour in high-tech needs to be viewed as a *resource* as opposed to a *cost*.

Building up this base of human resources is a pressing issue for some developing countries as they attempt to establish an expanding and sustainable local electronics industry and to use the new technologies efficiently. However, it is not clear that this is occurring since, as with scientific and technological expenditures, education expenditures are faltering in most developing countries.[39]

## 5. The Crisis of the 1980s: Trends in Science and Technology in the Developing World

It was emphasized that if developing countries are to appropriate the benefits associated with IT they have to take into account that the required skills have to be created; they do not necessarily build up from previously accumulated experience. In other words, developing countries will need an extensive upgrading in number and quality of human resources if they want to succeed in sharing the potential benefits of these revolutionary technological developments. Expenditures in science and technology become increasingly important for developing countries in the context of the new IT-based techno-economic paradigm.

However, the decade when these new technologies started to diffuse in the developed world also witnessed a widening disparity between North and South. In fact, during the 1980s the gap between rich countries and poor countries increased. For example, per-capita income in Latin America and Africa as a percentage of per-capita income in OECD countries diminished substantially in the decade. Also, the scientific and technological gap between the developed and the developing world widened during the 1980s. The period in which considerations regarding the strengthening of the scientific and technological infrastructure should have prevailed over short-run concerns was precisely the one in which such infrastructure underwent a rapid process of obsolescence.

Economic policies in most developing countries, particularly in Latin America and Africa, in the 1980s consisted of successive rounds of crisis management. The need to manage the external and internal debt and to control inflation dominated the government agenda in these countries. Pressing needs for short-term stabilization crowded out work on long-term economic and technological strategy. Concern with industrial strategy and science and technology policies existed, but there was not enough space to sustain such work. Science and technology expenditures have, as a consequence, fallen dramatically in these countries.

**TABLE 2.1** World R&D expenditures (estimated expenditure in US$ million by region, 1970–90)

|  | 1970 | 1975 | 1980 | 1985 | 1990 |
|---|---|---|---|---|---|
| USA and Canada | 27 620 | 38 382 | 66 796 | 115 882 | 193 721 |
| Europe | 15 739 | n/a | 70 712 | 65 540 | 104 956 |
| Asia | 4 540 | 12 304 | 31 726 | 47 188 | 91 218 |
| Latin America and Caribbean | 498 | 1 686 | 3 635 | 3 062 | 2 860 |
| Arab states | 115 | 334 | 3 824 | 3 465 | 3 078 |
| Africa | 105 | 300 | 784 | 620 | 746 |
| Oceania | 497 | n/a | 2 147 | 2 115 | 2 984 |

NOTE: USSR not included.
SOURCE: UNESCO, *Statistical Yearbook* (various issues).

Tables 2.1 and 2.2 show the evolution of world R&D total expenditures by region from 1970 to 1990, according to UNESCO statistics. The two decades present very different patterns regarding regional distribution of R&D expenditures. During the 1970s, although all regions increased their absolute R&D expenditures, the gap between more developed and less developed countries decreased. The share of the USA and Canada in the world's R&D expenditures (excluding USSR and Eastern Europe) fell from 56 per cent in 1970 to 38 per cent in 1980. In all remaining regions absolute growth in R&D expenditures was accompanied by a relative increase: in Europe from 32 per cent in 1970 to 40 per cent in 1980; in Asia from 9 per cent to 15 per cent; in Latin America and the Caribbean from 1 per cent to 2.2 per cent; and in Africa from 0.25 to 0.47 per cent.

In the 1980s, however, a very different trend was observed. North American and Asian countries experienced significant growth in absolute terms and increased their share in the world's R&D expenditures. The USA and Canada spent approximately US$193 billion in 1990 (48.8 per cent of the total) as compared to US$66 billion in 1980 (38 per cent). Asian countries almost trebled their R&D expenditures from US$31 billion in 1980 (16 per cent) to US$91 billion in 1990 (22 per cent). European countries increased their expenditures from US$70 billion in 1980 to US$105 billion in 1990, but their share of the world's efforts decreased from 40 per cent in 1980 to 26 per cent in 1990.

**TABLE 2.2** World R&D expenditures (% distribution by region, 1970–90)

|  | 1970 | 1975 | 1980 | 1985 | 1990 |
|---|---|---|---|---|---|
| USA and Canada | 56.258 | 42.337 | 37.988 | 49.363 | 48.803 |
| Europe | 31.985 | 40.201 | 40.118 | 27.926 | 26.454 |
| Asia | 9.229 | 13.568 | 15.976 | 18.772 | 22.349 |
| Latin America and Caribbean | 1.011 | 1.884 | 2.130 | 1.275 | 0.684 |
| Arab states | 0.253 | 0.377 | 2.130 | 1.506 | 0.798 |
| Africa | 0.253 | 0.377 | 0.473 | 0.232 | 0.228 |
| Oceania | 1.011 | 1.256 | 1.183 | 0.927 | 0.684 |

NOTE: USSR not included.
SOURCE: UNESCO, Statistical Yearbook (various issues).

So far as African countries and the Latin America and Caribbean region are concerned, both experienced a sharp decrease in relative terms. In Africa, the estimated R&D expenditure for 1990 was US$746 million (0.22 per cent), slightly less than the US$784 million spent in 1980 (0.47 per cent). As for Latin America and Caribbean countries, expenditures of US$3.6 billion in 1990 (0.7 per cent) compares with US$2.8 billion in 1980 (2.1 per cent).

With regard to policies for science and technology and R&D, the experience of developing countries during the 1980s differed from one region to another. In Latin America and Africa, crisis conditions brought diminishing commitments by governments to science and technology infrastructure. It is true that most Southeast Asian developing countries were able to increase significantly their R&D efforts both by government and private capital. In Korea, for example, the rate of R&D expenditures to GNP increased from 0.65 per cent in 1981 to 1.93 per cent in 1987, as a direct result of industrial and technological policies.[40] The augmenting of R&D expenditures is not, by itself, a sufficient condition for achieving success within the new techno-economic paradigm. The export-oriented strategy of Korea, for example, is running into problems that are linked to trade and technology. In the words of a Korean economist: 'We have to compete with China and South East Asian countries on price while our product quality and our technology seems to be far behind that of advanced nations'.[41]

To sum up, with the exception of some East Asian developing countries, the 1980s witnessed a trend towards decreasing (as in Latin America) or stagnating (as in Africa) expenditures in science and technology. But just as important is the trend in such countries towards a very low overall commitment by the private sector to R&D.

## 6. Conclusions

For four decades the international economic system has been governed by rules established by the dominant powers at the end of the Second World War. Such rules encompassed multilateral institutions and agreements covering most areas of trade (GATT), international financial management (Bretton Woods rules and the IMF), and certain areas of development (such as those under the jurisdiction of the World Bank). Other areas such as international business activities and technological transactions were not addressed at the multilateral level, but remained the preserve of bilateral arrangements constrained by individual national policies.

In the late 1980s, under the significant changes in the economic and political balance of power among developed countries, a new series of international norms were negotiated. Examples include not only new bilateral and regional initiatives (such as the various US–Japanese agreements and the EEC) but also attempts to change multilateral agreements (such as the Uruguay round of GATT). It is true that developing countries have participated more actively in these new multilateral agreements than they did in the immediate postwar arrangements. However, these parallel developments have been negotiated under two very distinct principles, which relate to the new technologies and trade. The North–North bilateral and regional agreements have been increasingly dictated by a concept of 'fair trade', whereby access to markets is dependent upon the effects it has on the economic structure of the recipient countries/regions. In North–South relations the old concept of 'free trade' has been pushed further and concentrated in areas related to the new technologies (such as intellectual property rights and services).

This dichotomy means that protectionism and market liberalization are treated unevenly in North–South relations. Very mature industries and core novel technologies are protected in the North while various forms of market access are sought in the South, which, in turn, faces a different horizon of protectionist preferences, including those technology-related. In essence, the South is again considered as purely a market for products/technologies developed in the North. Even when we consider the almost insignificant amount of aid 'given' by advanced countries, trade is prominent. A substantial amount of bilateral aid is tied. Recipient countries are required to buy goods and services from donor countries. Approximately two-thirds of all aid supplied by developed countries falls into this category.[42] Also, these countries prefer to finance physical capital installations that help their own firms and exporters, and they are reluctant to support the operating costs of aid-funding undertakings.

Throughout this paper, the following points have been stressed:

- Science and technology become increasingly important for developing countries in the context of the new techno-economic paradigm.
- The required skills for efficient use of the new information technologies have to be created; they do not necessarily build up from previously accumulated experience.
- However, the scientific and technological gap between the developed and the developing world widened during the 1980s. The period in which considerations regarding the strengthening of the scientific and technological infrastructure should have prevailed over short run concerns was precisely the one in which such infrastructure underwent a rapid process of obsolescence.
- Two of the main factors that have changed the environment for the development of new technologies – global interdependence in the economic and technological spheres, and growing involvement of governments in industrialized and industrializing countries in policies to promote technical change and innovation – are blurring the boundaries between technology and trade policies. In other words, trade policies, particularly in advanced countries, have been increasingly dictated by technology issues and vice versa.
- Significant changes are occurring in the composition of international trade. One consequence of these structural changes – which are part and parcel of the IT technological revolution – has been a significant reduction in the complementarity which existed between developed and developing countries.
- Reliance on foreign technology should not substitute for an intense, costly and long *internal* commitment to science and technology. More specifically, the existence of local scientific and technological infrastructure, R&D efforts by local firms, and 'networks' of innovating firms are preconditions not only for a national system of innovation but also for the efficient use of the new technologies.

- In order to exploit the potential benefits of new information technologies, developing countries should be able to develop capabilities to produce and use them.
- In the new techno-economic paradigm users of new technologies in the developing world are particularly important, not only as 'receivers' of technical progress generated in advanced economies. Traditional mechanisms for 'transferring' technology may be necessary conditions for attaining efficient levels of use of new technologies, but they are not sufficient. As users seem now to play a more active role in development and use of new technologies, optimum conditions should include strategies to give users a comprehensive knowledge about their technology and, whenever economic conditions permit, strategies to ensure a local production capacity of selected new technologies in order for user–production relations to emerge.
- Price and other conditions for technology transfer may change as a consequence of specific characteristics of new technologies. Technological solutions tend to be more 'firm-specific', which limits the scope for 'transferring' technology. Also some features of the innovation process of new technologies – particularly the growing importance of basic research and collaborative arrangements – may hinder the access of developing countries to imported technology.

We argued above that efficient utilization of the new technologies is dependent upon the capacity the recipient economy has to understand fully the various characteristics of new information technologies. If most advanced countries keep considering less developed ones only as potential markets, the benefits accruing from new technologies would barely spread beyond the developed world. The following is a tentative list of actions for an effective use of new technologies by developing countries. At the level of developing countries it is clear that a significant increase in overall expenditures in education and science and technology is more than desirable. It has to be recognized, however, that this is not a straightforward endeavour, particularly because most of these countries face severe problems of poverty and are engaged in short-term crisis management. In such an environment, long-run commitments tend to be relegated to the future and a vicious circle is perpetuated. International organizations have an important role to play here, and the same vigour they show in forcing developing countries to pursue policies for 'structural adjustments' should be used towards long-run policies for science and technology.

Among these policies, government support for technology infrastructure is important; but what is crucial are policies to encourage the private sector to increase R&D expenditures. The problem, however, is that these policies should include tax and other incentives which became less fashionable during the last decade.

Increase in expenditures, however, is just a small step towards an efficient use of new technologies by developing countries. Even if it can be achieved, as the example of South Asian countries shows, the result does not necessarily mean that new technologies will be used to solve basic problems of shelter, health, energy, and so on, or even to develop technologies and products more suitable to local conditions. On the contrary, using new technologies in developing countries with the basic aim of increasing exports and generating trade surpluses may bring further pressures on the environment.

The efficient use of new technologies by developing countries should be directed primarily towards the *exploitation of local and regional markets*. It should also be emphasized that such 'efficient use' bears no relation to the concept of 'appropriate technology'. It is worth recalling that, although fashionable among some Western institutions, the 'appropriate technology' movement failed to gather significant support among developing countries precisely because of its implicit association with low-grade, second-class technologies.

Under the idea that new technologies should be directed towards local markets, and bearing in mind that technology is mostly firm-specific, international organizations should change their priorities in order both to include science and technology in their agenda and to shift their emphasis from export-oriented growth to a more balanced approach whereby local markets matter. Institutions in advanced countries could contribute more specifically by pursuing several courses of action.

Among them one could single out the establishment of programmes for scientific and technological cooperation with developing countries in new technologies targeted to specific areas/problems. Distinct from those already in existence, which concentrate on the exchange of scientific personnel and joint research efforts, these programmes should include firms and be aimed at products/processes directed to local markets.

Finally, an essential part of any effort should be the establishment of financial mechanisms to provide long-run resources for firms to invest in technology programmes. This is particularly important given that most developing countries have very imperfect capital markets, and the non-availability of these resources is normally associated with lack of investment in technology in these countries.

## Notes

1. SPRU, 'Global Perspective 2010: Tasks for Science and Technology – A Synthesis Report on the FAST-EC Programme', Science Policy Research Unit, University of Sussex, Brighton, 1992, p. 4.
2. D. Mowery and N. Rosenberg, *Technology and the Pursuit of Economic Growth* (Cambridge: Cambridge University Press, 1989), pp. 274–89.
3. US Department of Commerce, *The Competitive Status of the US Electronics Sector*, Superintendent of Documents (Washington DC: US Printing Office, 1990), p. 8.
4. In 1991 the ten largest Asian firms showed a 46 per cent decrease in their net profits (*Datamation*, 1 September 1992, p. 80), while the 100 largest US firms witnessed a 72.3 per cent fall in their 1991 net profits. It was widely publicized that IBM, DEC and UNISYS sustained in 1991 a combined loss of US$5 billion on sales of US$85 billion (*Datamation*, 15 June 1992, p. 16). The three largest European firms (Siemens Nixdorf, Olivetti and Groupe Bull) experienced proportionally larger losses: US$1.4 billion on sales of US$19 billion.
5. The basic features of the sector during the crisis period are an increasing diffusion of open systems and standards, the growing importance of distributed data-processing systems, the interchangeability of small and large systems, so-called 'downsizing' and an increase in the degree of internationalization of the industry via direct investment and the setting up of joint ventures with local partners.
6. Although the delivered computing power in the US economy has increased by more than twofold in the past two decades, and productivity, especially in the service sector – by far the most important user of IT systems – seems to have stagnated. Disillusionment with IT is evident in statements such as that by Robert Solow: 'We see computers everywhere except in the productivity statistics.' As a result users are becoming increasingly selective in their automation strategies and are participating more actively in the design and development of IT systems.
7. A. Aksoy, 'Computers Are Not Dynamos – Frontiers in the Diffusion of Information Technologies', *Futures*, vol. 23 (1991), no. 4, p. 405.
8. C. Freeman, 'Networks of Innovators: A Synthesis of Research Issues', paper presented at the International Interdisciplinary Workshop on Network of Innovators' (Montreal: University of Quebec, 1990), p. 32.
9. See C. Freeman, 'The Economics of Technical Change', *Cambridge Journal of Economics* (1993), for a recent review.
10. M. Bell and J. Cassiolato, 'Technology Imports and the Dynamic Competitiveness of the Brazilian Industry: The Need For New Approaches to Management and Policy', a report for the Estudo da Competitividade da Indústria Brasileira, IE/UNICAMP and IEI/UFRJ, University of Campinas and Federal University of Rio de Janeiro, Campinas and Rio de Janeiro, 1993.
11. P. David, *Computer and the Dynamo: The Unclear Productivity Paradox in a Not Too Distant Mirror*, paper presented at the OECD Seminar on Science, Technology and Economic Growth, Paris, 1992.

12. N. Katz and C. Shapiro, 'Network Externalities, Competition and Compatibility', *Discussion Paper*, No. 54, Woodrow Wilson School, Princeton University, Princeton, 1988; D. Allen, 'New Telecommunications Services: Network Externalities and Critical Mass', *Telecommunications Policy*, September 1988.

13. O. Williamson, 'Technology and Transaction Cost Economics: A Reply', *Journal of Economic Behaviour and Organisation*, vol. 10 (1988).

14. R. Kaplinsky, 'Industrial Restructuring in LDCs: The Role of Information Technology', paper prepared for Conference of Technology Policy in the Americas, Stanford University, Stanford, 1988.

15. E. von Hippel, *The Sources of Innovation* (Oxford: Oxford University Press, 1988).

16. B. Carlsson and S. Jacobsson, 'Technological Systems and Economic Performance: The Diffusion of Factory Automation in Sweden', in D. Foray and C. Freeman, eds, *Technology and the Wealth of Nations: The Dynamics of Constructed Advantage* (London: Pinter Publishers, 1993).

17. F. Sercovitch, 'Domestic Learning, International Technology Flows and the World Market: New Perspectives for the Developing Countries', WEP-2-22/WP 189, World Employment Programme Research, ILO, Geneva, 1989, p. 5.

18. G. Dosi, 'Sources, Procedures and Microeconomic Effects of Innovation', *Journal of Economic Literature*, vol. XXVI (1988).

19. G. Dosi, K. Pavitt and L. Soete, *The Economics of Technical Change and International Trade* (Hemel Hampstead: Harvester Wheatsheaf, 1990).

20. J. Metcalfe, 'Technological Innovation and the Competitive Process', in P. Hall, ed., *Technology, Innovation and Economic Growth* (Southampton: Camelot Press, 1986).

21. B.-Å. Lundvall, 'Innovation, the Organised Market and the Productivity Slow-down', paper presented at the International Seminar on Science, Technology and Economic Growth, Paris: OECD, 1989. Equally interesting are studies that show what happens in the absence of user/producer interactions. When producers dominated users (or when users had a limited technical competence) there was a tendency towards 'hyperautomation'; that is, users were faced with design values different from their needs and with capital goods too complex and costly. In such cases, instead of attaining productivity gains, automation led to the upsurge of diseconomies (ibid., pp. 16–17). Empirical studies in wastewater technology and office technology in Denmark showed how a lack of local user competence had a negative effect upon the systems developed (ibid.).

22. D. McKendrick, 'Information Technology and Performance in Indonesian Commercial Banking', School of Management, The University of Texas, Dallas, 1991.

23. J. Cassiolato, 'The User–Producer Connection in Hi-Tech: A Case Study of Banking Automation in Brazil', in H. Schmitz and J. Cassiolato, eds, *Hi-Tech for Industrial Development* (London: Routledge, 1992).

24. L. Soete and G. Dosi, *Technology and Employment in the Electronics Industry* (London: Frances Pinter, 1983).

25. L. Soete, 'International Diffusion of Technology, Industrial Development and Technological Leapfrogging', *World Development*, vol. 13 (1985), no. 3. This

*José E. Cassiolato* 63

also tends to happen for biotechnology (A.W.F. Anciães and J.E. Cassiolato, *Os impactos da biotecnologia no Setor Industrial*, Brasilia: CNPq, 1985) and advanced materials (H. Lastres and J. Cassiolato, 'High Technologies and Developing Countries: The Case of Advanced Materials', *Materials and Society*, vol. 14 (1990), no. 1).

26. V. Walsh, 'Technology and the Competitiveness of Small Countries: A Review', in C. Freeman, and B-Å. Lundvall, eds, *Small Countries Facing the Technological Revolution* (London: Pinter Publishers, 1988).

27. A. Aksoy, 'Innovation and Diffusion Dynamics of the Information Technology Paradigm', *Working Paper*, No. 11 (1990), Centre for Information and Communications Technologies, Science Policy Research Unit, University of Sussex, Brighton

28. K. Hoffman, *Technological Advance and Organizational Innovation in the Engineering Industry: A New Perspective on the problems and Possibilities for Developing Countries*, Report submitted to the World Bank (Brighton: Sussex Research Associates, 1988).

29. J. Bessant and B. Haywood, 'The Introduction of Flexible Manufacturing Systems as an Example of Computer Integrated Manufacturing', *Operations Management Review*, Spring 1986.

30. These types of dynamic efficiencies are acquired through the development of technological skills and knowledge in order to attain and retain competitive ability. For example, it has been established that the development of supplier–subcontractor networks has been a major source of technology accumulation and diffusion first in Japan and more recently in developing countries like South Korea (A. Amsden, *Asia's Next Giant: South Korea and Late Industrialisation* [Oxford: Oxford University Press, 1989]; OECD, *Draft Background Report*, Chapter 11, Technology/Economy Programme [Paris: OECD, 1990]). That is precisely why the early stages of the development of an innovation are so important for industrializing economies according to neo-Schumpeterians (C. Freeman, 'Innovation and Long Cycles of Economic Development', paper presented at the International Seminar on Innovation and Development in the Industrial Sector', University of Campinas, Campinas, Brazil, August 1982).

31. C. Antonelli, 'The International Diffusion of New Information Technologies', *Research Policy*, vol. 3 (1986).

32. Allen, 'New Telecommunications Services'.

33. Kaplinsky, 'Industrial Restructuring in LDCs', pp. 7–8.

34. D. O'Connor, 'The Computer Industry in the Third World: Policy Options and Constraints', *World Development*, vol. 13 (1985), no. 3, p. 325.

35. M. Dertouzos, R. Lester and R. Solow, *Made in America – Regaining the Competitive Edge* (Cambridge, Mass.: MIT Press, 1989), p. 137.

36. J.C. Alexim, 'Las nuevas fronteras de la formación profesional', *Revista Critica & Comunicación* (Lima: OIT, 1992).

37. See Freeman, 'Networks of Innovators'; R. Kaplinsky, '"Technological Revolution" and the International Division of Labour in Manufacturing: A Place for the Third World?', EADI Conference on New Technologies and the Third World, Institute of Development Studies, University of Sussex, Brighton, 1987;

Soete, 'International Diffusion of Technology'; and D. Ernst, *Automation, Employment and the Third World: The Case of the Electronics Industry*, IDPAD Project on Microelectronics, Institute of Social Studies, The Hague, 1985.

38. T. Hewitt, 'Employment and Skills in the Electronics Industry: The Case of Brazil', D.Phil. thesis, University of Sussex, Brighton, 1988.

39. T. Whiston, 'Education and Employment for a Sustainable World', paper prepared for the FAST-EC programme 'Global perspective 2010: Tasks for Science and Technology ', Science Policy Research Unit, University of Sussex, Brighton, 1993.

40. M. Hobday, *The Needs and Possibilities for Cooperation between Selected Advanced Developing Countries and the Community in the Field of Science and Technology – Country Report on the Republic of Korea*, Strategic Analysis of Science and Technology, Commission of the European Communities, Brussels, 1991. It is important to emphasize that the current idea among international lending institutions and some development experts in Western societies that governments should leave market forces to deal with industry/technology problems runs completely counter to what is happening in their countries. A review of OECD policies indicates that 'priority sectors are chosen; winners are picked up; R&D subsidies are handed out; rescue operations are launched; orders are guaranteed; selective trade protection is granted ... leading Western European companies in frontier industries call on their governments for five to seven year protection periods on top of that received through industrial property rights and quantitative restrictions in order to enjoy enough "breathing space" to upgrade facilities, develop new products and get in a better position to compete' (Sercovitch, 'Domestic Learning, International Technology Flows and the World Market').

41. *Financial Times*, 25 February 1992, p. 4.

42. World Bank, *World Development Report 1991* (Washington DC: World Bank, 1991), p. 128.

# References

Aksoy, A., 'Innovation and Diffusion Dynamics of the Information Technology Paradigm', *Working Paper*, No. 11 (1990), Centre for Information and Communications Technologies, Science Policy Research Unit, University of Sussex, Brighton.

Aksoy, A., 'Computers Are Not Dynamos – Frontiers in the Diffusion of Information Technologies', *Futures*, vol. 23 (1991), no. 4, pp. 402–14.

Alexim, J.C., 'Las nuevas fronteras de la formación profesional', *Revista Crítica & Comunicación* (Lima: OIT, 1992).

Allen, D., 'New Telecommunications Services: Network Externalities and Critical Mass', *Telecommunications Policy*, September 1988, pp. 257–71.

Amsden, A., *Asia's Next Giant: South Korea and Late Industrialisation* (Oxford: Oxford University Press, 1989).

Anciães, A.W.F. and J.E. Cassiolato, *Os impactos da biotecnologia no Setor Industrial* (Brasilia: CNPq, 1985).

Antonelli, C., 'The International Diffusion of New Information Technologies', *Research Policy*, vol. 3 (1986), pp. 139–47.
Bell, M. and J. Cassiolato, 'Technology Imports and the Dynamic Competitiveness of the Brazilian Industry: The Need For New Approaches to Management and Policy', a report for the Estudo da Competitividade da Indústria Brasileira, IE/UNICAMP and IEI/UFRJ, University of Campinas and Federal University of Rio de Janeiro, Campinas and Rio de Janeiro, 1993.
Bessant, J. and B. Haywood, 'The Introduction of Flexible Manufacturing Systems as an Example of Computer Integrated Manufacturing', *Operations Management Review*, Spring 1986.
Carlsson, B. and S. Jacobsson, 'Technological Systems and Economic Performance: The Diffusion of Factory Automation in Sweden', in D. Foray and C. Freeman, eds, *Technology and the Wealth of Nations: The Dynamics of Constructed Advantage* (London: Pinter Publishers, 1993).
Cassiolato, J., 'The User–Producer Connection in Hi-Tech: A Case Study of Banking Automation in Brazil', in H. Schmitz and J. Cassiolato, eds, *Hi-Tech for Industrial Development* (London: Routledge, 1992).
David, P., *Computer and the Dynamo: The Unclear Productivity Paradox in a Not Too Distant Mirror*, paper presented at the OECD Seminar on Science, Technology and Economic Growth, Paris, 1992.
Dertouzos, M., R. Lester and R. Solow, *Made in America – Regaining the Competitive Edge* (Cambridge, Mass: MIT Press, 1989).
Dosi, G. 'Sources, Procedures and Microeconomic Effects of Innovation', *Journal of Economic Literature*, vol. XXVI (1988), pp. 1120–71.
Dosi, G., K. Pavitt and L. Soete, *The Economics of Technical Change and International Trade* (Hemel Hempstead: Harvester Wheatsheaf, 1990).
Elson, D., 'Transnational Corporations in the New International Division of Labour: A Critique of the "Cheap Labour" Hypothesis', *Manchester Papers on Development*, vol. 4 (1988), no. 3.
Ernst, D., *Automation, Employment and the Third World: The Case of the Electronics Industry*, IDPAD Project on Microelectronics, Institute of Social Studies, The Hague, 1985.
Freeman, C., 'Innovation and Long Cycles of Economic Development', paper presented at the International Seminar on Innovation and Development in the Industrial Sector', University of Campinas, Campinas, Brazil, August 1982.
Freeman, C., 'Networks of Innovators: A Synthesis of Research Issues', paper presented at the International Interdisciplinary Workshop on Network of Innovators', Montreal: University of Quebec, 1990.
Freeman, C., 'The Economics of Technical Change', *Cambridge Journal of Economics*, vol. 18 (1994), no. 6, pp. 467–514.
Frobel, F., J. Heinrichs and O. Kreye, *The New International Division of Labour* (Cambridge: Cambridge University Press, 1980).
Hewitt, T., 'Employment and Skills in the Electronics Industry: The Case of Brazil', D.Phil. thesis, University of Sussex, Brighton, 1988.
Hobday, M., *The Needs and Possibilities for Cooperation between Selected Advanced Developing Countries and the Community in the Field of Science and Technology* –

*Country Report on the Republic of Korea*, Strategic Analysis of Science and Technology, Commission of the European Communities, Brussels, 1991.

Hoffman, K., *Technological Advance and Organizational Innovation in the Engineering Industry: A New Perspective on the problems and Possibilities for Developing Countries*, Report submitted to the World Bank (Brighton: Sussex Research Associates, 1988).

Jenkins, R., *Transnational Corporations and Uneven Development: The Internationalisation of Capital and the Third World* (London: Methuen, 1987).

Kaplinsky, R., '"Technological Revolution" and the International Division of Labour in Manufacturing: A Place for the Third World?', EADI Conference on New Technologies and the Third World, Institute of Development Studies, University of Sussex, Brighton, 1987.

Kaplinsky, R., 'Industrial Restructuring in LDCs: The Role of Information Technology', paper prepared for Conference of Technology Policy in the Americas, Stanford University, Stanford, 1988.

Katz, N. and C. Shapiro, 'Network Externalities, Competition and Compatibility', *Discussion Paper*, No 54, Woodrow Wilson School, Princeton University, Princeton, 1988.

Lastres, H. and J. Cassiolato, 'High Technologies and Developing Countries: The Case of Advanced Materials', *Materials and Society*, vol. 14 (1990), no. 1.

Lundvall, B.-Å., 'Innovation, the Organised Market and the Productivity Slowdown', paper presented at the International Seminar on Science, Technology and Economic Growth, Paris: OECD, 1989.

McKendrick, D., 'Information Technology and Performance in Indonesian Commercial Banking', unpublished paper, School of Management, The University of Texas, Dallas, 1991.

Metcalfe, J., 'Technological Innovation and the Competitive Process', in P. Hall, ed., *Technology, Innovation and Economic Growth* (Southampton: Camelot Press, 1986).

Mowery, D. and N. Rosenberg, *Technology and the Pursuit of Economic Growth* (Cambridge: Cambridge University Press, 1989).

Nochteff, H., 'New Technologies and Dependency: Towards the Development of Interdependence-management Capacities', *Development*, special issue: Communication, Participation and Democratization of the Media and Development, Rome, 1990.

O'Connor, D., 'The Computer Industry in the Third World: Policy Options and Constraints', *World Development*, vol. 13 (1985), no 3.

OECD, *Draft Background Report*, Chapter 11, Technology/Economy Programme, Paris: OECD, 1990.

Scott, A., 'The Semiconductor Industry in South East Asia: Organisation, Location and the International Division of Labour', *Regional Studies*, vol. 21 (1987), no. 2.

Sercovitch, F., 'Domestic Learning, International Technology Flows and the World Market: New Perspectives for the Developing Countries', WEP-2-22/WP 189, World Employment Programme Research, ILO, Geneva, 1989.

Soete, L., 'International Diffusion of Technology, Industrial Development and Technological Leapfrogging', *World Development*, vol. 13 (1985), no. 3.

Soete, L. and G. Dosi, *Technology and Employment in the Electronics Industry* (London: Frances Pinter, 1983).

SPRU, 'Global Perspective 2010: Tasks for Science and Technology – A Synthesis Report on the FAST-EC Programme', Science Policy Research Unit, University of Sussex, Brighton, 1992.

US Department of Commerce, *The Competitive Status of the US Electronics Sector*, Superintendent of Documents (Washington DC: US Printing Office, 1990).

von Hippel, E, *The Sources of Innovation* (Oxford: Oxford University Press, 1988).

Walsh, V., 'Technology and the Competitiveness of Small Countries: A Review', in C. Freeman and B-Å. Lundvall, eds, *Small Countries Facing the Technological Revolution* (London: Pinter Publishers, 1988).

Whiston, T., 'Education and Employment for a Sustainable World', paper prepared for the FAST-EC programme 'Global perspective 2010: Tasks for Science and Technology ', Science Policy Research Unit, University of Sussex, Brighton, 1993.

Williamson, O., 'Technology and Transaction Cost Economics: A Reply', *Journal of Economic Behaviour and Organisation*, vol. 10 (1988), pp. 355–63.

World Bank, *World Development Report 1991* (Washington DC: World Bank, 1991).

# 3

# The Advanced Materials Revolution: Effects on Third World Development

*Helena M.M. Lastres*

## 1. Introduction

The group of advanced materials includes ceramics, polymers, metals and composites. These were introduced into the world economy on a significant scale in the second half of the twentieth century; they are characterized by a high degree of purity, high technical performance features, and by high value added in the production process.

Many differences have been pointed out in the literature between the resource-intensive bulk standardized materials and the advanced materials. In comparison with traditional materials, AMs are supposed to require less mineral input and can contribute to saving energy. At the present stage of their progress and diffusion into the world economy, one major characteristic of AMs is their customized, specific applications. The development and production of AMs are recognized as demanding a closer interaction with their market and a much more flexible form of industrial organization. Their markets are regarded as the fastest growing of the materials markets. R&D and also the production of AMs are activities carried out mainly by major potential users and some specialized firms rather than by traditional materials producers. The production of AMs, which typically is highly information-intensive, requires specific types of inputs, technology and organizational structures.[1]

The development of AMs has importantly influenced mineral and materials markets, and such changes are bound to affect the Third World in several ways. The basic objective of this chapter is to discuss and help understand the present changes and main impacts on developing countries. Section 2 analyses the recent trend in patterns of materials usage during the 1970s and 1980s, contrasting the break in the trend of production and consumption of major bulk materials,

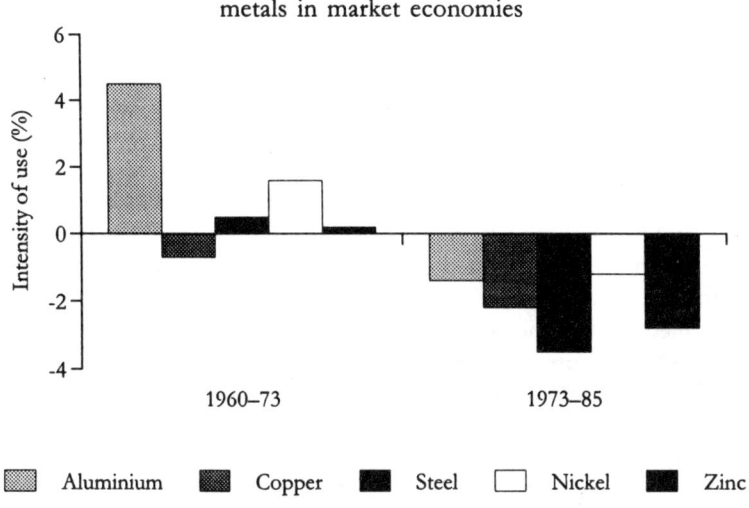

**FIGURE 3.1** Declining trend in the intensity of use of major metals in market economies

[Bar chart showing Intensity of use (%) for periods 1960–73 and 1973–85, with bars for Aluminium, Copper, Steel, Nickel, Zinc]

☐ Aluminium  ■ Copper  ■ Steel  ☐ Nickel  ■ Zinc

SOURCE: H.M.M. Lastres, *Advanced Materials Revolution and the Japanese System of Innovation* (London: Macmillan, 1994).

especially metals, with the rise of new materials. Section 3 summarizes the main ideas that underpin the debate about the so-called present materials revolution. Section 4 discusses the ways in which the recent changes in the materials sector may influence Third World development, while section 5 discusses recent Brazilian experience in defining S&T policies for the development of advanced materials.

## 2. Break in the Trend of the Demand and Supply of Materials

Among materials economists, there is no agreement about the nature, importance, causes and consequences of the changes that have occurred in the materials sector during the last two decades. However, there is a reasonable consensus that from the 1970s onwards a shift in the trend of the advanced countries' consumption of raw and traditional materials has taken place.[2]

The rather abrupt and general slowdown in the growth of metal demand has attracted considerable attention. However, as Auty pointed out,[3] it was ironically the concern in the early 1970s about the shortages of critical resources – which could result from the exhaustion of physical reserves or from the emergence of developing country cartels – that spawned several investigations into patterns of demand growth of materials inputs.

One argument produced by some analysts of these changes related to the link between these trends and the temporary difficulties the world economy faced at the beginning of the 1980s. However, as comparison between metals consumption and other different aggregate indices of economic performance indicated, such trends could not simply be considered as a conjunctural shift. In the last few years, several studies have aimed at measuring and explaining the recent changes by analysing the trend in the intensity of use of major metals – defined as the ratio of their consumption to gross domestic product (GDP).[4] As Figure 3.1 illustrates, in market economies, with the exception of copper, the annual rate of growth of intensity of use was rising before 1973; and has since been declining for all five metals considered – steel, aluminium, copper, nickel and zinc.

The main attempts made to explain this break in trend were based on the conclusion that, over time, intensity of metal use was declining because (i) consumer demand was shifting towards metal-saving services and industries; and (ii) the same goods were being produced with fewer or different materials. For instance, the amount of steel used in the car industry may decline due to the introduction of resource-saving new technologies or to the substitution of new engineering plastics and composites as alternative materials for steel.

The most important conclusions to emerge from a series of case studies analysing the changes in the material composition of products and services particularly stressed: (i) the importance of the increasing share of computers and other products which require far fewer pounds of basic materials per unit of added value than motor vehicles and other more traditional manufactured products; and (ii) the dramatic impact that polymers, advanced ceramics and composites have had on the material composition of many products consumed not only in the sectors with highly increasing shares in the economy – electronics, informatics and telecommunications – but also in others, such as the automobile industry, beverage container industry, and so on.[5]

Even relying on precarious data, in the late 1970s and early 1980s, a number of analyses revealed that the advanced materials were experiencing high rates of growth. In the mid-1980s, a proliferating number of reports expressed the same conclusion and also high expectations in terms of their future growth.[6]

**FIGURE 3.2** World consumption of engineering plastics

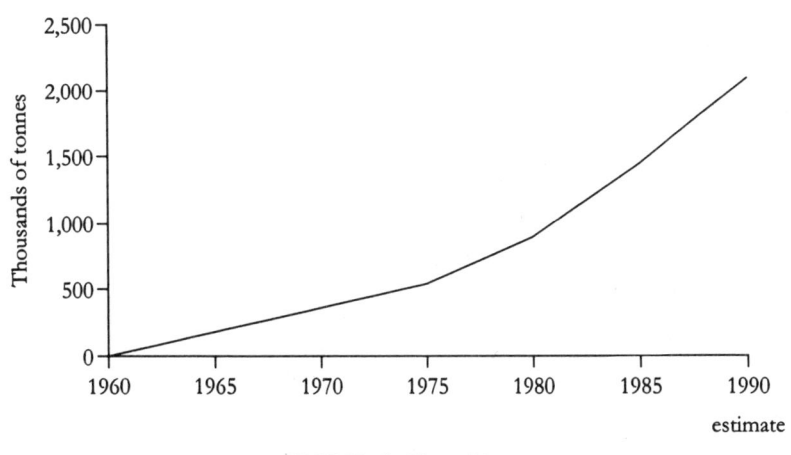

*[Line graph showing Thousands of tonnes on y-axis (0 to 2,500) versus years on x-axis (1960 to 1990 estimate). The curve rises from near 0 in 1960 to approximately 2,100 in 1990.]*

SOURCE: As Figure 3.1.

Already in 1979 in the USA, the volume of production of plastics in general surpassed that of steel. Among the former, engineering plastics present the highest rates of growth. Since their advent in the 1960s, their world consumption has increased significantly and was estimated to exceed 2 million tonnes by the beginning of the 1990s (see Figure 3.2). According to OECD, in 1986 the US was responsible for 37 per cent of the consumption of these materials, Western Europe for 35 per cent and Japan for 17 per cent.[7]

In terms of advanced ceramics, although most of the scientific breakthroughs which led to new materials and processes were produced in the USA and the UK years earlier, the beginning of the 1980s marked world-wide recognition of Japanese leadership in this area. In Japan, the sales of ceramic IC packages and substrates alone were estimated at ¥ 117 billion (US$59 million) in 1980. These sales – which represented less than 28 per cent of Japanese total sales of advanced ceramics at that time – were increasing at rates higher than 55 per cent per year at the turn of the decade.[8]

According to US sources, in 1988, the world market for advanced ceramics was valued at US$12 billion (Japan with 57 per cent of this market, the USA with 31 per cent and Western Europe with 12 per cent). Japanese sources have estimated a much higher market share for

**FIGURE 3.3** Advanced ceramics production in Japan

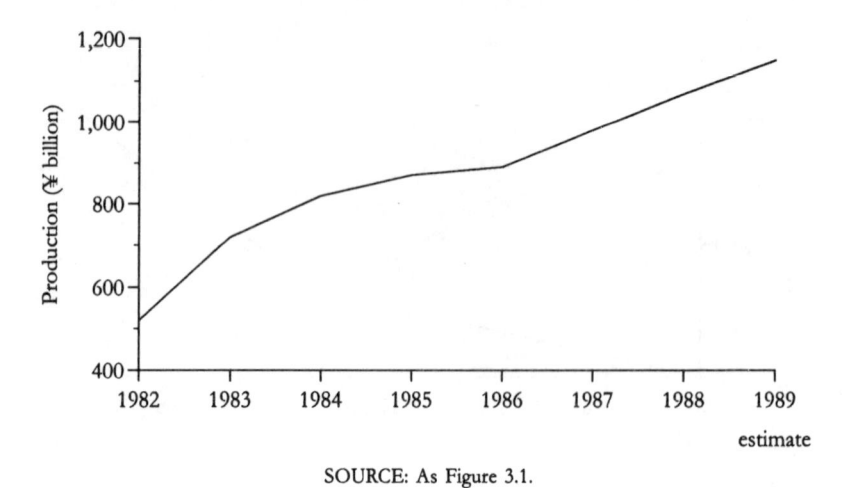

[Line chart: Production (¥ billion) vs year 1982–1989, rising from ~520 in 1982 to ~1,150 in 1989 (estimate).]

SOURCE: As Figure 3.1.

their country and also a bigger world-wide market for advanced ceramics. According to the Japan Fine Ceramics Association (JFCA), the value of Japanese production of advanced ceramics almost tripled from 1983 to 1989, to an estimated ¥ 1.2 trillion, as illustrated by Figure 3.3.

This set of figures suggests that a real shift in the intensity of different materials usage has been taking place during the two last decades. The trend in the previous growth of materials consumption has been broken and nowadays there is a more rapid rise of a particular group of materials, and a new term – advanced materials (AMs) – was coined to identify it.

## 3. Advanced Materials Revolution

The ability to manipulate raw material inputs has increased several times in the past. Different groups of materials have succeeded each other in forming the basis of successive techno-economic paradigms. The last two decades have witnessed the development of a particular group of materials closely linked to the development of information technology.[9]

With the advent of a new techno-economic paradigm a set of new requirements was imposed on the economy as a whole. Among the main changes the materials sector had to face were: demand for products with new properties and of better quality; larger variety; greater flexibility; new consumer standards; energy and mineral conservation; and compliance with environmental regulations. These new requirements had a catalytic effect on the research and industrialization of new materials developed between the 1950s and 1970s.[10]

One fact has been recognized as the most important among those leading to significant improvements and developments in materials analysis, design, processing, testing and applications. This fact is the deepening of the understanding of the general rules which determine how atoms and molecules combine together within matter, allowing the intervention at the molecular and atomic level of matter and the rearrangement of the microstructure of materials to create entirely new and synthetic materials displaying the desired combinations of properties and performance; and allowing the development of theoretically predictable materials, built atom by atom. Therefore the major difference between earlier booms in materials and the present so-called advanced materials revolution is that properties can now be designed. As one consequence, this materials revolution – which relies on advances in both organic and inorganic materials – presents a quantum leap in terms of number and range of materials offered in the market. In fact, what has been happening (especially during the last 20 years) is that, with the advanced materials revolution, a huge increase in variety, properties and range of applications of materials has become a reality.

Nowadays, new polymers and composites with sophisticated mechanical functions are competing more and more with metals in structural applications, especially those which require high strength combined with durability, elasticity, heat resistance and light weight. Advanced ceramics are displacing metals and are being used as cutting tools materials and also in parts of diesel engines. Other special ceramics and some new polymers can also compete with metals in applications which require thermal, optical, magnetic, electrical and electronic functions. The progress of metals itself has led to the development of new metallic materials – such as single crystal superalloys and superplastic alloys – with improved properties and functions, which make them a substitute for past generations of structural metals and, at the same time, competitors with other classes of advanced materials.

In every materials boom there is partial substitution of materials (as when iron was substituted for stone and wood, steel was substituted for iron, plastics were substituted for wood or steel, and so on) because these new materials have a better quality for some applications and sometimes lower prices. But there are also new properties and functions being offered. Therefore there is a combination of substitution and new functions. In the same way, the so-called present advanced materials revolution is introducing a great deal of substitution (for lower quality and lower-performance older materials). Also a substantial increase in the range of new functions is now offered by the development of materials such as:

- the advanced ceramics and polymers used in optoelectronics;
- electric conductive polymers used as electric wires and in batteries;
- histocompatible polymers, which are used to produce artificial bones and organs;
- selectively permeable polymeric membranes, which have introduced new separation processes based on dialysis and reverse osmosis, influencing the industries which rely on these processes (such as the separation of gases, purification of water, desalination of sea water, separation of enzymes from microbial broth, and so on), and have also allowed the development of drug delivery systems (which use a microporous membrane to introduce drugs through the skin);
- amorphous alloys (produced by rapid solidification processing) with their unique soft magnetic properties which allow their application in electronics, power distribution, motors and sensors;
- hydrogen-absorbing alloys which absorb and discharge hydrogen with changes in temperature and are the basis of the hydrogen car fuel storage system;
- shape memory alloys, which, after having been deformed, can revert to their original shape when submitted to a predetermined temperature, and which are being used in various kinds of valves (closing mechanisms for air conditioning, coffee makers, greenhouse windows, fire shutters, grill shutters, radiator fan switches, and so on) and in artificial joints.

In addition to the increase in the number of alternative materials displaying a variety of functions, the new substances and processes developed during recent years led to an *exponential* improvement in the properties offered by the AMs, compared to earlier materials. The result of the efforts of materials scientists and engineers over this period has also been illustrated by a number of examples given various indicators of engineering measurement. Among the most prominent examples of the exponential improvement in properties and functions of specialized materials are:

**FIGURE 3.4** Evolution of transition temperature of the best superconducting material since the discovery of the phenomenon

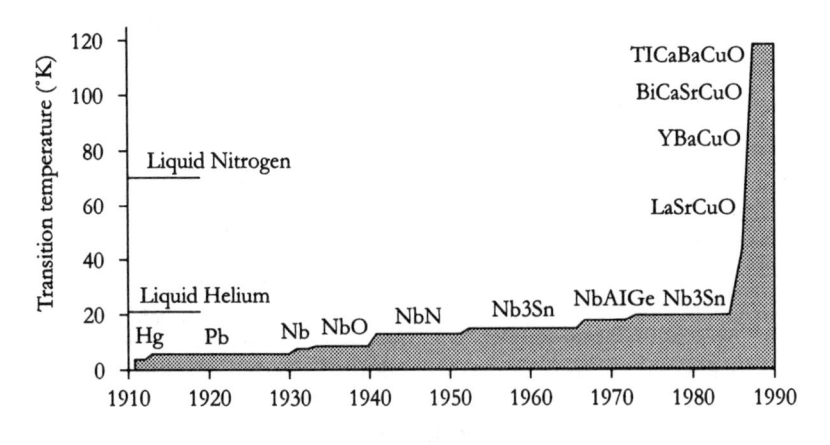

[Graph: Transition temperature (°K) vs. year 1910–1990. Materials labeled: Hg, Pb, Nb, NbO, NbN, Nb3Sn, NbAlGe Nb3Sn, LaSrCuO, YBaCuO, BiCaSrCuO, TlCaBaCuO. Reference lines for Liquid Helium and Liquid Nitrogen.]

SOURCE: As Figure 3.1.

- the increase in the transition temperature of superconductors from 23 K in the 70s to 125 K in the late 1980s. The discovery of the high temperature ceramic superconductors is promising cheaper energy (they can conduct electricity without resistance, i.e. losses), revolutionary means of transport (using electromagnetically levitated bullet trains and battery-powered cars with highly efficient electromotors), as well as important improvements in microelectronics, computers, telecommunications, medical diagnosis equipment, satellites and mineral processing, among others;
- the great improvement in the transparency of silica glass achieved from 1966 onwards with the development of optical fibres, which has revolutionized the telecommunications industry. These fibres are now about 100 orders of magnitude more transparent than they were in 1966. A single fibre 0.001 mm in diameter can transmit thousands of telephone conversations, i.e. many more than a thick conventional cable;
- the increase in the strength-to-density ratio of structural materials. As Figure 3.5 shows, advanced composites and polymer fibres represent a significant improvement on modern metals. The results of these advances are being experienced, for instance, in aeroplanes and automobiles that use less fuel and go faster;

**FIGURE 3.5** Progress in materials' strength-to-density ratio as a function of time

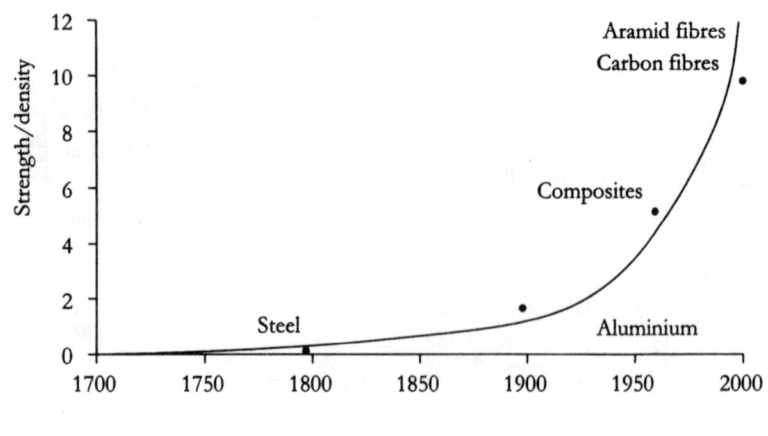

[Graph showing Strength/density (y-axis, 0 to 12) vs. year (x-axis, 1700 to 2000), with data points labeled: Steel, Aluminium, Composites, Aramid fibres, Carbon fibres]

SOURCE: As Figure 3.1.

- the growth in the efficiency with which heat energy is converted to mechanical or electrical energy in engines and power plants by the development of materials that are stronger at higher temperatures. Superalloys can now operate at temperatures of over 2000° F and advanced ceramics may allow engines to operate at temperatures up to 2500° F, The maximum theoretical efficiency of such engines is about 80 per cent, while the efficiency of conventional engines is limited to about 60 per cent. The foremost consequence is the possibility of a more efficient production of energy with a concomitant reduction in cost, fuel requirements and pollution.[11]

Thus, in the last two decades a formidable increase in variety, properties and range of applications of materials has become a reality, producing a radical change in the conceptual basis of materials production. As stressed above, a main qualitative difference of these changes is that it is now possible to produce synthetic materials displaying the required properties by rearranging the microstructure of the materials.

As a result of the recent advances, the focus of the new production systems tends to concentrate more and more on specifications to be met and capabilities to be realized, instead of type of materials (or mineral input) to be used. In other words, a given product no longer relies on a given material or on a given input. Instead, several materials compete to assume a given function (the concept of 'hyperchoix des matériaux'):

> overchoice means that for a given product not one but several materials can be adopted. In other words, the variety offered by new materials increases the number of degrees of freedom of economic choice regarding materials. An example illustrating this overchoice is the rivalry among aluminium, high yield point steels, reinforced plastic sandwiches in car body parts... The material, formerly a limitation, has become an optimizable variable which can be adapted to the most stringent requirements in a set of specifications.[12]

With the new paradigm, a real industrial changeover is occurring in a way which is both creative and destructive, and which requires a more flexible production system based on a wider variety of materials, properties and processing methods, and on higher levels of information. With AMs the economy has significantly expanded the variety of feasible alternatives as a response to these new market requirements.

Because of the close interdependence between information technology and advanced materials, the consolidation of the new paradigm depends on the emergence of the AMs, while the profitability of the latter depends on information (from their research to their production and commercialization, as well as throughout the life of the products which incorporate these materials). The rapid development and diffusion of advanced materials resulting from the pressures imposed by the emerging new techno-economic paradigm simultaneously became the instrument which made the new situation feasible and coherent.

## 4. Effect of Advanced Materials on Third World Development

The above-mentioned changes comprise a whole process of industrial restructuring, affecting patterns of investment, technological accumulation, industrial organization, employment and trade; they also involve rethinking intrafirm relations and those between the different actors. Therefore they have various consequences for the less developed countries (LDCs), especially those which are important producers of traditional materials and ores. The most visible one is the expected negative impact on the balance of trade of these countries.

As pointed out above, from the 1970s onwards a change in the trend of consumption of raw and traditional materials has been observed. In particular, the analysis of world consumption of the seven major metals shows that their yearly rate of growth turned negative after 1979.[13] As illustrated by Figure 3.6, the crisis that started in the early 1970s put an end to the high rates of growth experienced by most major metals during the 1950s and 1960s. This slowdown was further deepened by the recession that took place in the early 1980s. It is worth stressing that such a declining trend does not necessarily imply an absolute decline in the volume of metals consumed in the world. But it does mean that the consumption of such products is no longer increasing at the same rate as it did in the past. As stressed by Gonzales-Vigil,

> The significance of this fall is paramount indeed, as the seven metals [analysed in Figure 3.6] represent around three quarters or more of the value of all metal minerals in the world economy and, in particular, they accounted for three quarters of developing countries' exports of all non-fuel minerals in the mid-70s and together with manganese ore, for 53 per cent of the total non-fuel mineral export earnings received by developing countries in 1980.[14]

It is also expected that the materials industry's current international division of labour will continue for some time, but not for long. Developing countries are expected to be the fastest growing market for the major metal raw materials produced by themselves. It is also pointed out that the intensity of metal use is likely to rise only in those LDCs where per-capita income is growing rapidly, for in such countries the shift in demand toward metal-intensive goods will more than offset the tendency for material substitution and resource-saving technologies.[15] Thus, while in the past a greater rate of consumption of major metals used to express a greater level of industrial development within different countries, today it means the opposite. Low consumption growth of metals is now considered to be a characteristic of mature developed economies.

But if it is really true that the dynamic axis of industrial growth is moving away from the production of basic, traditional and raw materials, probably worse than this is the dramatic depression in price trend such materials are experiencing. As the World Bank has demonstrated, such a declining trend can be seen as part of the general tendency for commodity prices (excluding petroleum) in the last decades.

Figure 3.7 shows the weighted index of metal and mineral prices for the period 1950–2000. From it we notice that the price index of these commodities has been experiencing a remarkable decline. It is also worth emphasizing that, after the 1980s, short-run price increases were always followed by greater price decreases. An improvement of about

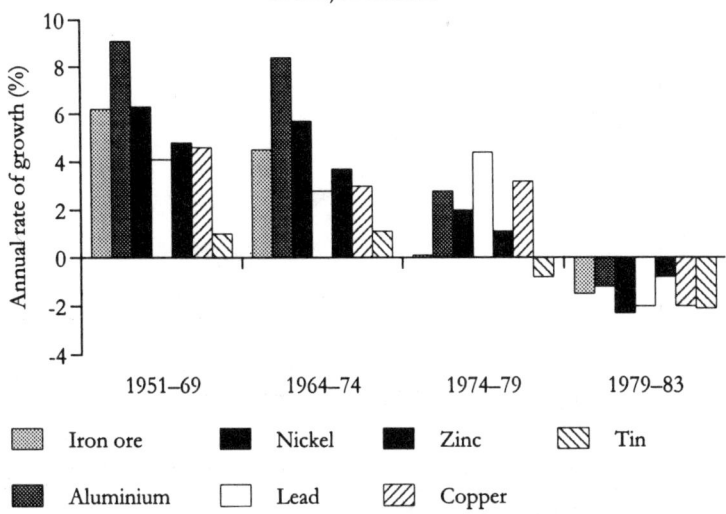

**FIGURE 3.6** Declining trend in world consumption of major metals

[Bar chart showing annual rate of growth (%) for periods 1951–69, 1964–74, 1974–79, 1979–83 for: Iron ore, Nickel, Zinc, Tin, Aluminium, Lead, Copper]

SOURCE: H.M.M. Lastres, *Advanced Materials Revolution and the Japanese System of Innovation* (London: Macmillan, 1994); original source of data, F. Gonzales-Vigil, 'New Technologies, Industrial Restructuring and Changing Patterns of Metal Consumption', *Raw Materials Report*, vol. 3 (1985), no. 3.

3 per cent is expected over the period 1990–1995, with the average price index for these six years representing less than half the average of the 1950s and 1960s.

From some materials producers' point of view, particularly developing countries, the so-called advanced materials revolution poses important challenges. Strategies based on producing ores and basic metals for export are becoming more and more vulnerable. A significant number of developing countries have had exports of minerals and basic metals as the core of their growth strategies. In 1989, various countries among the LDCs had more than 50 per cent of their exports originating in the mineral sector: Zambia (92 per cent), Zaire (85 per cent), Guinea (83 per cent), Bolivia (80 per cent), Congo (76 per cent), Chile (57 per cent), Peru (55 per cent), Papua New Guinea (54 per cent) and Togo (53 per cent). Even bigger countries, like Brazil and Australia, which have implemented more sophisticated economic structures have a significant share of their export revenues (15 per cent and 32 per cent respectively) derived from mineral production.[16]

**FIGURE 3.7** Weighted index of metal and mineral prices (constant 1985 US$)

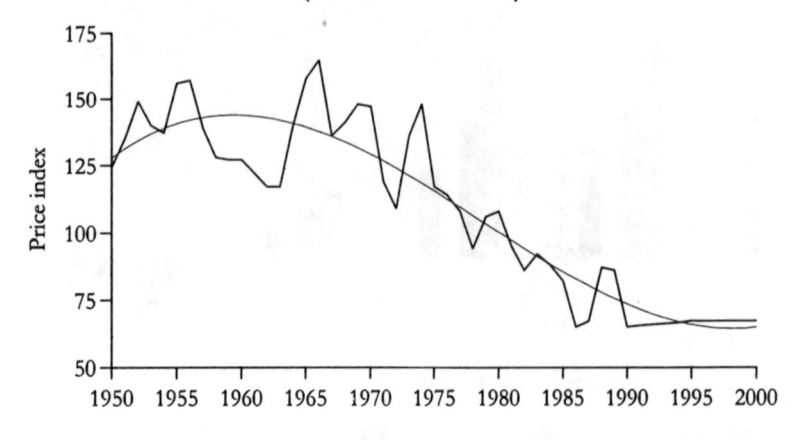

SOURCE: H.M.M. Lastres, *Advanced Materials Revolution and the Japanese System of Innovation* (London: Macmillan, 1994); original source of data, World Bank, *Price Prospects for Major Commodities, 1988–2000*, vol. 1 (Washington DC: 1992).

In the new situation, the importance of large and high-grade national mineral reserves, relatively cheap and abundant energy inputs, and a non-skilled labour force (even with extremely low levels of wages) is tending to diminish. Therefore the analysis of the consequences of such changes has necessarily to focus on the threat to developing countries' international market share and export earnings. However, there seem to be aspects even more important than those: such as the loss of attractiveness of most of the countries mentioned above to foreign investments and the need to redefine the new basis for partnership between less and more developed countries.

The analysis made by Takeuchi on the trend in Japan towards foreign direct investment (FDI) is particularly significant because of Japanese extreme dependence on foreign sources for its supply of raw materials and energy. Takeuchi's conclusions regarding such a trend are that: (i) since the 1960s the Japanese government has encouraged firms to invest in overseas production and primary processing of minerals and energy, reflecting the desire for secure and long-term supply of these goods; (ii) Japanese FDI in these sectors peaked in the early 1970s and remained relatively high until the early 1980s, when it dropped both relatively and in absolute terms, indicating that Japan's concern for the long-term security of its supply of raw materials and energy had begun to subside by that time.

An example of the corresponding trend in a Third World country is the case of the Carajás Project in Brazil. The project was designed in the early 1980s to exploit Latin America's most important mineral province in the Amazon. The Brazilian government provided the necessary infrastructure for the project, including ports, railroads, energy supply and various subsidies (ten year exemption of federal and regional taxes, energy prices below cost, etc.), aiming at attracting private and particularly foreign capital. Even with such subsidies, the project failed so far as its main objective was concerned. The 'failure' was mainly due to lack of interest by foreign investors (both financial and productive firms) in investing in Brazil in minerals and basic metals. It is interesting to note that similar models were considered as very 'successful' throughout the 1960s and 1970s.[17]

One related implication here is the expected change in terms of the location of materials production. Given (i) the relatively lesser importance of the availability of inputs and (ii) the greater importance of the linkages with their consumers, the processing plants of advanced materials tend to be located near the consuming and end-using industrial markets. Hence, the emerging geographical pattern of AM production is expected to be centred in those countries with better technological capabilities and strong markets for high-tech products. Therefore, it is expected that the plants located in LDCs will tend to serve local and regional markets only.[18]

The impact of the recent changes on developing countries' metal industry will, of course, depend on the degree and pace of the dissemination of AMs. For most metal producers, even in advanced countries, their attempts to remain in the market and make profits within such a difficult scenario involve some very difficult options. In fact, the combination of mature markets and low rates of innovation has far-reaching implications for economic planning even in the most advanced countries.[19]

But the discussion of the foreseen impacts of the introduction of AMs on LDCs, as well as the alternatives which can be open to these countries, transcends the limits of a change in the materials basis of the economy and the specificities of the AMs – such as, for instance, their sophisticated technological requirements. The range of aspects that have to be discussed includes also the difficult financial situation of most of the LDCs (and, in specific terms, of the metal producers in these countries), the new mode of the international division of labour, and the role of these countries in it.

Apart from all the other macroeconomic problems faced by the developing countries (high external debt, accelerating inflation, and political and institutional instability), together with the uncertainty surrounding minerals and basic metals, it has to be considered that the traditional means of articulation between LDCs and the advanced countries are experiencing major changes. In any case, a major issue here is that these recent changes have led to far more complex industrialization processes, where comparative advantages depend increasingly on innovation (both technical and organizational), rather than on purely physical factor endowment. The high requirements of sophisticated scientific and technological knowledge that the production of AMs implies can be seen as a handicap for those LDCs who intend to produce such materials.

On the other hand, the shift to the production of AMs also requires a new industrial and sectoral organization. Among all these requirements, the more flexible and agile style of production and the linkage between research, production and consumption of materials should be particularly stressed. But it is worth pointing out that, like any other profound historical transformation, the recent change brings with it heavy costs (especially for those strongly committed to traditional structures) as well as some important opportunities. What seems to be of crucial significance is to recognize that a fundamental shift is taking place and to understand the character of this change.

To make the best of the opportunities that the new paradigm presents, it is also extremely important to operate with sufficient agility and creativity in both public and private sectors. In this sense, Perez, when developing her concept of 'windows of opportunity', argues that much of the knowledge required to enter a technology system in its early phase is not as crystallized as it is in mature phases, most of it consisting in public knowledge available in universities. Her conclusion is that, 'given the availability of well-qualified university personnel, a window of opportunity opens for the relatively autonomous entry of lagging countries into new products in a new technology system in these early phases.'[20] At the same time it is stressed that firms (industries or countries) lacking the necessary educational, R&D and industrial capability may become even more seriously disadvantaged in the international competition. The existence of an adequate political and institutional framework – that is, of an active and efficient 'national system of innovation' – is also considered to be of particular importance, especially in moments when the diffusion of a new techno-economic paradigm is taking place.[21]

The case of optical fibres in Brazil could be considered as an example of successful exploitation of such opportunities of entering a new technology system in its early phases. The initial project was designed by the state company in charge of telecommunications, Telebrás, and matured in the Institute of Physics of the University of Campinas, which, since 1975, has been developing a research programme on optical fibres together with the Brazilian state telecommunications company – Telebrás's research centre. In 1984, the results of the research programme were transferred to a private national firm instituted to produce the fibres required for the development and renewal of the Brazilian telecommunications network. At that time an agreement was signed granting a five-year market reserve by Telebrás (more than 90 per cent of the Brazilian market for optical fibre).[22]

One result of such measures is that Brazil was at that time one of the few developing countries in the world to hold an autonomous research programme on optical fibres. In this case, a modern and agile governmental institution was able to take the opportunity to articulate internal political interest in designing strategic planning, promoting research–production–utilization linkages, and making effective use of its active market procurement policy to build up capabilities in a high-tech area, which was new even for advanced countries.

## 5. The Institutional Building of Policies for New Materials in Brazil

In 1985, a Ministry for Science and Technology was created in Brazil. The setting up of a specific ministry to deal with such activities represented a substantial improvement in the institutional building of S&T in Brazil since it upgraded the area within the country's policy-making strategy. One of the main tasks the new ministry was given was the designing and implementation of policies for the new areas of advanced technology, such as informatics, advanced materials, biotechnology and precision mechanics.[23]

During 1986, the Ministry of Science and Technology created the National Commission on New Materials (CONMAT) headed by the ministry's Secretary for Planning with senior representatives from all agencies of the ministry (especially CNPq and Finep, which are in charge of financing R&D activities in universities, research institutes and firms) and also from the Brazilian Special Secretariat for Informatics (SEI). Including SEI in the Commission on New Materials was due both to the importance of the IT market for the development of advanced materials and to the role of such materials in establishing a solid capability in microelectronics.

At the same time, a study and planning group was created to work on a full-time basis as the Executive Secretariat of the Commission. In its creation the Nucleus of Study and Planning in New Materials (NMAT) was charged with three immediate basic tasks: (i) to draw a preliminary set of short-term proposals and plans of action to be discussed and implemented by the National Commission on New Materials; (ii) to provide an assessment of the world-wide situation of new materials; and (iii) to evaluate the internal capabilities of universities, research institutes and firms. With the participation of leading Brazilian scientists in the areas of new metallic materials, advanced ceramics, new polymers, semiconductors and composites, the Nucleus defined a set of proposals and plans of action to be discussed by the Commission. In early 1987, an emergency two-year programme for new materials was proposed by the Commission and launched by the ministry. The basic document of the programme[24] stated that three main factors justified an urgent definition of a national policy for these materials:

- Brazil possessed important reserves of strategic minerals in a quasi-monopoly situation: quartz (95 per cent), niobium (86 per cent), titanium, beryllium, rare earth and others. As such inputs are fundamental for a series of strategic applications (in microelectronics, telecommunications, transport, space, etc.) the country could find itself in the future in the position of importing its own resources transformed into materials or final products if a strategy were not immediately devised and implemented.
- Brazil had already acquired reasonable scientific and technological capacity in some universities, research centres and firms which should be supported to be an effective basis where a global programme for new materials would be anchored.
- The Brazilian market was considered to be promising for high-tech sectors in the foreseeable future.

The main priorities of this programme concentrated on the research and development of new metals, new metallic alloys, semiconductors, quartz and optical fibres, accordingly to Brazilian advantages, capabilities and market. An extensive survey and evaluation of the internal capabilities in new materials and an assessment of the world-wide situation of new materials were completed in 1987 and early 1988 by

**TABLE 3.1** New materials in Brazil: number of institutions, researchers, research projects and patents granted, 1987

| Sub-sectors | New metallic materials | Ceramics | Quartz/ silicon | Polymers | Composites | Others | Total |
|---|---|---|---|---|---|---|---|
| Institutions | 57 | 63 | 23 | 41 | 19 | – | 102 |
| Researchers | 467 | 358 | 134 | 256 | 43 | 252 | 1322 |
| R&D projects | 177 | 153 | 22 | 66 | 38 | 36 | 492 |
| Patents | 33 | 28 | 15 | 16 | 7 | 12 | 111 |

SOURCE: Ministério da Ciência e Tecnologia, Brazil, *Pesquisa e Desenvolvimento de Novos Materiais no Brasil – Programa e Metas*, Secretaria de Novos Materiais (Brasília, 1990).

NMAT acting as the Executive Secretariat of the Commission. Both documents provided the basic information needed for further and complementary policy action.

According to the survey on Brazilian capabilities in new materials,[25] 102 institutions were performing R&D activities related to advanced materials in 1987 in Brazil (62 in advanced ceramics, 57 in new metallic materials, 41 in polymers, 23 in quartz and silicon, and 19 in composites). At that time, there were 1322 researchers participating in 492 R&D projects in these institutions, and the number of patents granted in the area amounted to 111 (Table 3.1).

It is worth emphasizing that even if, according to that study, universities were the main institutions developing research in the area in Brazil, there were, in 1987, 44 private and state-owned firms doing R&D in advanced materials and employing 429 researchers (Tables 3.2 and 3.3).

In January 1987, the two-year Programme of Scientific and Technological Development in New Materials was launched by the ministry and its agencies. As part of the programme the initial amount of approximately US$76 million were to be invested in 1987 and 1988 by the various agencies of the ministry in six main sub programmes:

- short-run opportunities in R&D (59%);
- human resources (25%);
- R&D infrastructure (9%);
- setting up of new productive units (6%);
- studies and planning (1%).

**TABLE 3.2** Number and type of institutions performing activities in new materials in Brazil, 1987

| Type of institution | No. | % |
|---|---|---|
| University | 44 | 43 |
| R&D centre | 14 | 14 |
| Firm | 44 | 43 |
| Total | 102 | 100 |

SOURCE: Ministério da Ciência e Tecnologia, Brazil, *Pesquisa e Desenvolvimento de Novos Materiais no Brasil – Programa e Metas*, Secretaria de Novos Materiais (Brasília, 1990).

In August 1987, secretariats for New Materials, Biotechnology and Precision Mechanics were established in the Ministry of Science and Technology with the task of designing a national policy for those sectors and implementing it. However, the creation of specific secretariats to design, implement and coordinate policies for high-tech sectors were almost immediately followed by the first of a series of political crises the new ministry started to face.

As the main political and institutional alliances of the new government which took over in 1985 were not sustained, and with the government being increasingly subjected to external pressures,[26] the measures regarding high-tech areas were gradually left out, and within five years the area of science and technology had four different ministers. With the inauguration of the Collor government in March 1990, the ministry was again converted into a secretariat now under the president's cabinet.[27] By that time the Secretariat for New Materials, as well as those relating to Biotechnology and Precision Mechanics, were dissolved and in 1991 even SEI was dissolved.

After the impeachment of President Collor in 1992 and the inauguration of the Itamar Franco government in 1993, the Secretariat of Science and Technology was once more reconverted into a ministry. However, no step has been taken to reinstall the policy structure concerning the new areas of high technology, as originally designed in 1985–87.

The various ups and downs of policy planning for new materials in Brazil in the last few years reflect the fragile political support for the establishment of a long-term strategy for the area. It is also true that this was not an isolated problem, since the crisis conditions of the 1980s and early 1990s hindered both the definition and implementation of a broader national development strategy in Brazil.

**TABLE 3.3** Distribution of researchers in new materials in Brazil by degree according to type of institution, 1987

| Type of institution | Level | | | |
|---|---|---|---|---|
| | PhD | MSc | BSc | Total |
| University | 368 | 171 | 77 | 616 |
| R&D centre | 51 | 84 | 142 | 277 |
| Firm | 35 | 139 | 255 | 429 |
| Total | 454 | 394 | 474 | 1 322 |

SOURCE: Ministério da Ciência e Tecnologia, Brazil, *Pesquisa e Desenvolvimento de Novos Materiais no Brasil – Programa e Metas*, Secretaria de Novos Materiais (Brasília, 1990).

The adoption of such policies is a very difficult task, especially for those LDCs that are facing important macroeconomic problems combined with political and institutional instability. However, it is also recognized that the definition of new forms of development is considered to be fundamental, particularly for those countries, as they face important challenges relating to their future chances of growth.

## 6. Concluding Remarks

The aim of this chapter has been twofold. First of all, it attempted to show that the present materials revolution is bringing about significant changes in the structure of production and consumption of materials. With the advanced materials revolution, a pronounced shift in the intensity of different materials usage has taken place in the last two decades. The previous growth of consumption of major bulk materials has been broken and nowadays there is a more rapid increase in advanced materials consumption. The new possibilities for designing novel synthetic materials, opened up by the change in the control over matter, plus the new requirements of industry, are the most important variables explaining the recent materials revolution.

The achievement of the mastery of the microstructure of matter, combined with the emphasis on applications research, the improvement in processing methods and the introduction of new research, design and materials processing technologies had numerous consequences. As argued above, with the present materials revolution a significant increase in variety, properties and range of applications of materials has become a reality. This has set the beginning of a radical change in the conceptual basis of materials production. One of the main qualitative differences of this revolution is that it is now possible to produce synthetic materials displaying the required properties by rearranging the microstructure of the materials. Properties can now be designed, not just discovered and selected.

With AMs, the economy has significantly expanded the variety of feasible alternative responses to the transition from an energy and resource-intensive mass-production style to an information-intensive production style. There is now a vast and proliferating number of materials and technical ways of manufacturing materials for specific requirements. A given product no longer relies so strongly on a given material. Instead, several materials now compete to assume a given function. This explains why the concept of advanced materials is not a closed one and includes not one but all families of materials: the homogeneous materials (ceramics, polymers and metallic materials) and also combinations of two or more materials (composites).

The recent shifts in materials evolution represent a combination of radical and incremental, technical and organizational innovations. These changes comprise a whole process of industrial restructuring – affecting patterns of investment, technological behaviour, industrial organization, relations between different actors, employment and trade, and involving a break in the previous institutional settings and previous competitive strengths. This changeover is occurring in a way which can be seen as a prerequisite for the introduction of a more flexible production system – based on a wider variety of materials, properties and processing methods, and on higher levels of information.

This chapter has also argued that these changes are bound particularly to influence countries which based their development strategies on the exploitation of natural resources. It is true that such far-reaching shifts will not dramatically affect – at least in the short run – the total consumption of basic materials. Nevertheless, they tend to reduce the strategic importance of minerals resources and will lead to a new international division of labour.

The second aim of the chapter has been to point out that such changes provide both threats and opportunities to developing countries. The opportunities depend on envisaging new development opportunities which take into account the exact significance of the new changes. As the Brazilian experience in optical fibres shows, it is not impossible successfully to design and implement strategies to niche markets. However, as the Brazilian experience also shows, such strategies are not free from pressures caused by the need to tackle short-run problems. Only a commitment to a long-run strategy seems to be a realistic option that developing countries should pursue in order eventually to cope with the challenges of the materials revolution.

## Notes

The author is indebted to Chris Freeman, José Eduardo Cassiolato, Adilson de Oliveira and João Lizardo de Araujo for their useful comments.

1. See H.M.M. Lastres, *Advanced Materials Revolution and the Japanese System of Innovation* (London: Macmillan, 1994).
2. See for instance, J.E. Tilton, 'Changing Trends in Metal Demand and the Decline of Mining and Mineral Processing in North America', *Resources Policy* (March 1989); E.D. Larson, M.H. Ross and R.H. Williams, 'Beyond the Era of Materials', in T. Forester, ed., *The Materials Revolution: Superconductors, New Materials, and the Japanese Challenge* (Oxford, 1988); W. Vogely, 'Proceedings of the Conference on Metals Demand', *Materials and Society*, vol. 10 (1986); L. Sousa, *Problems and Opportunities in Metals and Materials – an Integrated Perspective* (Washington DC: US Bureau of Mines, 1988); and P. Cohendet, M.J. Ledoux and E. Zuscovitch, *New Advanced Materials: Economic Dynamics and European Strategy*, ed. M. Ledoux (Berlin: Springer Verlag, 1988).
3. R. Auty, 'Materials Intensity of GDP – Research Issues on the Measurement and Explanation of Change', *Resources Policy* (December 1985).
4. These studies suggest an inverted U-shaped relationship between the intensity of metal use and per capita income. In low-income developing countries, rising per-capita income was expected to produce an increase in the intensity of metal use, while the opposite was expected in developed countries.
5. See Lastres, *Advanced Materials Revolution*.
6. See particularly those produced by or for the OECD, the Department of the Interior and of Commerce in the USA, MITI and STA in Japan, and CEE in Europe.
7. Organization for Economic Cooperation and Development (OECD), *Advanced Materials – Policies and Technological Challenges* (Paris, 1990).
8. H.M.M. Lastres, *Advanced Ceramics: Industrial Organization, Technological Features, Relevance of Government Policies and Expected Changes in the International Division of Labour*, paper prepared for the United Nations Development Programme in Brazil, Universidade de Campinas, Brazil, 1990.
9. Nowadays, there is widespread agreement that information and communications technology leads an exceptionally important cluster of innovations which are profoundly affecting technological competitiveness in virtually every branch of the world economy, and which are promoting organizational changes and restructuring.
10. See Lastres, *Advanced Materials Revolution*.
11. See also Chapter 7 of this volume, and Lastres, *Advanced Materials Revolution*.

12. Cohendet et al. *New Advanced Materials*, pp. 43, 73.

13. Analysts who consider this trend in the OECD economies conclude that this sharp break in the long-run growth rates of metal consumption took place earlier than that.

14. F. Gonzales-Vigil, 'New Technologies, Industrial Restructuring and Changing Patterns of Metal Consumption', *Raw Materials Report*, vol. 3 (1985), no. 3, p. 12.

15. See, for instance, K.H. Hwang, and J.E. Tilton, 'Leapfrogging, Consumer Preferences, International Trade and the Intensity of Metal Use in Less Developed Countries – The Case of Steel in Korea', *Resources Policy* (September 1990).

16. World Bank, *World Development Report 1991* (Washington DC, 1991).

17. See H.M.M. Lastres, *Programa Grande Carajás: Aspectos Políticos e Institucionais*, mimeo (Rio de Janeiro, 1982).

18. See for instance G. Gregory, 'New Materials in Japan', in T. Forester, ed., *The Materials Revolution: Superconductors, New Materials, and the Japanese Challenge* (Oxford, 1988).

19. As E.D. Larson et al. ('Beyond the Era of Materials', in T. Forester, ed., *The Materials Revolution: Superconductors, New Materials, and the Japanese Challenge*, Oxford, 1988), point out: 'Those who make economic policy in the industrial countries must recognize that reforms such as subsidizing the steel industry would not restore rapid growth, because they would have no effect on the underlying stagnation in demand. The materials industries cannot be sustained by protecting antiquated technology. Substantial innovation will be necessary to bring these industries into conformity with the present reality. In the past the rapid growth of demand was a spur to technological innovation, but that stimulus is gone' (p. 156).

20. C. Perez, 'New Technologies and Development', in C. Freeman and B. Lundvall, eds, *Small Countries Facing the Technological Revolution* (London, 1988), p. 92.

21. Networks of institutions in the public and private sectors whose activities and interactions initiate, import, modify and diffuse new technologies. See C. Freeman, *Technology Policy and Economic Performance: Lessons from Japan* (Brighton: SPRU, University of Sussex, 1988).

22. H.M.M. Lastres et al., *Diagnóstico dos Novos Materiais*, Núcleo de Estudos e Planejamento em Novos Materiais do Instituto Nacional de Tecnologia do Ministério da Ciência e Tecnologia, NMAT/INT/MCT, Rio de Janeiro, 1988. See especially the section on quartz and silicon by Lemos.

23. J.E. Cassiolato, 'Policies in Newly Industrialized Countries: the Case of Brazil', in *Advanced Technologies Alert System Bulletin* 5 (Washington DC: United Nations Organization, 1988).

24. Brazil, Ministério da Ciência e Tecnologia, 'O desafio dos Novos Materiais – Programa Brasileiro', *Série Brasil Ciência no. 2* (Brasília, 1987).

25. Lastres et al., *Diagnóstico dos Novos Materiais*. Data and information about the research centres that are developing activities in this area were obtained through direct interviews and include: characterization of the institutions (university, research institute or firms; private – national or foreign – or state-owned; etc.); share of R&D funds allocated to new materials and their importance in the total annual expenditure; main areas of activity in new materials; description

of laboratories and equipment available for research in the area; researchers developing activities in new materials (by degree and background); national and international programmes of technological and industrial collaboration; patents held or applied for; research projects (scope and objectives, type of material focused on, main technological routes investigated, main inputs and equipment used in the research, researchers involved, technical and industrial cooperation with other institutions, duration, funding, results, corresponding patents, industrial utilization of the results obtained, and transfer of knowledge); and agenda for R&D activities in the following couple of years.

26. See, for instance, M.I. Bastos, 'State Policies and Private Interest: the struggle over information technology policies in Brazil', in Schmitz and Cassiolato, eds, *Hi-Tech for Industrial Development – Lessons from the Brazilian Experience in Electronics and Automation* (London: Routledge, 1992).

27. Within the Brazilian governmental hierarchy, ministries are the highest institutional level below the president. Each ministry usually comprises several different secretariats.

## References

Auty, R., 'Materials Intensity of GDP – Research Issues on the Measurement and Explanation of Change', *Resources Policy* (December 1985), pp. 275-83.

Bastos, M.I., 'State Policies and Private Interest: the struggle over information technology policies in Brazil', in Schmitz and Cassiolato, eds, *Hi-Tech for Industrial Development – Lessons from the Brazilian Experience in Electronics and Automation* (London: Routledge, 1992).

Brazil, Ministério da Ciência e Tecnologia, 'O desafio dos Novos Materiais – Programa Brasileiro', *Série Brasil Ciência no. 2* (Brasília, 1987).

Brazil, Ministério da Ciência e Tecnologia, 'Programa de Formação de Recursos Humanos nas Areas Estratégicas – RHAE', *Série Brasil Ciência no.7* (Brasília, 1988).

Brazil, Ministério da Ciência e Tecnologia, *Pesquisa e Desenvolvimento de Novos Materiais no Brasil – Programa e Metas*, Secretaria de Novos Materiais (Brasília, 1990).

Cassiolato, J.E., 'Policies in Newly Industrialized Countries: the Case of Brazil', in *Advanced Technologies Alert System Bulletin* 5 (Washington: United Nations Organization, 1988).

Cohendet, P., M.J. Ledoux and E. Zuscovitch, *New Advanced Materials: Economic Dynamics and European Strategy*, ed. M. Ledoux (Berlin: Springer Verlag, 1988).

Freeman, C., *Technology Policy and Economic Performance: Lessons from Japan* (Brighton: SPRU, University of Sussex, 1988).

Gonzales-Vigil, F., 'New Technologies, Industrial Restructuring and Changing Patterns of Metal Consumption', *Raw Materials Report*, vol. 3 (1985), no. 3.

Gregory, G., 'New Materials in Japan', in T. Forester, ed., *The Materials Revolution: Superconductors, New Materials, and the Japanese Challenge* (Oxford, 1988).

Hwang, K.H. and J.E. Tilton, 'Leapfrogging, Consumer Preferences, International Trade and the Intensity of Metal Use in Less Developed Countries – The

Case of Steel in Korea', *Resources Policy* (September 1990), pp. 210–24.
Key, P.L. and T.D. Schlabach, 'Metals Demand in Telecommunication', *Materials and Society*, vol. 10 (1986), no. 3, pp. 241–3.
Kuroda, T., 'From Traditional Ceramics to Technical Ceramics – Examples of a Japanese Company', *Industrial Ceramics*, vol. 10 (1990), no. 2.
Larson, E.D., M.H. Ross and R.H. Williams, 'Beyond the Era of Materials', in T. Forester, ed., *The Materials Revolution: Superconductors, New Materials, and the Japanese Challenge* (Oxford, 1986).
Lastres, H.M.M., *Programa Grande Carajás: Aspectos Políticos e Institucionais*, mimeo (Rio de Janeiro: 1982).
Lastres, H.M.M., *Advanced Ceramics: Industrial Organization, Technological Features, Relevance of Government Policies and Expected Changes in the International Division of Labour*, paper prepared for the United Nations Development Programme in Brazil, Universidade de Campinas, Brazil, 1990.
Lastres, H.M.M., *Advanced Materials Revolution and the Japanese System of Innovation* (London: Macmillan, 1994).
Lastres, H.M.M., et al., *Diagnóstico dos Novos Materiais*, Núcleo de Estudos e Planejamento em Novos Materiais do Instituto Nacional de Tecnologia do Ministério da Ciência e Tecnologia, NMAT/INT/MCT, Rio de Janeiro, 1988.
Lastres, H.M.M., and J. Cassiolato, 'High Technology Sectors and Developing Countries: the Case of Advanced Materials', *Materials and Society*, vol. 41 (1990), no. 1, pp. 1–9.
Organization for Economic Cooperation and Development (OECD), *New Materials (Countries Annexes)* (Paris, October 1988).
Organization for Economic Cooperation and Development (OECD), *Advanced Materials – Policies and Technological Challenges* (Paris, 1990).
Perez, C., 'New Technologies and Development', in C. Freeman and B. Lundvall, eds, *Small Countries Facing the Technological Revolution* (London, 1988).
Sousa, L., *Problems and Opportunities in Metals and Materials – an Integrated Perspective* (Washington DC: US Bureau of Mines, 1988).
Takeuchi, K., 'Japan's Experience in Linking Foreign Direct Investment and Imports of Minerals', *Resources Policy* (December 1990), pp. 307–12.
Tilton, J.E., 'Changing Trends in Metal Demand and the Decline of Mining and Mineral Processing in North America', *Resources Policy* (March 1989), pp. 12–23.
United Nations Conference on Trade and Development (UNCTAD), *UNCTAD Commodity Yearbook – 1988* (New York, 1989).
US Department of the Interior, Bureau of Mines, *The New Materials Society – Challenges and Opportunities* (Washington DC, 1990).
US Congress, Office of Technology Assessment (OTA), *New Structural Materials Technologies – Opportunities for the Use of Advanced Ceramics and Composites: a Technical Memorandum* (Washington DC, 1988).
Vogely, W., 'Proceedings of the Conference on Metals Demand', *Materials and Society*, vol. 10 (1986).
World Bank, *World Development Report 1991* (Washington DC, 1991).
World Bank, *Price Prospects for Major Commodities, 1988–2000*, vol. 1 (Washington DC, 1992).

*Part II*

Case Studies of Developing Countries

# 4
# High Technology Programmes in China
## Jian Song

China still belongs to the developing world, and its per-capita income is less than US$400 per annum. After more than a decade of effort, it has managed to provide enough to eat and wear for its 1.2 billion people. Although about 5 per cent of the population has yet to cast off poverty, the average Chinese person has a secure food supply of about 400 kgs per annum and a modest income. At the present stage, the most important task in China is to advance to a better standard of living, to a level of US$1000 per capita per annum. The Chinese decision-makers believe that China should, in addition, embark upon high-tech development. China hopes to shorten the lengthy development road that the developed countries experienced and to employ high technology for national growth. The scientific community in China sees high-tech development as the key to seizing the opportunity for advancement.

It is the belief of scientific communities world-wide that the world has entered the age of high technology, and this is indeed now a powerful component of economic growth. Its development has become a must for today's nations, industrialized and developing alike. In its drive to achieve high-tech development, China possesses a number of valuable strengths that should work to its advantage:

- a 17-million-strong contingent of scientists and technicians, 1.8 million of whom are engaged in R&D activities, 1 million in research institutes, 0.5 million in universities and 0.3 million active as entrepreneurs;
- a fast growing economy that will generate huge demand for high technology;
- continuous government support in spurring on, and providing funding for, high-tech development programmes, that are set to follow global trends;

- an increasingly open attitude, favourable to wider international co-operation.

However, the country is encountering a number of problems in the development process. In the case of industrialized nations, high-tech is the outcome of an evolutionary process, the result of a long accumulation of technologies on a solid industrial base. But China has had to start from a self-sustained traditional economy with conventional industry, and a rather weak technological infrastructure that lacks proper linkages between science and economy. For a long time in China, as in most developing countries, scientific research and commodity production were divorced from each other. Starting from 1985, China launched reforms in its science and technology (S&T) management systems in order to orient scientific research towards national economic development. It is now stressed that economic growth must be S&T-driven.

In line with this principle, China has formulated an S&T strategy of 'holding fast to one end [of basic research] and leaving open the whole field [of R&D and market orientation]'. It is popularly known as a three-tier S&T development strategy.

In the first tier, most of the S&T capabilities and resources have been mobilized to serve the country's economic development. This is the main battlefield, involving roughly two-thirds of the country's scientific and technical resources. They are to work at the forefront of the production sector to ensure technological progress and raise productivity.

In the second tier, a considerable amount of S&T capabilities and resources are being employed to follow the global trends in high technology, and to work on programmes of national priority.

In the third tier, a creative scientific contingent works on the frontiers of basic research. The objective here is to contribute to the building of scientific strength in the long term, without being constrained in any way by short-term goals of achieving tangible results.

In the following, we will limit ourselves to describing briefly the salient points in three government-sponsored high-tech programmes: the '863 Programme', the 'Torch Programme' and the 'Key S&T Achievements Dissemination Programme'.

The *863 Programme* was launched in March 1986. It has mobilized an elite S&T task force to work on several selected frontier areas. Twenty national R&D centres and networks, with an R&D team of 15 000 people, form the core of the programme. Its objectives are to follow the latest international trends in high-tech, pursue its own innovations, narrow the gaps between Chinese and foreign technologies, to break through in areas where China has development potential, disseminate research results to industry and create a new generation of high-tech talent. It is focused on biotechnology, space technology, information technology, lasers, automation, energy technology and new materials.

Since its inception, the 863 Programme has worked on about 550 projects, achieving success in many important areas such as computer software in Chinese characters, new lasers, powerful new space carriers, large-scale separation and purification of hepatitis B vaccine, interferon, interleukin, new high-yielding varieties of hybrid rice, *in vitro* bovine fertilization, artificial crystals, robotics and computer-integrated manufacturing (CIM). About ten business firms are to implement CIM programmes in various industries, including machine building, appliances, garments, aircraft and automobiles.

Under the aegis of the 863 Programme, research institutions have been created for management information systems, computer aided design, computer aided manufacturing, numerically controlled machine tools, equipment for material distribution and management, and so on.

The *Torch Programme*, initiated in 1988, is aimed at promoting the commercialization, industrialization and internationalization of China-developed high technologies and products in a market-driven fashion. Its central objective is to establish 52 science parks of high-tech industries in the national and provincial capitals. With the provision of loans and risk capital as incentive, scientists and industrialists are to be encouraged to commercialize research results. The science parks will have advanced infrastructure and management systems, with telecommunications and transport, banking and insurance, as well as enjoying favourable tax policies. Here, technology-intensive investment, both domestic and foreign, will be made welcome.

The Torch Programme is also aimed at orienting scientist-turned-entrepreneurs towards the domestic and international markets, combining good technology development skills with sound business management.

The *S&T Dissemination Programme*, begun in January 1990, has the task of disseminating advanced and appropriate technologies to producers situated country-wide. The dissemination occurs radially from centres scattered across the nation. So far, this Programme has succeeded in propagating about 500 new technologies for use in agriculture and industry.

In all high-tech programmes, environmental protection is a major consideration. An integrated approach to global environmental issues calls for, among other things, technology transfer from developed to developing countries in a spirit of global partnership. Developing countries are badly in need of technology transfer.

In the past ten years, in what has been the prelude to its modernization, China has benefited enormously from world-wide cooperation. And it will continue to work to strengthen international cooperation in a spirit of partnership.

# 5
# Development of New Generic Technologies in India
## P.K.B. Menon

The growth of scientific and technological culture in India has played an important role in attracting world attention to indigenous capabilities. Today, India has a vast S&T infrastructure with more than 200 national laboratories, over 190 universities including institutes of technology, and about 1200 in-house R&D units in the industrial sector. Indigenous efforts have enabled the country to make considerable progress not only in developing local capabilities but also in adopting and adapting exotic technologies. The president of the Institute for Scientific Information (USA) has referred to India as a Third World 'research superpower'. India is the only 'developing country' to figure as high as eighth (above Italy and Australia) in a list of the ten top countries contributing to world literature in science. In a Science Citation Index (SCI) coverage of 5000 journals, India was found to publish more than 10 000 papers a year during 1981–85.[1]

The impact of the changing economic environment on the S&T system in India is quite evident today. From a situation of total dependence on government funding for R&D, gradual shifts are taking place in the direction of partially self-supported efforts, academic research attracting sponsored support from industry, and the growth of multi-institutional research programmes involving industry. The trend towards a market-driven economy and globalization is creating a new environment for R&D, forcing institutions and industries to become more competitive and productive. In what follows, three areas have been chosen to provide a scenario of developmental efforts in new generic technologies. These are: biotechnology, new materials and information technology.

## Biotechnology

Biotechnological applications have great potential for developing countries: for increasing agricultural production and industrial production, and producing adequate healthcare facilities. The sector offers the possibility of creating new jobs through value-added products and processes. The development of non-polluting and eco-friendly technologies through biotech research has become vital for the survival of modern man. A concerted effort has been made in India to promote the development of biotechnology through a strong research base, academic excellence, manpower development, and product and process development. As a result, research capabilities exist today in the areas of industrial biotechnology, medical biotechnology, microbial biotechnology, plant molecular biology and agricultural biotechnology, biological pest control, aquaculture and marine biotechnology, animal biotechnology including veterinary sciences, animal husbandry and leather biotechnology, biochemical engineering, downstream processing, biomedical technology and bioinformation. Several Indian Institutions have emerged as centres of excellence, specializing in immunology, DNA fingerprinting techniques, oligonucleotide synthesis, protein engineering, tissue culture, hybridoma techniques and production of vaccines. A DNA probe for DNA fingerprinting has been developed. This is the second such effort in the world. A process for efficient conversion of molasses to alcohol using an osmotolerant and ethanol tolerant strain of yeast is another significant achievement. Impressive developments in plant tissue culture include *in vitro* flowering of Bamboo (the first time in the world) and micro-propagation techniques for teak, eucalyptus, sugar-cane, and cardamom. Microbes have been cultured for the degradation of oil spills and desulphurization of crude oil and natural gas.

Research work in biotechnology is supported mainly by the Department of Biotechnology (DBT), the Council of Scientific and Industrial Research (CSIR), the Indian Council of Medical Research (ICMR), the Indian Council of Agriculture Research (ICAR), the Department of Science and Technology (DST) and the University Grants Commission (UGC). A large number of private-sector companies have also shown keen interest in developmental activities in biotech products and processes. However, financial support amounts to less than $30 million, with industry's contribution to R&D perhaps not exceeding $0.5 million, annually. The areas of R&D activity include production of certain cell cultured vaccines, immuno-diagnostics and certain primary and

**TABLE 5.1** Selected research projects in biotechnology commercialized in India

| Sl no. | Technology | Institute/ company | Applications |
|---|---|---|---|
| 1. | ELISA | Lupin Labs | Pregnancy detection |
| 2. | r-DNA molecule immunology | CDRI | Cholera (field trials) |
| 3. | Using antigen of E.histolitica | CDRI | Amoebiasis |
| 4. | ELISA, Hybridoma | CDRI | Filariasis |
| 5. | ELISA, Hybridoma | CDRI | Lechmaniasis |
| 6. | RIA | CDRI | Reproduction of steroidal set of hormones |
| 7. | Culturing and harvesting of virus | Bio-vaccine | T.T. |
| 8. | r-DNA & G.E. Tech's | Indian Immunologicals | virus animal vaccines |
| 9. | r-DNA & G.E. Tech's | Serum Institute | DPT, IT, DT, measles |
| 10. | Fermentation culture | Haffkine, BioPharma | Cholera Typhoid |
| 11. | Tissue culture | Haffkine, BioPharma | Polio (Oral) |
| 12. | r-DNA | NIV | J-encephalitis |
| 13. | RIA, IAHA, RPH | NIV | Hepatitis-B |
| 14. | RIA, IAHA, RPH | NIV | Yellow fever, diarrhoea, dysentery |
| 15. | ELISA | HAL | Pregnancy |
| 16. | Microbial culture | BIOCON | Industrial enzymes |
| 17. | Continuous fermentation | VDSI | Alcohol |
| 18. | Microbial enzyme | IMTECH | Ethanol |
| 19. | Submerged fermentation | CFTRI | Glycerol |
| 20. | Horticulture & Floriculture | Several institutions | Varietal improvement |

NOTE: CDRI: Central Drug Research Institute, Lucknow; NIV: National Institute of Virology, Pune; VDSI: Vallab Dada Sugar Institute, Pune; IMTECH: Institute of Microbial Technology, Chandigarh: CFTRI: Central Food Technological Research Institute, Mysore; RIA: Radio Immuno Assay, ELISA: Enzyme Linked Immuno Sorbant Assay; RPH: Reverse Passive Haemagluttination; IAHA: Immune Adherence Haemagluttination Assay.

SOURCE: Technology Information Forecasting and Assessment Council: *TMS in Biotechnology: Fermentation, Tissue Cuture, Medicine etc.*, TIFAC: TMS 0:45 (1992).

secondary metabolites produced by fermentation. An R&D-cum-manufacturing unit, M/s Bharat Immuno Biologicals Co. Ltd (BIBCOL), with an annual production capacity of 100 million doses of oral polio vaccine, has been established. Another manufacturing unit is being set up for the production of viral vaccines, including injectable polio, hyperattenuated measles and quadruple DPTP (diptheria–pertussis–tetanus–polio) vaccines.

Most of the research effort in fermentation is focused on the improvement of the current fermentation process – the conversion from batch to continuous processes, and the use of immobilized microorganisms in fermentation. Research is also under way to develop microbial enzymes and to improve the existing process. The focus of research in the private sector has been on the production of microbial enzymes using fermentation processes. In the field of horticulture and floriculture, efforts are concentrated on varietal improvement and development of micropropagation. In the area of health care, the emphasis is on the development of vaccines and diagnostic kits using the r-DNA and MAb techniques. Some of the most significant research projects, the results of which have been commercialized, are shown in Table 5.1. In addition, a DNA diagnostic probe for the detection of malaria due to *Plasmodium falciparum* was developed at IISc, Bangalore. DNA probes for the detection of M. tuberculosis, hepatitis, and others, are also under development in the country. Other products include biotech applications in effluent management, composting, microbial leaching and beneficiation of ores, forest biotechnology, synthesis of oligonucleotides, restriction endonucleases, gels, polymers and special materials used in biotechnology research and industry.

There has been a degree of success in transferring the research knowledge for use in production. Monoclonal antibody (MAb) technology has helped in the development of diagnostic kits. The r-DNA and MAb technologies have also made it possible to develop vaccines for many of the communicable diseases. The pharmaceutical industry has been one of the major beneficiaries of biotechnology research. The new biotechnology methods have enabled development of new pharmaceuticals and improved efficiency in the production of those currently available.

There has also been a major shift in the pharmaceutical industry towards exploiting this emerging area. In 1985, biotechnology-based products contributed only 0.4 per cent of the pharmaceutical industry's output. The conservative market prediction for output in the year 2000 is 21.8 per cent. This scale of increase is already evident, since in a

**TABLE 5.2** Selected biotech products and processes in India

| Sl. no. | Product/technology | Institute/company | Industry |
|---|---|---|---|
| **Fermentation technology** | | | |
| 1. | Immobilized yeast | NCL, Pune | VDSI, Dhanpur Sugar Mills |
| 2. | Microbial enzymes for alcohol | IMTECH, Chandigarh | VMSRF, UB |
| **Tissue culture** | | | |
| 3. | Mungbean, black gram, pigeon pea, oil seeds, rice, groundnut | BARC, Bombay | SSC |
| 4. | Wheat, soybean, grapes | MACS, Pune | SSC |
| 5. | Cardamom, sugarcane, rubber | NCL, Pune | AV Thomas, EID Parry |
| 6. | Banana, cardamom papaya, sugarcane | Lupin, Bhopal | Lupin Labs Ltd |
| 7. | Tomato | MKU | – |
| **Diagnostic kits** | | | |
| 8. | Filariasis detection kit | MGIMS, Wardha | Cadila Labs Ltd |
| 9. | Pregnancy slide test | NII, New Delhi | Ranbaxy Labs Ltd |
| 10. | Dot ELISA pregnancy test | NII, New Delhi | Ranbaxy Labs Ltd |
| 11. | Typhoid fever detection kit | NII, New Delhi; AIIMS, New Delhi | Lupin Labs Ltd, Ranbaxy Labs Ltd |
| 12. | Amoebic liver abcess detection | NII, New Delhi | Cadila Labs Ltd |
| 13. | Blood grouping | NII, New Delhi | Cadila Labs Ltd |
| 14. | Detection of hepatitis-B | NII, New Delhi | Lupin Lab Ltd |
| 15. | Leprosy immuno-modulater | NII, New Delhi | Cadila Labs Ltd |

NOTE: NCL: National Chemical Laboratory, Pune; VMSRF: Vittal Mallaya Science Research Foundation, Bangalore; UB: Union Biotech, Hyderabad; BARC: Bhabha Atomic Research Centre, Bombay; SSC: State Seed Corporation; MKU: Madurai Kamaraj University, Madurai; NII: National Institute of Immunology, New Delhi; MGIMS: Mahatma Gandhi Institute of Medical Sciences, Wardha; AIIMS: All India Institute of Medical Sciences, New Delhi.

SOURCE: Technology Information Forecasting and Assessment Council: *TMS in Biotechnology: Fermentation, Tissue Cuture, Medicine etc.*, TIFAC: TMS 0:45 (1992).

**TABLE 5.3** Demand for biotech products by the year 2000

| Product category | Estimated market by 2000 ($ million) |
|---|---|
| 1. Human and animal health, including diagnostics, vaccines, antibiotics by fermentation, bio-active protein, targeted drugs, blood products etc. | 1178 |
| 2. Agriculture, including plant tissue culture, hybrid and artificial seeds, biofertilizers, biopesticides, poultry feed additives, aquaculture, embryo transfer etc. | 128 |
| 3. Industrial products by fermentation, including industrial enzymes, organic acids, amino acids, improved alcohol strain etc. | 500 |
| 4. Other biotech products | 43 |
| Total | 1849 |

SOURCE: P.K. Ghosh, 'Biotechnology in India', *Australian Journal of Biotechnology*, vol. 3 (1994), no. 4, pp. 214–22.

short span of time the number of new drug applications through biotechnology has overtaken conventional new drug applications. Considering the great opportunities that biotechnology will offer in the near future for the discovery and development of new vaccines, therapeutic agents and immunodiagnostics, Indian industry should move rapidly into this area. An initiative has indeed already been made in this direction, as is reflected in the fact that a number of Indian companies have developed and marketed immunodiagnostic reagents and kits based on indigenously produced antibodies.

Selected biotech products and processes that have helped the development of industry are shown in Table 5.2. The demand for biotech products in India is on the increase. The estimated demand for different categories of products by the year 2000 is shown in Table 5.3. The highest demand will be for health-care and agricultural products. The major share in pharmaceutical products is likely to be accounted for by the cytokine market. Another area that is likely to play a major role is cell-based therapeutics, especially stem-cell-based therapeutics. It is envisaged that through this route, gene therapy will become a reality. The annual demand for industrial products is estimated at around $500

**TABLE 5.4** Biosensor technologies transferred to industry

| Sl. no. | Sensor type | Biomaterial | Application | Institute |
|---|---|---|---|---|
| 1. | Glucose (enzymatic) | Glucose oxidase | Glucose level in blood | NPL, New Delhi |
| 2. | BOD (bacterial) | Trichosporon beigilli | Pollution level of water, industrial effluents | IACS, Calcutta |
| 3. | Ethyl alcohol (liquid bacterial) | Trichosporon brassicae | For food, beverages and fermentation processes | IACS, Calcutta |
| 4. | Urea (enzymatic) | Urease | For clinical purposes | NPL, New Delhi |
| 5. | Biomaterials (glucose oxidase) | Aspergillus niger | For glucose sensor, canning and textile industry | CBT, Delhi |
| 6. | Biomaterials (urease) | Jack bean | For urea sensor | CBT, Delhi |
| 7. | Lactate | Lactate oxidase | For sports persons | NPL, New Delhi |
| 8. | Fish freshness | Oxidase nitrifying bacterial | For detection of freshness of fish | IACS, Calcutta |

NOTE: NPL: National Physical Laboratory; IACS: Indian Academy for the Cultivation of Science; CBT: Centre for Biotechnology.
SOURCE: Department of Science and Technology, unpublished information.

million. Technologies for effluent treatment, composting and microbial leaching would also be in great demand.

Biosensors (biologically based sensors) have been in use for several years, but design problems have limited their acceptance. Enzymes and monoclonal antibodies are particularly suited for use as sensors because of their high specificity for given substances. Sensors using biomolecules will be more sensitive than traditional ones. Some of the biosensor technologies developed and transferred to industry are shown in Table 5.4. Biosensors using enzymes have also been used to detect the presence of various organic compounds.

## Materials

Materials play an important role in every system associated with modern technologies, and are encountered in a wide variety of sectors like heavy engineering, aerospace, microelectronics, bioengineering, power generation and transportation. In all these sectors technological progress was possible only through the development of critical materials with improved performance capabilities. In fact, some of the most exciting scientific advances are taking place in the realm of materials. The big difference in R&D effort in this area from that of the past is the attempt to produce special materials to suit special requirements rather than adopting naturally occurring materials. It has become increasingly possible to predict the properties of materials before they are even made, and to modify them to suit a particular application.

The advanced materials programme of India was formulated not only with the future materials needs of our technologies in mind; also taken into account were the scope of national resources and indigenous materials processing capabilities. India's advantages in the area of S&T are: (i) a significant resource base, (ii) an excellent R&D infrastructure, (iii) engineering and manufacturing capabilities, (iv) a large domestic market, and (v) highly competent manpower. The present status of advanced materials research can be summarized as follows:

*Semiconductor materials*

In India, the main integrated circuit facility based on silicon is located at Semi Conductor Complex Ltd (SCL), Chandigarh. Recently, major facilities like silicon MBE, reduced vapour phase epitaxy and electron beam lithography have been set up within DRDO. Mettur Chemicals have established a plant to produce electronic-grade silicon.

A major programme on semiconductor III-V compounds is under way at Solid State Physics Laboratory (SPL) – where facilities for the synthesis and growth of single crystals, epitaxial processing and metal-organic chemical vapour deposition (MOCVD) devices and electron beam micro fabrication equipment have been established. Several sophisticated diagnostic facilities have also been installed. At the Central Electrical and Electronic Research Institute (CEERI), Pilani, the emphasis is on device design and development.

India has the potential to become a major source of gallium in the world, as this is a by-product of aluminium production from bauxite, and, in terms of gallium content, Indian bauxite is among the best in the world. Central Electro Chemical Research Institute (CECRI) and

Bhaba Atomic Research Centre (BARC) have developed processes for gallium recovery. Nuclear Fuel Complex (NFC) have established purification techniques to produce 4–6N electronic-grade gallium.

## Optical fibres

Production equipment to draw optical fibres from preforms made by using modified chemical vapour deposition (MCVD) technology is available at the Madhya Pradesh State Electronics Development Corporation and Hindustan Cables Limited.

## Superconductors

A National Superconductivity Programme (NSP) was initiated in 1988 as a multi-institutional project. Under this project, some 66 programmes, distributed in among 37 R&D institutions in the country, are being supported. As a consequence, India today has several laboratories equipped with facilities for materials preparation and characterization. Specialized facilities are also available for thin film depositions by various techniques. The R&D achievements in this area include: (i) synthesis, characterization, modification and fabrication of yttrium, bismuth, lanthanium, thalium and mercury-based high temperature superconducting materials (HTSM) and LTSC ternary intermetallics; (ii) bulk processing techniques; (iii) various preformed shapes; (iv) sputtering target materials; (v) multiple film growth techniques; (vi) low temperature superconductivity (at LQD He) applications; (vii) development of SQUIDS; and (viii) superconductivity power generator and high gradient magnetic separator using low Tc superconducting materials.

## Magnetic materials

India has one of the world's richest rare-earth mineral resources. Processing into rare-earth salts is undertaken by Indian Rare Earths Ltd. Several major programmes are under way for extraction of rare-earth metals and alloys for various applications, as well as development of rare-earth magnets. These magnets, which are primarily of samarium cobalt and neodymium – iron – boron are the most volume-efficient high-energy materials produced through the sophisticated technology of powder metallurgy coupled with magnetic field alignment and compaction.

These magnets are highly efficient for device miniaturization in high-tech applications such as defence, aerospace, nuclear, medicine, electronics, and so on. Intense R&D efforts at Defence Metallurgical Research Laboratory (DMRL) have led to the establishment of powder metallurgical process technology. This has been upgraded from laboratory to semi-commercial scale, and prototype magnets have been produced for various applications, such as telephone, solar energy motors, microwave devices, MRI, magnetic separators. Market sensitization and application engineering for various end-users of magnets is currently under way.

*Oxidic and other single crystalline materials*

Laboratory-scale facilities for growing certain crystals and for their characterization exist at IISc, BARC and Anna University. Defence Service Centre (DSC) and SPL have facilities for pilot-scale production of YAG and GaAs crystals. Production of silicon crystals is underway at Super Semi-conductors (Calcutta), Silitronics (Hosur) and Mettur Chemicals (Tamil Nadu).

*Titanium*

Industrial-scale plants for beneficiation of titanium deposits and manufacture of titanium tetrachloride have been set up in Kerala and Orissa. A demonstration plant for the extraction of titanium sponge from titanium tetrachloride has been working at Defence Metallurgical Research Laboratory, Hyderabad. Advanced processing facilities are available at Midhani, Hyderabad. Facilities for making condenser tubes from titanium have now been established. A modern condenser tube plant with state-of-the-art testing facilities has recently been set up at Mishra Dhatu Nigam Ltd (MIDHANI). The capability to design a steam condenser exists at Bharat Heavy Electricals Limited. The metal, its alloys and coated products are used in power generation and many modern chemical process industries. There is a growing move towards large-scale use of the metal in building construction and marine-based operations.

A Titanium Development Advisory Committee (TDAC) was set up in 1991, comprising eminent scientists and industrialists, to provide direction to the developmental activities of titanium and its downstream products.

*Composites*

The increase in usage of composites is likely to reduce considerably the consumption of conventional materials – metals, wood, and so on. The use of composites in sectors like automotive, railways, sports and leisure would lead to some 10 per cent reduction in power/energy consumption. India occupies sixth position in the world composite (fibre reinforced plastics) output, according to 1992 data. The average growth rate of the glass fibre and glass-fibre reinforced plastics (GFRP) industry in India, especially over the last few years, has been about 15 per cent. The main reason for a growth rate higher than in the developed countries, is the continued efforts that have been made with development and the identification of new areas of application (roads, railways, defence, renewable energy sectors, and so on).

The global position of composite technology is highly advanced due to the availability of advanced fibre-resin systems and fabrication processes. Indian advances in the field, however, have not kept pace with global developments. Nevertheless, India has a fairly good understanding of expensive polyesters, superior epoxies, phenolics that are difficult to process, and glass-fibre reinforced composites, which are all now in use in the country. Efforts to indigenize production of carbon and 'Kevlar' fibres are under way at the National Aeronautical Laboratory, the Indian Space Research Organization, the National Physical Laboratory, IIT-Delhi, and Indian Petrochemicals Corporation Ltd. There are also industries capable of producing commercial products from carbon fibres. The efforts of DRDO in producing missile bodies and rocket-launcher tubes has given further impetus to the application of high-performance composites and fabrication techniques. The Light Combat Aircraft (LCA) project has brought together several R&D institutions to work together in the use and further development of advanced composites. The role played by private enterprise in indigenous manufacture and supply of fibres and resins to industry, which accounts for about 80 per cent of total composite activity in the country, is highly commendable.

A Composite Product Centre (COMPROC) has been established to pass on the benefits of defence technologies to civilian sectors. Progress is being made on a number of projects of national importance: these include the design and development of composite rotor blades for wind turbine generators, and the production of prosthetic and orthotic appliances. Composite products identified for design and development during the next two years include: string musical instruments, golf club shafts and fishing rods, medical appliances (radiography beds, wheelchairs, headrests), aircraft interiors, refill cylinders for CNG, wave guides for microwave communications, carbon composite brake discs, railway components, automotive components, bicycle components.

*'Home-grown' materials technologies*

An initiative has begun to identify 'home-grown' technologies that have been developed (fully or partially) at laboratories and in industry in India. Out of 400 such technologies, ten proposals on new materials

**TABLE 5.5** Development projects in 'home-grown' technologies

| Sl. No. | Name of agency | Project | Industry/ institute |
|---|---|---|---|
| 1. | IICT, Hyderabad | CFC substitute | M/s. Navin Fluorides |
| 2. | IICT, Hyderabad | Carbon-based chemicals | M/s. Lupin Labs Ltd |
| 3. | NFTDC, Hyderabad | High energy rare-earth magnets | DMRL |
| 4. | NAL, Bangalore | 64-bit parallel computer | WIPRO |
| 5. | Hindustan Zinc Ltd, Udaipur | Cobalt recovery | – |
| 6. | ATIRA, Ahmedabad | High-performance synthetic substitutes of kerosene in pigment printing | – |
| 7. | Wayana Cromates (P) Ltd, Hyderabad | Ferromagnetic $CrO_2$ powder | RCTC, APIDC, HAL |
| 8. | Precision Graphite India Ltd, Bombay | Carbon fibre | NPL, NRDC |
| 9. | DCM Shriram Rayons | Aramid fibre | NAL |
| 10. | IICT, Hyderabad | Vitamin A based on synthetic route | Glaxo |

NOTE: IICT: Indian Institute of Chemical Technology; NAL: National Aeronautical Laboratory; DMRL: Defence Metallurgical Research Laboratory; HAL: Hindustan Aeronautics Ltd; NRDC: National Research Development Corporation; RCTC: Risk Capital Technology Corporation; APDIC: Andhra Pradesh Industrial Development Corporation; NFTDC: Non-Ferrous Technology Development Centre; ATIRA: Ahmedabad Textile Industries Research Association.
SOURCE: Department of Science and Technology, unpublished information.

are being advanced for commercialization in the immediate phase (see Table 5.5). The endeavour is likely not only to provide a fillip to the development of indigenous technologies but also to give grounds for considerable confidence among Indian technologists in some of the 'niche' areas.

**TABLE 5.6** Trends and implications in information technology

| | Trends | Implications |
|---|---|---|
| 1. | Micro-miniaturization | Smaller, portable, high-capacity systems possible |
| 2. | Interconnectivity | Networking of various equipment |
| 3. | Intelligent systems | Sensors, databases, instruments, warheads with decision-making capacity |
| 4. | Cordless systems | Wireless interfaces |
| 5. | Distributed systems | Parallel and distributed processing and programming possible |
| 6. | Systems integration | Diverse equipment and systems can be integrated and controlled centrally |

SOURCE: TIFAC, *Information Technology: State-of-the-Art and Imperatives for India*, Study 40 (TIFAC, 1993).

## Information Technology

The six major areas of change shaping the 'digital future' are shown in Table 5.6. These are likely to induce rapid diffusion of advanced information technology into many application areas. So far as India is concerned, the aim is to sharpen focus so as to become a major player in this area of technology.

### Computers

A large number of firms in India are producing PCs and workstations of different configurations. HCL-HP Ltd manufactures micro-CAD systems and higher-end workstations. The technological strength of DIEL is in using a common architecture for all its systems, through the use of mainframes; for instance, the performance linking of departmental computers consisting of VMS, Unix Systems, PCs and workstations into one cohesive integrated system. The R&D work on Intel chips has given Godrej and Boyce in-depth knowledge of the architecture of 8-bit and 32-bit microprocessors and microcontrollers. Godrej's 860 RISC (Reduced Instruction Set Computer) station, the world's first implementation of the 860 processor, has placed the firm in the global picture for hardware architectural design. Other firms such as ICIM, Modi Olivetti, Pertech Computers, Rolta India Limited,

Tata Unisys Limited, WIPRO Infotech Co., Zenith Computers, also provide special facilities to meet user demand.

The current estimated demand for PCs is one million per year. A growth rate of 50 per cent is expected in the hardware sector of the IT industry. The Indian information industry has graduated to a level where several foreign vendors and buyers are now interested in this market.

India has developed three supercomputer systems: (i) the 'Flosolver' of the National Aeronautical Laboratory (NAL) for aeronautical and fluid dynamic computations; (ii) the 'PARAM' by C-DAC; and (iii) the 'PACE' of the Defence Research Development Organization (DRDO).

A major initiative taken by India is towards Integration of Computational Facilities for Scientific and Engineering Research. Under this programme, major R&D institutions as well as academic institutes will gain direct access through networking to a few centres of excellence possessing supercomputers. Already around six such centres have been established. A systems Engineering Group (SEG) has been constituted to take stock of existing computational facilities and to recommend advanced system configurations based on the needs of the various institutions.

*Communication and networking*

The recent trend in India has been to integrate communication networking and transmission (see Figure 5.1). The technologies emerging in this field are: universal mobile telecommunication systems, multimedia, teleconferencing, data communications and gigabit networks.

The National Informatics Centre (NIC) has a satellite-based nationwide communication network called NICNET, with over 600 nodes, connecting the national capital with all the state capitals and district headquarters. The Centre has now linked all the important commercial centres through a high-speed Ku-band network of 64 Kbp trunk. This high-speed network is overlaid on the existing C-band network and is interconnected via packet switches. This 'information highway' enhances data communication speed, provides multimedia transmission, enables video-conferencing and transmission of images, and offers electronic data transmission, among scores of other facilities.

*Information on technology and manufacturing*

Only a limited number of organizations are working in this area. The major organizations involved in technology development are: the Centre for Artificial Intelligence and Robotics, BARC, HMT, Sieflex Ltd, IISc, IITs, Central Machine Tools Institute.

```
┌─────────────────────────┐                    ┌─────────────────┐
│   Data communication    │                    │     Cable       │
│         UPT             │                    │   Fibre optics  │
│  Mobile communication   │                    │    Satellite    │
│      Multimedia         │                    │    Wireless     │
│    Video conferencing   │                    │                 │
└─────────────────────────┘                    └─────────────────┘
```

   ( Communication  ⟲  Transmission )
     technology        technology

         ( Networking
           technology )

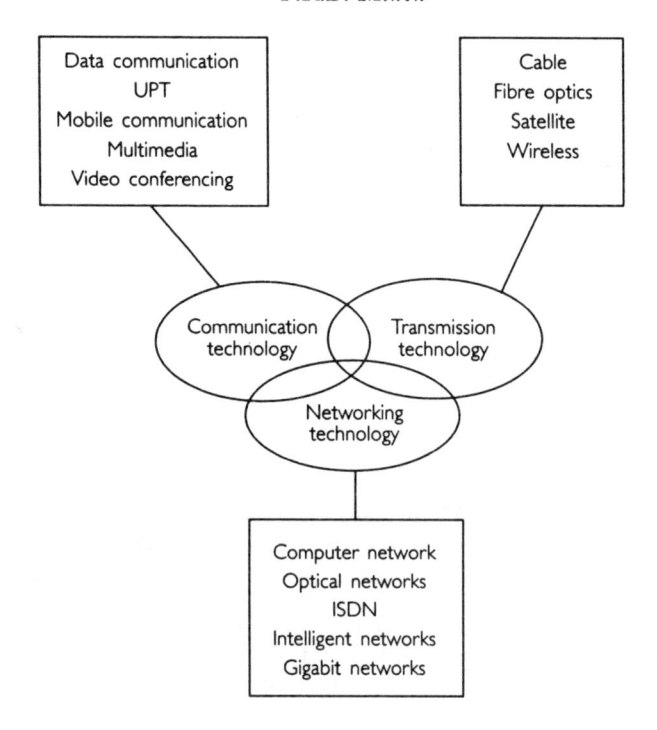

```
┌─────────────────────┐
│  Computer network   │
│  Optical networks   │
│        ISDN         │
│ Intelligent networks│
│  Gigabit networks   │
└─────────────────────┘
```

**FIGURE 5.1** Integration of communication, transmission and networking

SOURCE: TIFAC, *Information Technology: State-of-the-Art and Imperatives for India* (TIFAC, 1993), p. 407.

## Software

India's software industry output is only 2 per cent of world output. However, in the periods 1986–87 and 1991–92 the total output of the Indian software industry increased in value from $50 million to $245 million. In recent years the growth rate of the software industry has been much higher compared to that of the hardware sector. In 1991–92 software exports reached $135 million from $83 million in 1990–91. In 1993–94 software exports reached $330 million. A World Bank-sponsored study has predicted that Indian exports could have reached $1000 million by 1996.[2]

India has emerged as a major player in the supply of software for the global information highways. The latest initiative is to connect India, the United States, Canada and the United Arab Emirates through a satellite link. It is envisaged that packaged software for special construction and integrated engineering solutions and extended reverse engineering services will be supplied through the global satellite network.

## Notes

1. *Social and Public Policy*, vol. 10 (July 1983), no. 3.
2. 'International Software Studies – India's Software Services: Export Potential and Strategies', Department of Electronics (July 1992).

## References

Bhatnagar, D., 'Materials – TIFAC Initiatives' (unpublished document, 1993).

Chandrasekhar, S., 'Funding of Research and Promotion in Biotechnology: India and the Third World', Biotechnology Monographs Series 1, No. 1 (May 1993), pp. 75–84.

Ghosh, P.K., *Promotion of Biotechnologies in India: Growth of Biotechnology in India* (New Delhi: Narosa Publishing House, 1992), pp. 10–24.

Ghosh, P.K., 'Biotechnology in India', *Australian Journal of Biotechnology*, vol. 3 (1994), no. 4, pp. 214–22.

Ghosh, P.K., 'Investment Opportunities in Biotechnology-Based Indian Pharmaceutical Industry', Biotechnology Monographs Series 1, no. 1 (May 1993), pp. 7–21.

Government of India, Annual Report 1993–94, Department of Biotechnology, New Delhi.

*Perspectives in Science and Technology*. Technical Report by the Scientific Advisory Council to Prime Minister, Volume 1, Department of Science and Technology, New Delhi, 1990.

Ramachandran, S. and S. Natesh, 'Agriculture and Biotechnology Challenges and Opportunities in India', AISI Italian Association for International Development Workshop on Advanced Technologies for Increased Agriculture Production, September 1988.

TIFAC, *TMS in Biotechnology: Fermentation, Tissue Cuture, Medicine etc.* (TIFAC, 1992).

TIFAC, *Information Technology: State-of-the-Art and Imperatives for India*, (TIFAC, 1993).

# 6

# Public Policy and New Generic Technologies: The Case of Biotechnology in Sub-Saharan Africa

*Calestous Juma and John Mugabe*

**Introduction**

In recent years African countries have shown an increasing interest in biotechnology.[1] This interest has been enhanced by the growth in awareness of the subject generated through the recent negotiations for the Convention on Biological Diversity. Biotechnology has been viewed largely as a technical issue. However, the ability of the African countries to derive significant benefits from biotechnology will depend largely on the degree to which they reform their national policies to facilitate the acquisition and adoption of capabilities associated with biotechnology.

This chapter outlines some of the key policy research issues related to biotechnology and the prospects for more effective international assistance and cooperation. The first two sections deal with global and regional trends in biotechnology development. Much of the attention here is devoted to institutional issues. Section 3 stresses the links between biotechnology and biodiversity. Section 4 assesses the scope for international biotechnology cooperation. Section 5 presents six biotechnology policy issues that need to be considered when discussing development assistance. These represent the context of policy research in Africa: capacity-building, monitoring of international technology trends, policy studies, consultations and information dissemination. The chapter concludes by emphasizing the importance of innovations in modes of support for biotechnology policy research in Africa.

## 1. Global Biotechnology Developments

Biotechnology has experienced major advances in recent years and opened up a wide range of opportunities for application in the developing countries. The methods of undertaking genetic manipulation are being simplified and new areas of research such as sequencing the human genome are also advancing. It is difficult to establish the state of art in this field because of the rapid pace of technological innovation.

The majority of biotechnology research is still in the human health sector, followed by agriculture and biochemical processing. New areas of interest include environmental management as well as mining. Judging from the distribution of corporate activities among dedicated biotechnology companies in the USA, most of the R&D effort is directed to medical biotechnology.

As shown in Table 6.1, human therapeutics has received the most attention, followed by diagnostics, chemicals, plant agriculture, animal agriculture and reagents. It is notable that in Africa the emphasis has been on agriculture, and other areas such as therapeutics have received less attention. The direction of biotechnology research in Africa is influenced by the traditional research agenda and not the possibilities and prospects that exist in this new sector. While it is true that African agriculture requires special attention, it is not necessarily true that significant breakthroughs will also result from the same sector.

Furthermore, it is not necessarily true that the highest economic returns on research investment will also result from the same sectors. To the contrary, it can be argued that in most agricultural research areas, traditional breeding techniques have already pushed yields to relatively high levels, and the returns to research investment are not likely to be as high as in areas that have hitherto received less attention. It should be noted that the risks in non-agricultural biotechnology in Africa are relatively higher than in agriculture, and this may explain to a certain extent the current emphasis on agricultural biotechnology.

Other factors such as the line-up of research institutions also influence the priorities adopted in Africa. There are more research links between African and international organizations in agriculture than there are in other sectors, and therefore agricultural biotechnology developed in the industrialized countries finds its way to Africa more readily through these research channels. The direction of biotechnology research in Africa does not therefore reflect the global trajectory of research and the areas with highest potential for long-term participa-

**TABLE 6.1** Research focus of US biotechnology firms

| Research area | Biotech firms | % share | Large firms | % share |
| --- | --- | --- | --- | --- |
| Human therapeutics | 63 | 21 | 14 | 26 |
| Diagnostics | 52 | 18 | 6 | 11 |
| Chemicals | 20 | 7 | 11 | 21 |
| Plant agriculture | 24 | 8 | 7 | 13 |
| Animal agriculture | 19 | 6 | 4 | 18 |
| Reagents | 34 | 12 | 2 | 4 |
| Waste disposal/treatment | 3 | 1 | 1 | 2 |
| Equipment | 12 | 4 | 1 | 2 |
| Cell culture | 5 | 2 | 1 | 2 |
| Diversified | 13 | 4 | 6 | 11 |
| Other | 31 | 18 | 0 | 0 |

SOURCE: Office of Technology Assessment, *Biotechnology in a Global Economy* (Washington DC: US Congress, 1991).

tion in the global economy. This point is important considering the fact that Africa's traditional cash crops are currently threatened by biotechnology products, and the continent must prepare to adjust its production to new crops and industrial products.

Major biotechnology initiatives affecting cacao and sugar-cane do not augur well for those countries in Africa that depend on these crops for their foreign-exchange earnings. Pennsylvania State University, with support from the US chocolate manufacturers, has taken the initiative to improve the quality of cacao plants through genetic engineering. Should this initiative succeed in producing commercially viable cacao hybrid varieties, the organization of production will most likely favour large, capital-intensive farms. This would certainly have adverse financial consequences for the major cocoa-producing countries in Africa.[2] Major food companies in the USA, such as Nestlé and Hershey, have teamed up with biotechnology R&D firms and Cornell University and other universities to produce cocoa butter in the laboratory through tissue culture techniques. The aim is to 'dissociate cocoa butter production both from cocoa imports and from all land constraints, since cells could be grown in fermentation vats.'[3]

The other initiative, which is in the sugar industry, is proving successful. Sugar substitutes such as high fructose corn syrup (HFCS) – a sweetener from corn (maize) using immobilized enzymes – has begun to compete with refined sugar. For instance, the per-capita consumption of refined sugar in the USA in 1970 stood at 46.3 kg and that of sugar substitutes at 12.1 kg. If the present trend in the quite dramatic fall in consumption of natural sugar in North America and Western Europe continues, the livelihood of millions of people in the developing countries will be at stake.[4]

The competitive advantage of nations in international trade is currently being defined by the technological competitiveness of the leading firms in these countries. African countries can only seek real economic transformation in the acquisition and development of technological skills. Africa is still at a disadvantage on this front and therefore any choice of technology or priorities for technological development needs to take into account these long-term competitive considerations.

The current focus is not in itself a disadvantage given the fact that Africa has genuine agricultural problems that can be solved by biotechnology. The problem, however, is that Africa may be pursuing a route that has more limited opportunities in terms of overall biotechnology development. By paying less attention to other areas of biotechnology, Africa may be limiting its ability to draw from fundamental advances in these sectors.

Advances in tissue culture have been of particular interest to the African countries because of their potential to solve some of the persistent problems of African agriculture. The potential benefits of tissue culture include rapid plant multiplication, development of disease-free plantlets, production of uniform plants, year-round propagation of plants, rapid development of improved varieties and better conditions for exchange and storage of genetic material; as well as the related large-scale production for the mass market.

One of the key features of agricultural biotechnology research is the growing collaboration between the International Agricultural Research Centres (IARCs) and the institutions in the industrialized countries. This collaboration has made it possible for the centres to have access to the latest techniques developed by specialized research institutes.

As pointed out above the emphasis in agricultural biotechnology research has been on traditional crops. This is not unusual since these crops already have a market, and research investments are expected to improve the economics of producing the crop. Many of the crops that are receiving biotechnology attention are also major crops in the region. These include banana, cassava, cocoa, coconut, coffee, potato, pyrethrum, rice and wheat. The most advanced application of biotechnology to these crops is diagnostics. This is followed by rapid propagation. Less work has been done in transformation and regeneration systems. Except for crops such as potato, rapeseed and rice, the application of biotechnology is still in the medium range (5–10 years) and long range (over 10 years). This projection, however, could change considerably depending on advances in technological innovation. Already, recently developed techniques such as restriction fragment length polymorphism (RFLP) are revolutionizing genetic engineering. RFLP makes it possible to undertake breeding activities without first identifying the relevant genes or understanding their biochemical processes.

Transformation and regeneration systems are available for other crops. These include maize, rye, cotton, flax, soybean, carrot, cauliflower, celery, cucumber, potato, tobacco, tomato, lucerne, stylosanthus, poplar and walnut. The range of new products is expected to increase as the techniques for genetic manipulation become more refined.

## 2. Biotechnology Research in Eastern Africa

### Developments in biotechnology research

Biotechnology research in eastern Africa is scattered across the institutional terrain with varying levels of activity, funding, expertise and experience. The work is carried out mainly in international research institutions located or operating in the region, national research bodies and universities. The private sector is becoming interested in the prospects of biotechnology and is starting to fund local research institutions to conduct certain types of biotechnology research. These are mainly large corporations with financial resources but limited research capability. But, on the whole, the role of the private sector is still limited.

There are no institutions expressly charged with the mandate to conduct research on biotechnology. However, feasibility studies have been carried out in countries such as Kenya and Burundi to establish the viability of setting up biotechnology centres. The studies aim at consolidating biotechnology under currently existing institutions.

Biotechnology research and development (R&D) in most African countries have evolved from prior agricultural research activities. Except for Zimbabwe, Kenya and Ethiopia, which have attempted to outline priority research areas in biotechnology, most countries in the region are yet to identify research areas that relate to their socioeconomic needs. Biotechnology research in these countries has built on individual initiatives with the support of a few donor agencies. Zimbabwe's National Biotechnology Programme has been developed by the Research Council of Zimbabwe. The Programme sets agricultural biotechnology as the major priority area for research. It puts emphasis on the development and improvement of crops and livestock, the development of high-yielding crop varieties, the application of new techniques in food processing and preservation, and the conservation of indigenous genetic resources.

Biotechnology is still treated within the general framework of the national science and technology policy. However, a committee comprising several policy-makers and scientists from the University of Zimbabwe has been set up to draft a national biotechnology policy. The committee is expected to cover issues such as priority research activities, funding, intellectual property protection, information acquisition, training, the conservation of biodiversity and development of university–industry links.

In Kenya a national conference on biotechnology was held in March 1990. The aim of the conference was twofold: to establish ways of creating a national biotechnology institute and to identify crucial issues of biotechnology that should be articulated in the national biotechnology policy. The conference recommended that agricultural biotechnology (the development of crop varieties adapted to most of the ecological zones of Kenya) and livestock husbandry should be treated as priority research areas. It also addressed biosafety and intellectual property protection issues. Following this conference, the then Ministry for Research, Science and Technology established a national Biotechnology Advisory Committee composed of the directors of the seven national research institutes.

The Committee has been mandated to include representatives from the private sector with the purpose of advising the ministry on policy and institutional issues in the use of biotechnology across institutional sectors in Kenya. From its inception it was envisaged that the Committee would be responsible for such policy issues as patents, biosafety, ethics, commercialization of inventions and discoveries, and the viability of and prospects for establishing a high-level national biotechnology centre. The National Council for Science and Technology (NCST) has also set up a unit to look into intellectual property issues and intends to establish guidelines on biosafety.

However, since the conference little else has been done to implement the findings and resolutions. The country does not have a policy to guide technological and institutional developments in the regime of biotechnology. Most of the research activities are conducted in the public institutes without much understanding of the needs of the majority of the population: research priorities are essentially set by the researchers.

Ethiopia has made some attempts to define specific priority areas for research in biotechnology. The areas identified include environmental rehabilitation and protection, food production, health and industrial food processing. The priority areas are laid down in the overall science and technology policy document.

The Ethiopian Science and Technology Commission (ESTC) is in the process of developing a biotechnology policy. The Commission plans to establish a coordinating office in biotechnology. An *ad hoc* committee has been established to identify and study crucial issues in biotechnology which should be articulated in the national biotechnology policy. ESTC could go about this task by first carrying out a preliminary study to evaluate the nation's biotechnology assessment capability. This could be followed by sending out a national task force to various countries to study the methodologies used on the issues considered in the formulation of a national biotechnology policy. The results obtained could then be appropriately applied at home.

In Cameroon four of the country's five research institutes and existing facilities at three universities are in a position to assume a leading role in biotechnology if fundamental restructuring of the curriculum is undertaken. The creation of the University of Yaoundé Biotechnology Centre (BTC) at Nkolbisson, the university centres at Dschang and Ngaoundere, together with the activities of the Institute of Agriculture and Forestry Research (IRA), Institute of Medical Research and Medicinal Plant Studies and Research (IMPM), Institute of Zootechnical Research (IRZ) and the Institute of Geological Mining (IRGM), can in different ways contribute to the economic renewal programme and industrialization of the country through their different science technology, and R&D activities.

Most other African countries have not yet identified and clearly defined areas of priority in biotechnology. Cameroon, Tanzania and Uganda have recognized the role played by the technology in sustainable socioeconomic development and are already engaged in various biotechnology research activities. However, they have not started the process of defining priority areas of research and formulating national biotechnology policies. They continue to treat biotechnology issues within the broader science and technology policy framework.

## Institutional issues

The development of biotechnology in the region has moved faster than the countries have been able to formulate new institutional arrangements for coordination of research activities. The development of biotechnology in Zimbabwe is guided by RCZ, which is a government agency charged with the responsibility of promoting scientific and technological research and development activities. It advises the government on research priority areas and on ways of ensuring that research activities relate to national needs and capabilities. To enhance biotechnology activities in the country, efforts are under way to establish the Biotechnology Research Institute at the Scientific and Industrial Research and Development Centre (SIRDIC). The institute will have the responsibility of identifying and coordinating R&D activities in biotechnology. It will also be the organ for formulating national biotechnology policies that will direct Zimbabwe's aspirations in biotechnology to meet sustainable development needs. The Institute will bring together scientists from the University of Zimbabwe and other institutions engaged in biotechnology research activities to form a national team of researchers. This will enhance the exchange of scientific and technological knowledge.

In Ethiopia biotechnology research activities are carried out by a number of institutions. The Institute of Agricultural Research (IAR) is responsible for formulating a national policy for agricultural research within the framework of overall national development planning. The IAR has established an *ad hoc* committee to identify and formulate a national agricultural biotechnology strategy which articulates the main agricultural research priority areas for the country. Currently IAR is running five major agricultural biotechnology related projects: cereals, legumes, coffee, horticultural, livestock and alternative energy sources. The Institute is in the process of developing training programmes that will be geared to biotechnology research capability in the country.

Other institutions that spearhead biotechnology activities in Ethiopia include international organizations such as the International Livestock Centre for Africa (ILCA), and the Armauer Hansen Research Institute (AHRI), which does work on leishmaniasis and leprosy. Research in biotechnology in Kenya is carried out by a number of public institutions in conjunction with the NCST and the MRST. Examples of these institutions are the Kenya Agricultural Research Institute (KARI), the Kenya Forestry Research Institute (KEFRI), the Kenya Medical Research Institute (KEMRI), the Microbial Resource Centre (MIRCEN) based at the University of Nairobi's Department of Soil Sciences, the Kenya Trypanosomiasis Research Institute (KETRI), the International Laboratory for Animal Diseases (ILRAD), the International Centre for Research on Agroforestry (ICRAF), and the International Centre for Insect Physiology and Ecology (ICIPE).

Biotechnology R&D activities in Cameroon are currently coordinated by the Ministry of Higher Education, Computer Services and Scientific Research. The Ministry is charged with the responsibility of overseeing that R&D activities being conducted at the IRA, IRZ, IRGM, IMPM, University of Yaoundé and the Institute of Human Sciences (ISH) are well coordinated and relate to national development needs and capabilities.

In Uganda, there is no specific institutional mechanism to coordinate biotechnology activities. Currently most of the biotechnology-related research is done at Makerere University and a number of the public research institutes. The activities are evolving from the Green Revolution institutional set-up. There is no institutional mechanism to guide the evolution of biotechnology R&D activities so that they reflect national research capability and development needs.

Biotechnology R&D has not yet received much attention in Tanzania, but some work is being initiated at various institutions. An example of this is the Applied Microbiology Unit at the University of Dar es Salaam. This is the product of a technical assistance arrangement between the University of Dar es Salaam and the University of Nijmegen in the Netherlands.

## 3. Biotechnology and Biodiversity

The development of biotechnology is closely linked to the conservation of biological diversity.[5] It has taken a long time for this link to become explicit. Even now, only a small number of developing countries have formulated strategies to derive biotechnology benefits from genetic resources in their territories. A number of policy issues affect the ability of African countries to benefit from utilizing their resources. First, institutions that safeguard intellectual property rights in the form of technological innovations are larger and stronger than those that protect the interests of local communities involved in conservation efforts. Indeed, most possess scientific, technical and legal expertise available to few conservation groups. This issue has been extensively discussed in recent years and informed many of the provisions of the Convention on Biological Diversity signed in 1992 in Rio de Janeiro. It is still being discussed in the context of the follow-up to the convention.

More generally, African countries are not likely to benefit fully from the provisions of the biodiversity convention or other transactions involving genetic resources until new legal and institutional regimes are established. Accordingly, African countries should establish public agencies to address intellectual property concerns within the framework of national legislation and use them as a basis for international negotiation. But so far, little research is being carried out in these countries on alternatives to the existing intellectual property rights systems.

Second, African countries have limited national institutional and legal capacity for regulating access to genetic resources. Currently only the assertion of national sovereignty can be used in negotiations over access to biodiversity and biotechnology. During the negotiations for the biodiversity convention, for example, the question of sovereignty arose during discussion of a proposed global list of habitats and species threatened with loss. By arguing that the publication of such a list would impinge upon national sovereignty, developing countries were able to keep it from being included as a key element in the Convention. But unless legal and institutional regimes are established to assert and enforce that sovereignty, how can African countries benefit fully from the provisions of the Convention?

Third, African countries have tended to separate institutions working on biodiversity management from institutions involved in biotechnology. In fact, ratification of the Convention could stimulate the development of effective biodiversity prospecting and biotechnology development institutions by providing the policy, legal and institutional bases for linking biodiversity and biotechnology. Today, few of the key principles of this Convention are reflected in the character and organization of national and international biodiversity institutions.

The development of biotechnology represents the convergence of a wider range of skills and knowledge than any single institution or country is likely to possess. Thus, institutional cooperation will be critical to biodiversity prospecting and biotechnology development success. Clearly, the coordination of biotechnology activities, as part of the larger enterprise of science and technology, needs to be given legitimacy and impetus at the highest level of government. Unfortunately, the growing awareness of the role of science and technology in many African countries is rarely accompanied by measures aimed at putting this vital issue at the core of national development planning and economic liberalization activities.

Many African countries have set up stand-alone institutions to promote biotechnology or biodiversity conservation. But such units – often established in anticipation of funding rather than out of genuine interest to promote the use and conservation of biodiversity – are unlikely to yield long-term benefits unless they are part of a broader institutional and policy framework. New institutional arrangements aimed at promoting biotechnology development should be constructed with national goals in mind, rather than the politics and aspirations of donors.

Institutional cooperation should be extended beyond national boundaries – difficult for most African countries since their institutional arrangements for technological cooperation are limited. There are two main ways of dealing with this issue. One is through multilateral arrangements. For instance, African countries can use facilities of international institutions such as United Nations agencies, the International Centre for Genetic Engineering and Biotechnology (ICGEB), the Scientific Committee of Genetic Experimentation COGENE) of the International Council of Scientific Unions (ICSU), and the Consultative Group on International Agricultural Research (CGIAR) to acquire technological skills. Another is through bilateral arrangements provided by institutions such as the US Agency for International Cooperation (AID), the Directorate General of Development Cooperation (DGIS) of the Netherlands, and the Swedish International Development Authority (SIDA).

Fourth, harmonizing the rights of the state and local people to a nation's biodiversity is difficult. A government's interest in asserting national sovereignty over genetic resources in its territories should not override individual and human rights, especially those of local communities and indigenous peoples. Similarly, international prospecting institutions may erode more meaningful rights already held by local communities and undermine farmers' attempts to conserve biological diversity. For this reason, any international biodiversity prospecting and biotechnology development activities should be carried out in concert with national institutions that will make sure that the rights of local communities are recognized and enforced.

The Convention recognizes the contribution of indigenous knowledge and technology to sustainable development (see Preamble, Article 8(j) and Article 10(c) and (d)). This is an important step, but still needed are more specific and justifiable measures that ensure that the rights of innovators at all levels are recognized and protected. Without them, local communities will not be convinced of the benefits of conserving genetic resources. Unfortunately, the Convention does not provide a strong basis for such measures.

Finally, attempts to link biodiversity and biotechnology in international negotiations have so far tended to place excessive emphasis on monetary compensation and enforcement of intellectual property rights. Although these concerns are important, biodiversity prospecting and biotechnology development will not contribute much to the African countries unless they help them acquire and build up technological capacity. To ensure that national biotechnology policies enhance biodiversity prospecting, far more attention needs to be paid to human resource development, technological innovation, legal and institutional reforms, biotechnology regulation, and intellectual property management. These should be key elements in efforts to build policy-making capacity.

## 4. International Biotechnology Cooperation

### Technological capability

There are varied perceptions among donors, researchers and institutions as to whether support for physical infrastructural development should be regarded as part of the support necessary for R&D especially in the non-governmental sector. The subject is gradually becoming a major policy consideration in the region. It is notable that the issue is not normally articulated in this way but instead often in the context of 'access to biotechnology'. The issue has been extensively discussed in the framework of international negotiations for a convention on biological diversity. The region, led by Kenya, has strongly argued for linking access to genetic resources by the industrialized countries to access to biotechnology by the developing countries. Unfortunately, the industrialized countries tend to view the issue of access to biotechnology as similar to the transfer of other technologies.

Unlike mechanical technologies, biotechnology is science-intensive and its transfer or diffusion is often mediated through training and information flow and not necessarily the movement of equipment across borders. Issues such as training and information exchange are not being seriously discussed because they are not perceived as major channels for the transfer of biotechnology. Much of the discussion deals with intellectual property issues and not with access to public domain technologies. It is necessary to explain to policy-makers the various media for technology transfer and acquisition and how they differ from one form of technology to another.

The issue of access to biotechnology should be seen in a broader context of the acquisition of knowledge and expertise. Much of this is generic and therefore not restricted to a specific category of biotechnology.

## Technology transfer and acquisition

Technology transfer describes a complex process whereby the scale and type of flow of equipment, knowledge and expertise among nations changes over time. As used in international debates, the term has come to mean the transfer of systematic knowledge for the manufacture of a product, for the application of a process or for the rendering of a service, and not the simple sale of goods and services.

Implicit in this definition is the notion that the recipients in the transaction already have a critical minimum capacity to assimilate and operationalize the new technology – what Chapter 37 of Agenda 21 (the programme of work from the United Nations Conference on Environment and Development) calls human, scientific, technological, organizational, institutional and resource capabilities. Effective technology transfer also requires prospective recipients to negotiate – to articulate their goals and choose wisely among alternatives. In the same chapter of Agenda 21, further prerequisites are outlined: 'Skills, knowledge and technical know-how at the individual and institutional levels are necessary for institution-building, policy analysis and development management, including the assessment of alternative courses of action with a view to enhancing access to and transfer of technology and promoting economic development.' This list boils down to the possession of a basic infrastructure and a good command of the issues involved.

In practice, though, most African countries treat technology transfer as a process initiated and controlled by the holders of technologies for their own ends. Developing countries often contend that strong intellectual property rights in developed countries are the main obstacle to securing access to biotechnology and that simply removing certain barriers will considerably increase the flow of technology to the African countries.

Although abiding by the strictures of intellectual property rights does increase the transaction costs of getting access to biotechnology, the intellectual property and licensing restrictions are more scapegoats than major obstacles, and relaxing them will not necessarily lead to technology transfer. Developing countries have failed to utilize fully the technological information that is in the public domain – or that for which the patents have expired. Indeed, most of the biotechnologies that developing countries need are in the public domain and are based on conventional practices. Moreover, countries that cannot utilize public domain technologies are not likely to use patented ones.

A genuine and effective technology partnership will require willingness to transfer the relevant technology. But technology cooperation must also involve efforts by developing countries to *acquire* technology. The measure of how well such a partnership works will be reflected in the African countries' capacity to absorb and assimilate technologies, as well as develop local technological capability. To make technology appropriate for local circumstances, indigenous contribution to technological development is crucial, and this depends on indigenous technological capability.

Key elements of strategy to foster biotechnology transfer are training and access to information. Many African countries have severely restricted the international flow of technical information or have failed to give the local scientific community incentives to exploit the information available internationally. If the communications infrastructure is not improved and the research environment not liberalized, African countries will be hard pressed to enter the field of biotechnology.

Better pricing of biodiversity goods and services is also essential. Royalties received on patented technologies represent only a small proportion of the total cost of acquiring and operating imported technologies. Technology searches, negotiation, and user training cost far more – as does complying with the restrictive clauses in licensing agreements on the use of technology. This is not to say that intellectual property rights are not important, especially since they are sometimes used as non-tariff trade barriers. Nor is there any implication that additional resources to finance payments for proprietary technologies are unnecessary. But it is a mistake to think that biotechnology development can be boiled down to the issue of intellectual property rights.

Indeed, for African countries, most of the challenge in fostering the mutual growth of biotechnology and conservation of biodiversity lies in creating a political, economic and institutional environment in which the provisions of the Convention can be fully realized. African countries will not be able to develop biotechnological competence. Among other things, 'technological capability' means being able to manage a new technology deployed in the economy.

The acquisition of technological production capacity is associated with the flow of different kinds of knowledge and expertise. The first category includes the know-how needed to transfer and set up production facilities and all the various operational services associated with any investment project. The second category includes the expertise needed to maintain the new system once it has been installed – both the codified knowledge in manuals, schedules, charts and diagrams and the 'people-embodied know-how' fostered through training, information services and on-the-job learning. The third category includes the knowledge and expertise needed to implement technical change: an understanding of how technological systems work and the techno-managerial capabilities needed to evaluate and transform technologies already operating to meet new conditions.

In genuine technology transfer, developed countries help developing countries build up all three types of capacity, starting with the first and working up to the third. The third category is what triggers technological dynamism and helps countries fully utilize both public domain and proprietary technologies; policies that merely help countries adapt to international trends won't reach critical mass in this regard. The key policy elements are integrating legal provisions in technology transfer contracts (for example, on the local content) that emphasize the development and utilization of local resources – for instance, using national engineering, environmental and legal consultancy companies at all stages of project planning; proving national facilities for research and technological services; and providing an adequate national training system for scientific, technical and managerial manpower.

The most important factor in enhancing competitiveness in biotechnology is a country's ability to bring available knowledge and expertise to bear on the development of specific products and processes. The entry barriers for biotechnology – mastering traditional techniques, such as tissue culture – are lower than in other frontier technologies, such as microelectronics, so developing countries have unique opportunities to enter the field. Moreover, such precedents as the development of diagnostic kits for tropical diseases in Africa confirm that a small group of well-trained scientists in the South can contribute heavily to biotechnology development.

But African countries cannot fully benefit from the new access to biotechnologies provided for in the Convention unless they build a broad base of knowledge and expertise in complementary fields and make their institutions more amenable to research cooperation. Consider in this light the broad knowledge base needed to become a player in biotechnology. Article 2 of the Convention defines biotechnology as 'any technological application that uses biological systems, living organisms, or derivatives thereof, to make or modify products or processes for specific uses.' Not only does developing such applications require a deep grasp of molecular biology; commercializing them requires knowledge of a wide range of other technologies and processes – such as fermentation technology and process engineering – based on biological systems. Thus, training to meet the challenges of the knowledge-intensive industries of biodiversity exploration and biotechnology development must encompass genetics, taxonomy, molecular biology, biochemistry and process engineering. Both basic bioscience training and that in the 'applied' disciplines, such as biochemical engineering skills from biochemistry and microbiology, should be emphasized.

Concerns over financing, access to genetic resources, and access to and transfer of technology, tended to dominate negotiations for the Convention. But because biotechnologies are science-intensive and their economic application relies largely on the available technological capacity, the Articles promoting training, information exchange and technical cooperation are just as crucial to the transfer of biotechnology and the consequent improvement of African states' ability to conserve and sustainably use biodiversity.

How successful a country is in negotiating for favourable contractual arrangements with foreign firms or governments often determines its ability to promote its technology development programmes. Further, countries that explicitly recognize technology development as part of their long-term goals are less likely to be blinded by financial assistance concerns.

Developing countries often seek to legislate the terms of technology transfer and assume that such legal provisions will protect them against unfair business practices. Some, following the trend toward increasing economic liberalization, have also reduced restrictive regulations. But developed-country firms entering into technology transfer agreements may include a wide range of clauses that restrict the importing countries' ability to acquire the relevant technological competence.

In some biotechnology transfer negotiations, approaches formulated for other forms of technology, such as those embodied in the 1987 Montreal Protocol on Substances that Deplete the Ozone Layer and the 1990 London Amendment to the protocol, have been used. But two categories of technologies covered in these agreements differ fundamentally from biotechnology. First, ozone-related technologies are clearly identifiable products and processes; in contrast, biotechnologies represent a set of techniques that cannot be applied without expertise in a wide range of other sectors. Also, the scope of application of the ozone-related technologies is fairly obvious and easy to determine before they are employed, unlike that of generic biotechnologies. Negotiations on biotechnology partnerships should therefore be based on a detailed understanding of the technical characteristics of biotechnology, international technology trends, and the changing corporate strategies employed in market competition. But dealing with such issues requires capacity in policy analysis that is currently missing in most African countries.

## 5. Policy Research Issues

### Policy and policy research

In light of the above assessment of options, it is recommended that efforts to strengthen policy-making capacity in biotechnology be viewed as part of a process to support institutional development in the field of science and technology policy issues. The first step in this direction would be to reshape existing institutions in the various countries to bring science and technology to the centre of national development efforts. This can be done through a number of measures, depending on national realities. A recommended suggestion would be to set up a national council or commission for sustainable development which could incorporate key functions such as economic policy reforms, science and technology, human resource development, and environmental management. Such national efforts need to be complemented by regional institutions that provide policy research support, organize policy training programmes, and coordinate policy consultations.

It has been stressed that achieving sustainable development will require major policy changes in African countries. Such statements are often made without the necessary backing of an appeal for studies that indicate the kinds of changes that need to be introduced. The tendency has been for the industrialized countries to propose the extension of their policy measures to the African countries. This has often been possible because the discipline of policy research is nascent in developing countries and virtually absent in Africa. Pronouncements of a few people familiar with the policy needs for specific changes have often been confused with policy research. This problem is compounded by the fact that scientists have often believed that they should speak to power directly and that their research results should form the rational basis for policy-making.

This view, however, ignores the fact that interaction between decision-making and reality is not always rational and depends on a wide range of factors, which are the subject matter for policy analysts. The need for policy research and training has increased in recent years, especially given the changing world order and demands for democratic reforms in Africa. The traditional patterns of international relations are being redefined at a time when the pace of technological innovation is quickening. The introduction of multi-party politics and political reforms in Africa is changing the locus of policy decisions. In addition, the reduction in public sector participation is also altering the location of power in decision-making on technology issues.

These changes are occurring at a time when scholars are starting to question the foundations of past attempts to form regional markets in Africa. Emphasis is increasingly being placed on production-oriented approaches instead of trade-related measures. There is a sense of urgency in Africa to rethink development policies and programmes. This exercise requires competence in the field of policy research.

The few examples of policy reforms arising from research show clearly that efforts to promote policy research require the establishment of institutions specializing in this activity. There is more to the process of policy reform than the mere presentation of the results of studies at conferences. Realization of the reforms involves extensive follow-up and refinement of options to levels that can only be achieved through dedication. For instance, it took no more than three months to undertake a policy study on industrial property protection in Kenya. However, it took about a year to move from the identification of policy options to the enactment of legislation. It took another six months before the Kenya Industrial Property Office (KIPO) was established.

In situations where there is no research to back up policy-making, it is not unusual for policy reforms to take more that five years. In many other cases, laws are passed but not implemented because of the lack of ideas on how to implement them. Some laws become dormant because the research necessary to identify conflicts and overlaps with other laws is not done. Even where legislation and policies are being implemented, crises may occur because of lack of further research to evaluate their implementation. Such problems could be reduced through effective policy research and feedback between such research and the policy-implementation agencies.

So far, policy science is not taught in any of the sub-Saharan African universities as an independent field, but is often treated as part of economics, political science, government or other social-science disciplines.[6] Even more critical is the absence of training opportunities in science and technology policy analysis. Therefore, a starting point would be to strengthen biotechnology policy research capacity in the context of overall policy sciences.

## Capacity building

The building of capacity in the field of science and technology policy in general, and biotechnology in particular, cannot be supported unless considered in the context of specific activities. It should be noted that numerous donors have supported a wide range of projects in the name of 'capacity building', but such support has not been resulted in any important change in the configuration of institutions in the region. The term 'capacity building' is increasingly becoming a general term that is used to mean the same types of isolated projects that donors have supported in the last three decades in Africa. In many cases, such projects have contributed to the erosion of institutional capacity, or the projects were designed under the assumption that the institutional capacity to implement them already existed.

In the last three decades much of the support for institutional development has gone to international institutions conceived elsewhere but located in Africa. While these institutions have played an important role in their areas of competence, they have often been isolated from the policy-making process, and their impact has been minimal. In some countries, policy-makers have tended to avoid obtaining policy advice from such institutions because they view them as external and pursuing their own agenda. International institutions located in Africa are often viewed by host governments as places which provide local employment and bring foreign exchange into their countries, and not necessarily as potential sources of ideas for development and policy-making.

It is recommended that the strengthening of policy-making capacity in these countries be linked to the development of specialized institutions that focus on research, training and information dissemination, and that promote interaction and policy dialogue among representatives from governmental agencies, NGOs, research institutes and the private sector. Such institutions could enter into collaborative arrangements and organize joint activities. However, networking should be seen as an activity that emerges from the strengthening of certain efforts, and is not created merely to coordinate ongoing activities.

In view of the decline in institutional capacity in many African countries, it is important to place emphasis on training. This, however, involves a wide range of activities from university education to specialized courses. In most cases, training has been perceived as a means to provide individuals with skills instead of building institutions. It is recommended that training programmes be viewed in the context of long-term institutional development. The kinds of arrangements that are entered into between universities to support staff development need to be extended to research non-governmental organizations (NGOs).

Most donors insist on providing scholarships through bilateral programmes; and therefore government agencies are given first priority whenever training opportunities emerge. There are few cases where government agencies have extended these opportunities to NGOs. These organizations are often associated with short, non-degree courses, a factor which perpetuates the myth that NGOs are low-level institutions. As noted elsewhere, highly skilled researchers are starting to move to NGOs, and some of these institutions now have more trained personnel than many university or government departments. It is therefore important to design training packages, especially for postgraduate education, for NGOs.

## *Monitoring global technology trends*

The ability of African countries to formulate and institute coherent policies to guide the evolution of biotechnology largely depends on their capacity to monitor international developments in biotechnology that have an impact on the local economies. Currently there are no national programmes to study and analyse the impacts of technological developments in biotechnology and changes in international policy regimes on African economies. The few impact studies that have been generated in Africa do not discuss what policy measures African countries should adopt to address any negative impacts arising from developments in biotechnology.

One way around this problem is to develop an institutional set-up that will inform African leaders and policy-makers on international trends or changes in biotechnology. One way is for African institutions to establish links with policy research institutions in the North. This may offer them access to information on biotechnology policy development. The other way is to have Southern technology policy analysts situated in the North with the responsibility of studying and analysing biotechnological developments that have socioeconomic implications for local economies.

One of the main obstacles to policy-making in Africa is the absence of information on technological developments in other countries, as well as details of policy trends in other parts of the world. Most of the

African countries make policies under conditions of ignorance and often fail to take into account important developments which may influence their ability to implement the proposed actions. For example, very few African countries have made efforts to study the implications of the single European market for their own development strategies. They have often relied on UN agencies to provide information on these developments. However, in many cases, by the time such information is released, it is already out of date. The same concerns apply to access to scientific and technological information which is not readily available in these countries.

There are numerous ways to enable African countries to have access to the latest information necessary for decision-making. These include conventional needs such as strengthening library facilities, supporting efforts in the industrialized countries to provide such information, and encouraging the participation of African researchers at international scientific and policy conferences.

One area that requires concerted effort is access to the numerous databases available in the industrialized countries. For some African countries it is now possible to support on-line database facilities. However, for most of them, it is necessary to supply off-line facilities. It is also possible in selected African countries to establish satellite database facilities which can be used for searches. The provision of off-line facilities should be a starting point in the transition towards connecting Africa to the world network of scientific and technological information. In the past, such facilities have often been located in government agencies. It may be necessary to diversify the availability of such facilities and locate them in NGOs as well.

Activities have often been viewed as projects instead of being part of the effort to support the building of the relevant institutions. Such support should be linked to the building of special subject libraries at institutions specializing in science and technology policy studies. Such libraries, supported by on-line or off-line database facilities, could become important referral facilities for researchers and policy-makers. The main strategy for such support should be to strengthen institutional capacity to provide these services instead of merely providing support for projects in these areas.

## Policy studies

Much of the policy research relevant to biotechnology has been carried out in the form of 'networks' involving individuals. The establishment of networks often assumes that the institutional basis to sustain the research exists. However, recent experience has shown that networks have often been loose associations of individuals, not institutions; therefore the research activities tend to die as soon as support from the original donor stops. It is thus felt that an intellectual locus is necessary to support policy research. Support for individual projects should be accompanied by support for institutional development in the field of technology policy in general and biotechnology in particular. While many of the researchers from NGOs and government agencies felt that such support should be provided through universities, researchers from NGOs felt that such research should be conducted through institutions that are impartial.

The view here is that a certain measure of diversity should be maintained so that different institutions can undertake policy research that is relevant to their objective. For example, non-partisan institutions may focus on training and identification of policy options. However, partisan and governmental agencies as well as private-sector institutions may be interested in carrying out research that adheres to their stated policies.

In view of the fact that not much technology policy research is conducted in the region, the researchers and policy-makers interviewed did no more than provide a checklist of policy concerns. Many of these concerns relate to the broader problems of technological development. Not much attention has been given to biotechnology policy research except in the context of agricultural development.

## *Consultations*

The intense debate on issues relating to the exchange of genetic material and transfer of protected technologies will not be resolved through international conventions or UN fora. The issues seem to be so complex that they require a well-developed institutional environment that enhances dialogue between the different parties. For African countries the situation is even worse since most of those involved in such international negotiations have limited knowledge or experience in technology, environment and public policy issues. Moreover, most of the negotiations take place without much prior consultation on different policy issues; hence the space available for rigorous and informed discussion is limited.

The way out is to institute an activity that will allow African policy researchers and policy-makers on issues of science, technology and biodiversity to have informed consultations with their counterparts in other regions. This has the advantage of ensuring that those African representatives to UN discussions on technology transfer and biodiversity issues have some information on some of the important policy issues. Consultations should be continuous in order to enhance the capacity of African countries to negotiate for new technologies and resolve some of the crucial issues of property rights in genetic materials.

The organization of consultations on emerging policy issues is becoming an important activity, especially in light of the growing complexity of issues related to biotechnology. It is notable that African representatives at various negotiation fora are often unprepared and tend to rely on outdated views and positions on international issues. The tendency has often been for a few countries to sell their own agenda to the rest of the countries under the banner of collective action. Such consultations could be important in clarifying key issues as well as helping the African countries to formulate common positions on issues of regional concern. Such consultation should be based on well-prepared studies and should not be a substitute for conferences.

It is recommended that such consultations be aimed at high-level policy-makers, and also be organized in the context of overall institutional development. It is important for policy-makers in the region to identify a few institutions as sources of ideas for policy-making and a focal point for consultation. The organization of consultations requires access to information and the ability to mobilize experts in various fields to contribute to the process. In this regard, it is important that such an activity be organized as part of institutional development.

### Information dissemination

One of the key problems in African research institutions is the dissemination of information through publications and meetings. Very often donor agencies do not provide adequate support to ensure that research results are published and disseminated. A number of publishing houses have emerged in Africa which aim at publishing local research material. However, numerous obstacles stand in the way of such efforts. Significant advances have been made in ensuring that books published in Africa are distributed in the rest of the world.

### 5. Funding Options

While it is important to seek support from donors for the activities outlined above, opportunities for mobilizing local support for research have not been fully explored. There are two critical aspects to the lack of financial resources to support research in Africa. The first relates to the limited availability of financial resources for research. But this limitation is compounded by a second factor, which is the lack of domestic incentives that facilitate the mobilization of local revenue for development activities. Most African countries do not have the incentives that would enable individuals and private enterprises to provide support for local research. Reforms in tax laws, for example, would provide a much-needed inducement to such local support for research.

In a specialized field such as 'new technologies', it is necessary to provide a package of incentives that facilitate technology acquisition and adoption. Such incentives, for example, could include tax exemptions for *local foundations* that support the development of new technologies. Given the declining ability of most African governments to support R&D activities, it is critical that such incentives be provided as part of overall efforts to introduce technology-related economic reforms. Other measures that facilitate technology development need to be identified and introduced as part of the ongoing policy reform effort in Africa.

## Notes

1. This is based on C. Juma, J. Mugabe and P. Kameri-Mbote, eds, *Coming to Life: Biotechnology in African Economic Recovery* (London: Zed Books and Nairobi: Acts Press, 1994). This volume is the result of a study supported by the Swedish Agency for Research Cooperation with Developing Countries (SAREC) and the Swedish International Development Authority (SIDA).

2. C. Juma, *The Gene Hunters: Biotechnology and the Scramble for Seeds* (Princeton, N.J.: Princeton University Press and London: Zed Books, 1989), pp. 139–40.

3. F.C. Sercovich and M. Leopold, *Developing Countries and the New Biotechnology* (Ottawa: IDRC, 1991), p. 20.

4. Ibid., p. 19.

5. This section is based on W. Reid et al., *Biodiversity Prospecting: Using Genetic Resources for Sustainable Development* (Washington DC: World Resources Institute, 1993).

6. See C. Juma, *Policy Research in Sub-Saharan Africa* (Nairobi: African Centre for Technology Studies, 1994).

# References

Clark, N. and C. Juma, *Biotechnology for Sustainable Development* (Nairobi: Acts Press, 1991).
Enos, J.L. and W.H. Park, *Adoption and Diffusion of Imported Technology: The Case of Korea* (London: Croom Helm, 1988).
Juma, C., *The Gene Hunters: Biotechnology and the Scramble for Seeds* (Princeton, N.J.: Princeton University Press and London: Zed Books, 1989).
Juma, C., *Policy Research in Sub-Saharan Africa* (Nairobi: African Centre for Technology Studies, 1994).
Juma, C., J. Mugabe and P. Kameri-Mbote, eds, *Coming to Life; Biotechnology in African Economic Recovery* (London: Zed Books and Nairobi: Acts Press, 1994).
Kayumbo, H., 'Biodiversity: What's in it for Tanzania?', *Sunday News* (Tanzania), 2 August 1992, p. 5.
Lall, S., *Learning to Industrialize: The Acquisition of Technological Capabilities by India* (London: Macmillan, 1987).
Nyiira, Z.M., *Research Resources in National Research Institutes in Eastern and Southern Africa* (Ottawa: IDRC, 1990).
Office of Technology Assessment (OTA), *Technologies to Maintain Biological Diversity* (Washington DC: US Congress, 1987).
Office of Technology Assessment, *Biotechnology in a Global Economy* (Washington DC: US Congress, 1991).
Opio-Odongo, J.M.A., *Higher Education and Research in Uganda* (Nairobi: Acts Press, 1993).
Reid, W., et al., *Biodiversity Prospecting: Using Genetic Resources for Sustainable Development* (Washington DC: World Resources Institute, 1993).
Sanchez, V. and C. Juma, eds, *Biodiplomacy: Genetic Resources and International Relations* (Nairobi: Acts Press, forthcoming).
Sasson, A., *Biotechnology and Natural Products: Prospects for Commercial Production* (Nairobi: Acts Press, 1992).
Sercovich, F.C. and M. Leopold, *Developing Countries and the New Biotechnology* (Ottawa: IDRC, 1991).
Tiffin, S., F. Osotimehin, with R. Saunders, *New Technologies and Enterprise Development in Africa* (Paris: Organization for Economic Co-operation and Development, 1992).
United Nations Environment Programme (UNEP), *Convention on Biological Diversity: Nairobi*, United Nations Environment Programme, 1992.
United Nations, *Report of the United Nations Conference on Environment and Development* (New York: United Nations, 1992).
Walsh, V., 'Demand, Public Markets and Innovation in Biotechnology', *Science and Public Policy*, vol. 20 (1993), no. 3, pp. 138–56.

# 7

## New Materials Technology in Developing Countries

*Hugo F. Lopez and P.K. Rohatgi*

### 1. Introduction

Activity in materials technology is age-old, starting from the use of agricultural materials, stone, bronze, iron, clays and ceramics. However, most of the developing countries have to varying degrees missed out on the scientific and industrial revolution of the last 300 years, including the revolution in materials technology. This has resulted in a reduced availability of materials per capita in developing countries in terms of both quality and quantity, as compared to the advanced countries.[1] The developed world is currently undergoing yet another revolution in materials, an example of which is high-temperature superconductors. It is imperative that the developing world is adequately prepared to exploit the opportunities opened up by these new materials. The priority is to build capability in the developing world to exploit the opportunities from the new revolution in materials while meeting its challenges. This will involve multidimensional activity, including building capacity for technology forecasting, technology assessment, formulating materials policy, education, training, research, development, manufacturing, testing and standardization.

The developing countries are at different stages of the materials cycle and at present represent a range of capabilities to deal with the new materials revolution. In view of this, uniform prescriptions cannot be offered. In this chapter, only broad trajectories are discussed; individual countries' priorities must emerge by way of country-specific analyses. The specific cases of Mexico and India are discussed in some detail.[2]

The current availability of manufactured materials and energy per capita in the developing world is up to one hundred times less than

□ STEEL
△ CEMENT
■ REFRACTORY
● GLASS

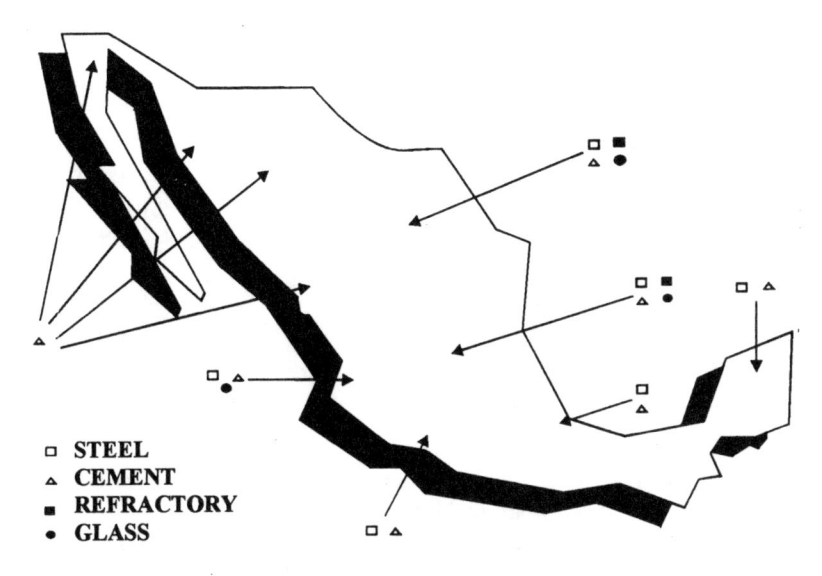

|  | Refractories | Steel | Cement | Glass |
|---|---|---|---|---|
| Plants | 14 | 18 | 27 | 14 |
| Production (*million tonnes*) | 0.39 | 8 | 32 | 3 |

**FIGURE 7.1** Location and production of refractory, steel, cement and glass products in Mexico

SOURCE: L. Martinez, 'Panel on Materials Science and Engineering in Mexico', Ixtapa State of Guerrero, Mexico (January 1993).

that in the advanced countries (see Tables 7.1 and 7.2). In addition, the present costs of materials in relation to incomes in developing countries are very high, and this results in further inequitable distributions of even these small quantities of materials. A considerable proportion of modern manufactured materials is in the possession of the rich elite in these countries, leaving the poor even more impoverished in terms of availability of materials. The abundant supply of materials and the services produced using materials, such as food, drinking water, housing,

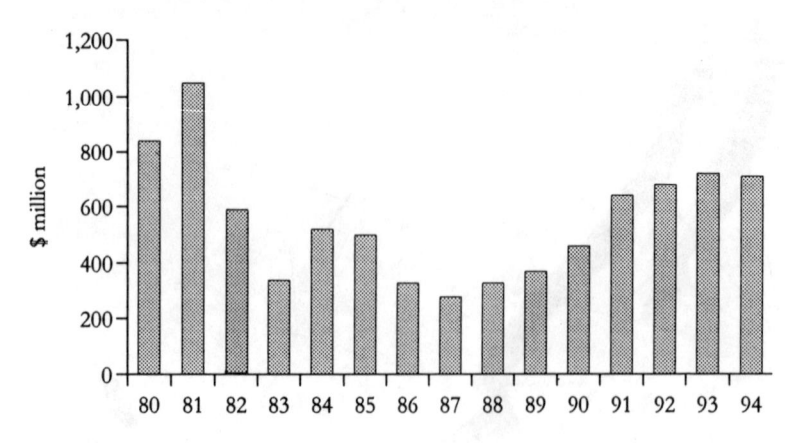

**FIGURE 7.2** Evolution of federal expenditure in science and technology in Mexico ($US millions)

SOURCE: M. Yacaman, *Indicators of the Scientific and Technological Activities in Mexico* (Mexico City: National Council of Science and Technology, 1993).

energy, health and clothing, could go a long way to reducing human misery in the developing world, and even in poorer sections of populations in the advanced countries.

## 2. Materials Technology in Mexico

Mexico, traditionally, has been a country whose economy is strongly linked to the export of oil and raw materials. In particular, Mexico is among the largest producers of silver, lead, zinc, mercury, and other metals (see Figure 7.8). It also possesses important petroleum and hydrocarbon reserves. Yet the development of technologies for the manufacture of new materials has been a relatively slow process. At present, the major manufacturing operations have been focused on steel (Mexico developed a direct reduction process known as HYL which was expected to have produced 750,000 tonmes per year by 1994), cement (CEMEX is the fourth largest producer in the world), and glass (VITRO is the fourth largest producer in the world). Figure 7.1 shows the location and production of these materials including refractories. Most of the materials thus produced are conventional, and

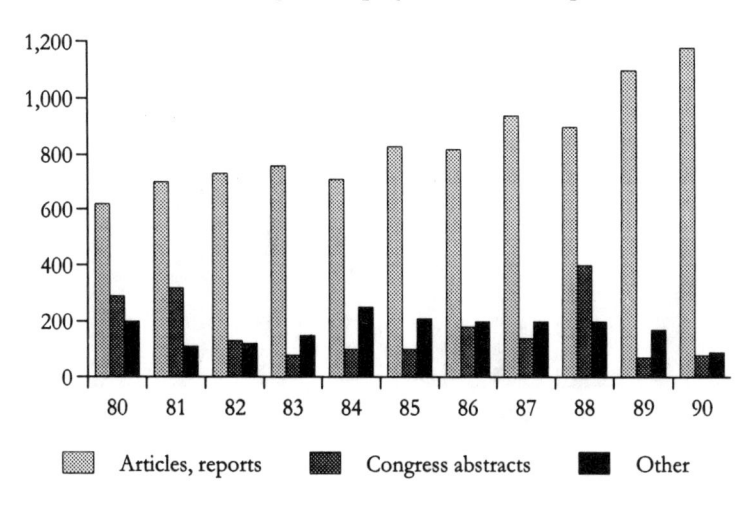

**FIGURE 7.3** Published works in science and technology by Mexican scientists, 1980–93

SOURCE: As Figure 7.2.

there are areas where there is a strong dependence on materials technology. For example, advanced steels with extremely small impurity contents, or advanced alloys including nickel-based superalloys, are not made in Mexico on a significant scale.

The materials-related areas where Mexico is strongly dependent on foreign technology are: (i) advanced or speciality alloys, (ii) structural ceramics, (iii) composites, and (iv) ceramics for electronic components such as microchips. The absence of these technologies in Mexico has been a major obstacle to a more harmonious utilization of its natural resources. Consequently, most of these materials are imported, resulting in considerable expenditures.

## 3. Research and Development

In order to reduce, or eliminate, the materials technological gap, the Mexican government has provided significant support to research and development (R&D) programmes through the national council of research and technology (CONACYT) (see Figure 7.2).[3] As a result, there has been an expansion in the number of research activities in univer-

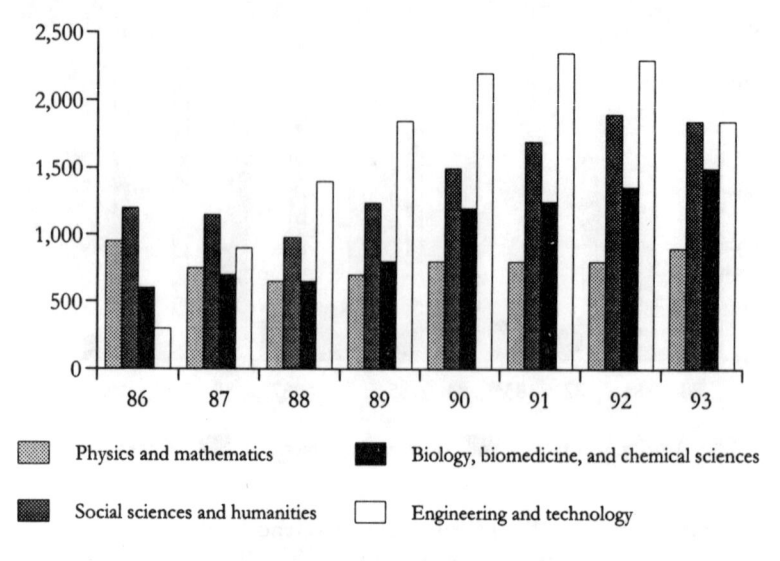

[Bar chart showing members by area from 1986-93, with categories: Physics and mathematics; Biology, biomedicine, and chemical sciences; Social sciences and humanities; Engineering and technology]

**FIGURE 7.4** Members of the National System of Mexican Scientists by area, 1986–93

SOURCE: As Figure 7.2.

sities and research institutes. Also, the number of graduate programmes in the field of materials science and engineering has been increased. Figure 7.3 shows the publication trends of Mexican researchers from 1980 to 1990, as well as the number of patents awarded in the years 1990–93. The R&D programmes included the creation of a national system of Mexican researchers with over 6000 members (see Figure 7.4). Nevertheless, the amount of support used in R&D activities accounts for only 0.3 per cent of GDP (see Figure 7.5). The contribution from the industrial sector was estimated to be in the order of 31 per cent in 1991 (see Figure 7.6).

However, the expected benefits from the research activities have not yet been achieved. A large manufacturing sector in Mexico is still involved in the production of relatively low quality material products. Furthermore, there is little or no link between industry and the R&D efforts. Since the physical and human infrastructure already existing in Mexico is large enough to provide the answers to its technological development, this poses important challenges. Traditionally, technology transfer has been the key factor in developing new materials technologies

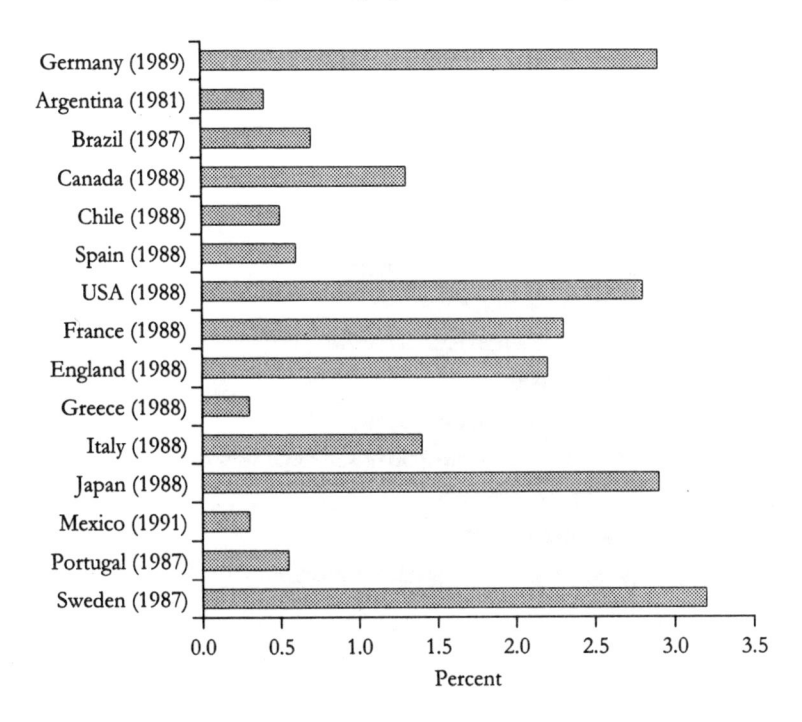

**FIGURE 7.5** National expenditure on R&D activities by country, as % of GDP

SOURCE: As Figure 7.2.

in Mexico. This will probably continue to be the case for the next decade. Technology transfer has the advantage of allowing rapid implementation of advanced technologies, and is low cost in comparison with the development of new technologies. Nevertheless, these types of activities give rise to a strong dependence on foreign know-how, and discourage the development of domestic technology.

Perhaps the most important challenge faced is the issue of quality. With the incorporation of Mexico into the North American Free Trade Agreement (NAFTA) with the USA and Canada, strong competition based on quality and low prices will promote the development of new technologies in Mexico as well as highly skilled personnel training. Unfortunately, a large number of small Mexican corporations will be unable to cope with this challenge. Hence, it is expected that most will be sold to their counterparts in the USA or Canada.

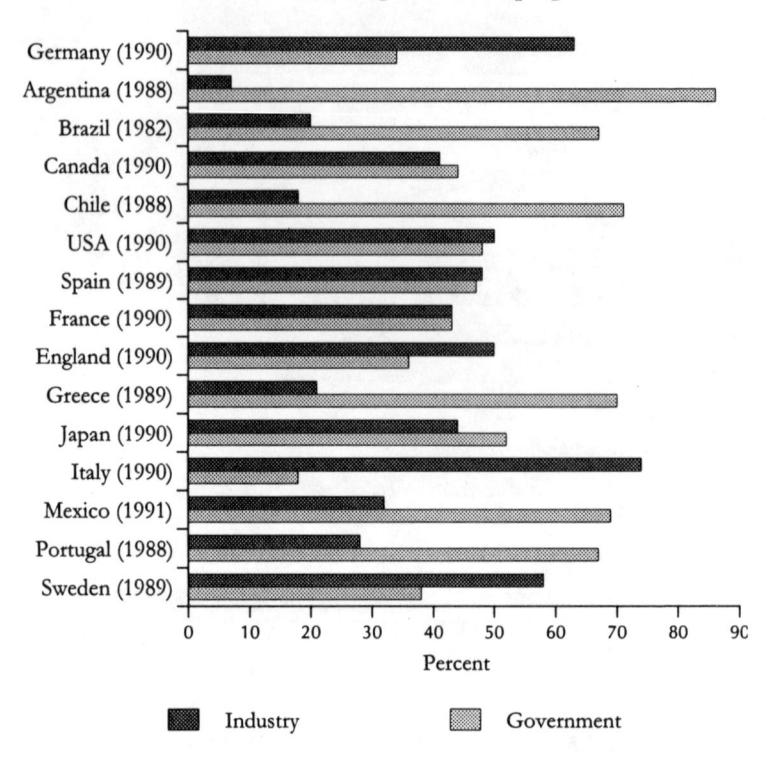

**FIGURE 7.6** Main sources of support for R&D activities by country

SOURCE: As Figure 7.2.

Another important challenge is the development of integrated manufacturing processes aimed at a better utilization of the natural resources of the country. The implementation of technologies which will make an integrated use of mining subproducts, among others, is badly needed for market diversification. However, at present there are no technological indicators pointing in this direction.

The materials field is undergoing a revolution with the emergence of advanced materials, engineered without the restraints of thermodynamic equilibrium, to meet specific needs;[4] these materials can now be tailored to meet property or performance targets. Hence, technological developments should be driven to meet the needs of developing countries. This, in turn, requires close monitoring and forecasting of materials science and technology developments in order to identify areas where a competitive edge exists in the production of new materials.[5]

Hence, it is imperative to establish suitable policies aimed at a better utilization of the raw materials existing in the country. Efforts should focus on producing high-quality materials at a competitive price which can be afforded by the different sectors of the population. Also, the importance of recycling and the effects of the degradation of the environment should be incorporated in the policy decisions to be made.

## 4. Challenges for Materials Technology Development in India

### Housing

Housing remains one of the most important problems of development due to lack of availability of materials, and this is an area where miniaturization cannot be applied beyond a certain point. One of the major challenges is to reduce the cost of materials for housing and to increase the performance of construction materials, particularly those based on local renewable or abundant resources. The highest priorities here will include a greater attention to materials science and the technology of alumino-silicates, earth, stone, laterite and clay-based products, which can be readily made everywhere. There is a need to improve the performance of bricks from common clay, and to develop biomass or solar-energy sources to fire them, or to develop low-temperature binders and sintering agents. The other area is application of modern materials science and technology to renewable resources, particularly locally available plant-based resources. Some examples of these resources for construction materials include bamboo; *Ipomoea carnea*; fibres from plants like coconut, sisal, banana, sunhemp, grasses; and large agricultural wastes like paddy straw and wheat straw.[6]

The shortage of cement and its high price due to high energy requirements in manufacture are great barriers to increasing the supply of housing in developing countries. Greater attention must be paid to using rice husk ash, fly ash and mineral-waste type materials to increase the volume of cement, and to bring down the cost of new high-tech cements such as zero-defect cement, rapid-setting cement, chemically bonded cement, and fibre-reinforced cement – the price of which is at present beyond the reach of the poor. Millions of people in the developing world use plant-based materials like coconut thatch for roofing, which do not provide adequate protection from nature and require replacement every year. Inputs of modern materials science and technology are required to increase the life and performance of these plant-based materials for housing and to make them more resistant to the elements and fire. It is absolutely essential that emerging materials technology should make a significant impact on materials for shelter.

## Infrastructure and communications

Macroeconomic development of all countries is directly correlatable to their infrastructure and communications systems. This infrastructure consists of large systems such as roadways, bridges, railroads, waterways and ocean, freight and air services, as well as telephones, wire services, and radio and television linkages. Often, the number of telephones or miles of railway tracks per capita are directly bound up with the progress of a country.

The key materials to meet these demands, fortunately, are well within the current production capabilities of most developing nations, including India. Because transportation is energy-intensive and the developing nations are energy-poor, it is imperative that significant material technology development resources should be immediately directed to meet these challenges. For advancement in communications, the basic raw materials – for example, fibre-optic glass materials, aluminium/copper conductors, and plastics from coal – are readily available.[7]

## Bio-processing of materials

A commitment to advanced research in genetic engineering would be likely to achieve an increase in the strength of wood and fibres available from fast-growing trees, in addition to the present focus on increasing the yield. The development of plants which take nitrogen from the air will reduce the considerable pressure in developing countries to manufacture fertilizers from minerals.

Biological routes to the production and preservation of materials[8] require greater attention in the context of development. One area that requires research is the microbiological extraction of metals from ores and ocean nodules, and the removal of sulphur and silica from minerals like coal and bauxite. Newer methods of microbial technology can be used to extract fibres and ultra-fine ceramic powders, at low energy costs, from agricultural products and wastes.

The problem of moisture absorption in these natural fibres must be solved by techniques such as acetylation. Greater attention needs to be paid to making silicon carbide whisker-type high-performance materials from agricultural resources other than rice husk, with reduced inputs of energy. A high priority should be to perfect modern microbiological techniques to extract fibres and ultra-fine powders of silica and other minerals from rice husk and other similar plant-based materials to produce advanced ceramics, and silicon for solar cells. Attention should be given to the possibility of controlled production of high-strength ceramic fibres for advanced composites by pyrolysis of natural fibres. The opportunity exists to produce polymeric materials from large quantities of agricultural resources and agricultural wastes. For instance, an agricultural waste like cashew nut shell liquid can be converted into very high performance polymers for composites.

## Strategic materials

Many developing countries do not have resources of strategic metals like nickel, cobalt, tungsten and chromium,[9] and it is necessary to synthesize high-performance materials from abundant elements like aluminium, silicon, oxygen, nitrogen and carbon. The synthesis of structural ceramics like silicon carbide and silicon nitride, and composites like aluminium-silicon carbide and aluminium-graphite can eliminate the need for special metals, which are in short supply in many countries. Greater emphasis must be placed on these kinds of ceramics and composites, in particular to reduce their production cost and to improve their performance, especially with regard to toughness. In terms of its use of coarse ceramics, the developing world remains very much in the stone age, as compared to the metals age in advanced countries; it needs to leapfrog into the world of advanced ceramics and composites without having to go through the stage of high-performance alloys, which will eventually be replaced anyway.

## Utilities

The need to develop inexpensive membranes and filters out of ceramics, composites and polymers for use in purifying and desalinating water is an important requirement in the developing world, and should be a priority of modern materials science and technology.

In the context of development, the advances made in the understanding of the structure and processing of newly emerging materials need to be applied to materials utilized in food production, transport and storage. Development of lighter, more durable and yet inexpensive clothing material, as well as recyclable paper and other materials to meet the demands of increasing literacy in the developing world, are important imperatives of materials technology. The demand for new inert and bioactive materials for transplants and health care will be much greater in the highly populated developing world. It is essential that the costs of these materials come down very considerably so as to be accessible to the poor. Some of the health-care materials of the future will be smart structures in the form of composites with embedded sensors, actuators and microprocessors, and their costs need to be reduced.

## Recycling

In the context of development, advances in materials science and technology related to recyclable materials,[10] and to materials that do not degrade or can be maintained by inputs of human labour, are extremely important. This is because the availability of resources for materials and energy will be a major constraint; regeneration of new material by recycling consumes much less material and energy than extracting it from source. There is a need for alloys and components that lend themselves to recycling and multifunctional uses, and that can be used in a succession of progressively downgraded applications. Increased understanding of surfaces and interfaces from a basic atomic and electronic perspective is necessary to produce surfaces that resist corrosion, oxidation, wear and fatigue, and thereby extend the life of materials.

## Energy

Materials for energy generation and transmission, and materials that can be made using decreasing amounts of energy, remain major imperatives for development. There is a correlation worldwide between per-capita income and per-capita energy consumption.[11] The impact of electricity on the villages of a developing nation, where a majority of the population lives rurally, is directly related to improvement in GDP (see Figure 7.7). In view of this, ceramics and composites that lead to higher efficiencies in energy conversion systems, to higher performance, and to lower-cost materials for solar energy and fusion energy production are important for development. New optoelectronic materials for transmitting energy and information will help overcome the constraints in these critical areas. Technologies for direct reduction of iron and

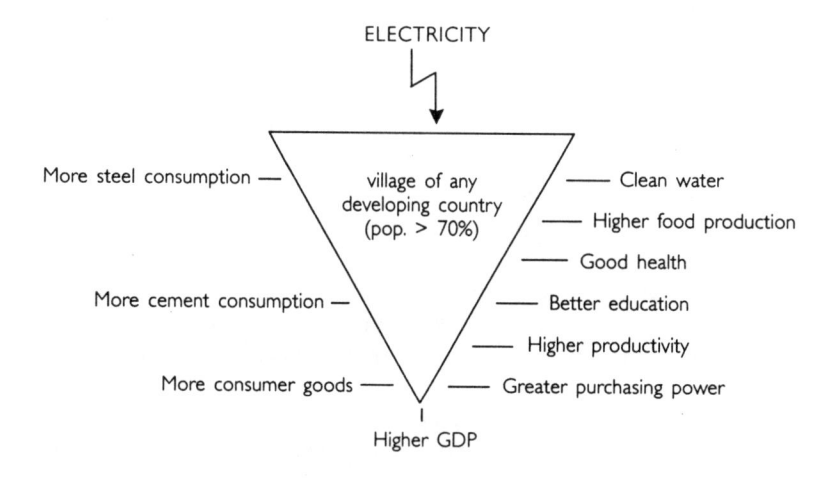

```
                    ELECTRICITY
                         ↓
More steel consumption ─╲  village of any   ╱─── Clean water
                         ╲ developing country╱
                          ╲  (pop. > 70%)  ╱──── Higher food production
                           ╲              ╱
                            ╲            ╱───── Good health
More cement consumption ────╲          ╱
                             ╲        ╱──────── Better education
                              ╲      ╱
                               ╲    ╱────────── Higher productivity
More consumer goods ───────────╲  ╱──────────── Greater purchasing power
                                ╲╱
                                 │
                              Higher GDP
```

**FIGURE 7.7** Impact of electricity on villages of a developing nation where the majority of the population live in rural areas

aluminium, and those that can make use of solar and biomass energies (e.g. plant-based reductants and bio-fuels) are also very important.

It is obvious that production of these primary materials will increasingly shift to the developing world to take advantage of mineral resources, lower labour costs and the current absence of pollution problems and stringent regulations. To acquire the science and technology to produce these conventional materials with low energy inputs, in small plants with low capital and high labour inputs, is a key imperative for development. Materials with high-temperature superconducting properties that have been discovered recently could have large implications for the developing world.[12]

## Materials processing

Micro-structural and more energy-efficient processes are called for in the processing of materials. With computer-aided design and simulation, it should be possible to minimize the amounts of materials used in various branches of industry (including construction), without lowering safety standards. In view of their labour-shedding character, the use of high degrees of automation and robotization in materials processing ought to be very sparing, and should be resorted to only where very high levels of quality and reliability are essential.

The information input that goes into materials processing should be as high as possible, but the actual process should be as simple a technology as possible, in order that it can be maintained in the most primitive developing environments. Materials in the context of development should as far as possible be made using local resources and manpower. It is worth upgrading the many traditional materials and processes that have long been used in the developing world by inputs of modern materials science and technology. A new trend in materials technology, parts consolidation, which deploys fewer parts as a result of single-step moulding of complex shapes, could prove to be very important.

The economies of several developing countries are very heavily linked to the export of a given mineral.[13] For instance, several countries in Africa depend upon the export of copper. In light of the development of glass communication cables, it is vital that materials science and technology is directed to find new uses for resources like copper otherwise the economies of these countries will collapse, and a global advance in materials science will be locally counterproductive in terms of development.

## 5. Objectives and Goals of Materials Science and Technology Institutions in Developing Countries

It is becoming quite apparent that the standard of living in any country is strongly dependent on the availability of materials per capita. The per-capita demand for materials in developing countries, which is at present very low compared to that in developed countries, will increase very rapidly as the standard of living of the large and growing masses of the developing world rises. This will place enormous pressure on the resources for materials, as well as create environmental problems related to the increased exploitation and use of the latter. Under certain conditions this issue could lead to international conflicts, which clearly must be avoided at all costs. In view of this, the objectives and goals of institutions in materials science and technology in developing countries should be:

- To engage in research, development and production related to materials to ensure adequate supplies of materials at the lowest

possible price, compatible with other national goals such as security, clean environment and resource conservation.
- To ensure that the increasing use of materials does not lead to undue pressure on resources for materials and degradation of the environment. To attempt to synthesize most materials from the most abundant elements of the earth's crust.
- In the context of developing countries, greater emphasis should be placed on materials related to the more basic of human needs; for instance, energy, shelter, food, water, infrastructure, education and health care. Normative exercises are needed to derive challenges for modern materials science from these human needs, to attempt to meet these needs quickly and as far as possible through local resources, local skills and local infrastructure.
- To maintain standards for materials which are compatible with international standards, and to develop the capacity to test locally produced and imported materials at local institutions.
- In the context of developing countries, greater emphasis should be placed on materials that can be produced from local resources, especially those which are renewable and abundant, to meet local needs while generating maximum local employment.
- To train interdisciplinary manpower in materials science and technology to cater for the emerging needs of the production, research, development, standardization and testing institutions, with special grounding in material problems related to local resources, needs and infrastructure. The institutions would begin at undergraduate level, but aim eventually to establish programmes for doctoral research degrees in materials science, materials technology and materials policy.
- To evolve a national materials policy based on technology forecasting and assessment; to develop a coherent long-term strategy for materials, and the capability to participate in the evolution of international materials policy. The materials policy would deal with issues relating to the import and export of raw materials or finished materials, stockpiling, materials substitution, international trade, research, development, production, distribution, pricing, conservation and recycling of materials. In addition to a national materials policy, objectives would include development of regional and local materials policies to complement national and international priorities in the light of regional and local resources and needs.
- To participate in international R&D efforts, to tackle common problems relating to resources, standards, environments, materials and legislation dealing with materials; to help develop international

materials policy, and form an international federation of professional societies concerned with materials within the country and worldwide. To set up mechanisms to promote the international exchange of information, experts and technology.
- To conduct technology forecasting and assessment from local and regional viewpoints, particularly in relation to the utilization of local resources, and make judgments relating to local manufacture and transfer of technology within the country, the region and the locality.
- To build the capacity to monitor the progress and maintenance of the above-mentioned institutions relating to materials technology. (Quite frequently in developing countries the right institutions are established, but a commitment to their maintenance is lacking and consequently they deteriorate.)
- To build the capacity to carry out environmental-impact assessments of present and proposed materials industries, and to take steps to prevent environmental degradation resulting from the manufacturing process, including the use and recycling of materials (in particular, stringent precautions are called for in respect of atomic energy development).

The various stages of the materials cycle require different kinds of institutions. In countries where materials activity is just starting, a few multifunctional institutions will be the right way to begin. These can then multiply and become more specialized.

## 6. The Linkage of Material Science Institutions and Industry

It is imperative that the institutions of materials science and technology[14] have links with industry and the government in order to influence national materials policy through the following mechanisms:

- The founding of materials science and technology institutions within the industry should be encouraged; these can then readily build bridges to those outside.
- Consultancy, sponsored research work, subcontracting, joint appointments, exchange of materials personnel between industry, government and academia should all be encouraged.
- Consortia of industries in a given field should be encouraged to sponsor collective research in academic institutions and government laboratories.

- The materials personnel in industry and academics should help in the formation of National Materials Advisory Boards and other materials policy bodies, and serve on these bodies charged with formulating and implementing policies.
- Academic institutions should offer customized seminars and short courses for personnel in the industry and in the government, in addition to their regular teaching responsibilities. Personnel working in both industry and government should be encouraged to register externally for higher degrees.
- Subcontracting of work between the industry and university should be encouraged.

## 7. The Structures of Materials Institutions in Developing Countries, with Special Reference to India

The structure of materials institutions in developing countries is very much dependent on the stage of development of the given country. Some of the relatively advanced developing countries like India have world-class teaching and research institutions in materials, whereas some of the countries just starting only have teaching facilities within materials-related departments. In general, institutions have evolved along the following lines:

- Establishment of an organized system of materials production (for instance, timber, cement and steel), and of a testings and quality-control institution within the factory, followed later by a research section. For example, in India the Tata Iron and Steel Company started production of steel and then set up a materials R&D section soon thereafter.
- Introduction of teaching in geology, mining, metallurgy, ceramics and materials (in that sequence) in materials-related departments of existing colleges. After this initial stage, teaching would be organized at departmental level or even in colleges dedicated to specific sectors. For instance, the teaching of geology, and subsequently of mining and metallurgy, began in chemistry departments; this was followed by the establishment of departments, and some colleges, of mining, metallurgy, ceramics and, recently, materials. Degrees in materials science have been offered only very recently, and only in the more advanced developing countries like India, China, Singapore and South Korea.

- Some of the materials-related departments located in a suitable research environment gradually take on research activity in materials. While some developing countries like India, Korea and China have recently developed undergraduate teaching that counts among the best in the world, the level of graduate education and postgraduate research in general does not compare favourably with that in the highly developed nations.
- Establishment of government-sponsored research, development and testing laboratories related to major materials; for instance, laboratories dedicated to building materials, cement, jute, metals, ceramics, glass and polymers. Some materials-related work was also started in government-sponsored laboratories established for research in physics, chemistry and aeronautics.
- Formation of government-sponsored institutions responsible for standards and quality control related to materials production. These institutions work to international standards; in addition, they maintain a local standard which is directly related to local conditions.
- Establishment of materials institutions sponsored by associations of industries – for example, the textile research institutions in India.
- Convening of working groups to deal with specific materials in the government ministries. For instance, the ministries of Mines and Steel in India set up groups to monitor and frame policy in the aluminium and steel sectors. In several countries, including India, the ministerial groups, including the National Planning Commission, steer policies related to materials through committees; these often involve experts from the academy, research laboratories and industry.
- Establishment of local and regional research laboratories by the government to exploit local resources to meet urgent local needs. These laboratories have sometimes been very effective in developing new materials using local resources. In fact, in the context of developing countries, the local and regional institutions dealing with local resources to meet local needs will be more important than the national or international institutions of science.
- Establishment of sections in ministries to determine which materials are to be manufactured domestically and which to be imported. Likewise, the setting up of groups with the capability and responsibility to decide what know-how is to be developed by the home industry and what must be imported, assimilated and modified to suit local needs. These groups operate through licensing procedures.

## 8. The Role of Materials Institutions in the Development of New Materials, Testing and Quality Control

The materials science and technology institutions in developing countries have been involved in technology assessment, utilization of local materials, development of new materials, testing and quality control, to varying degrees depending upon their relative stages of development. In most of the countries, the institutions have been reasonably successful in testing and quality control, and to a lesser extent in the utilization of local materials. There has been little development of new materials; and technology assessment has been used indirectly, while licenses have been granted for manufacture and import. No special emphasis has been placed on the development of innovative ideas.

These institutions in developing countries must, first and foremost, build the capacity for technology assessment in order to determine which current and future materials technologies are most relevant. They should determine the best directions for research, development and production of materials to provide maximum benefits to people by fulfilling basic needs; this will involve the use of quantitative techniques of technology assessment, sensitive to local needs and priorities. The cost of such assessments is quite low compared to the damage that can arise as a result of taking the wrong direction by making *ad hoc* judgements in the absence of technology assessments. The expertise required to carry out technology assessments from a local perspective can be found fairly easily, and can be supplemented from outside the country. The assessment should not, however, be left wholly to foreign experts, who may not fully understand local priorities and hence fail to do a comprehensive job of assessment from the viewpoint of the developing country. In fact, the lack of such proper assessments is one of the major reasons that, while developing countries have been able, after a time lag, to reproduce the materials research or production of developed countries, innovative development of new materials using local resources for local needs has hardly taken place. Key ingredients for fostering innovation and entrepreneurial activity are missing in the developing nations. This situation needs to be rectified through objective technology assessment if the materials scientists and technologists are to solve the problems of food, housing, health and energy that exist in the developing world.

The institutions of materials science have sometimes played an important role in the utilization of local materials – including minerals, coals, clays, woods and forest products. This role should be intensified in the future, and attempts must be made to apply the latest materials science and technology information to local materials to make products now being developed elsewhere, since this know-how will not be readily available from other countries. The international institutions such as UNDP have at times helped in this effort and can be expected to do so in the future if specific needs are identified.

As mentioned earlier, the development of new materials by developing country institutions has been less than adequate. The reasons include a preoccupation with reproducing what has been done in developed countries, lack of creativity, a perceived technological colonialism, and a lack of initiative in utilizing local resources to meet local needs. If the developing country institutions concentrate on local resources and local needs, as has been done in some national and regional research laboratories, the probability of developing new materials will increase.

The most readily feasible role for materials science and technology institutions in developing countries has so far been in testing and quality control. These functions have proved relatively straightforward to establish; yet effecting improvements in quality has not been so easy. With the growing emphasis on reliability and quality (ISO 9000–9004), the modern instrumentation and procedures for testing and total quality control will be increasingly demanding and complex. Acquiring the training and skills as well as the acquisition, maintenance and continual updating of equipment present a challenge to developing countries. It may be necessary to build up the existing infrastructure in teaching and research institutions to provide national central facilities for materials testing.

In monitoring and maintaining standards, care will have to be taken to ensure that these institutions do not become instruments of delay. In several instances, the instruments of policy, such as licensing authorities, lead to inordinate delays, and thus act as brakes on progress instead of promoting advancement and quality. A time limit should be set within which these institutions can raise an objection or deny a permission; if nothing is heard from them during this period the process should go ahead.

## 9. Materials Science and Technology Institutions: The Experience of India

In India, early materials science institutions very much followed the British pattern in terms of geological survey, national test houses and teaching within departments of applied chemistry. These latter in time became departments and colleges of mining, then of mining and metallurgy, then of metallurgy; and now, some of the teaching departments are being redesignated Materials Science and Engineering, following the recent pattern in the USA. Materials, in addition to being taught in dedicated departments, are also taught in materials-related departments such as physics, chemistry, mechanical engineering, chemical engineering, civil engineering and aeronautical engineering.

The undergraduate education in metallurgy and materials science provided at the Indian institutes of technology compares with the best in the world. Several universities, regional engineering colleges and institutes have also started doctoral and postdoctoral research programmes in the last thirty years. Although some of these programmes are quite good, they do not on the whole compare with the best in the world, as the undergraduate courses do. Some of the universities as well as the institutes have established advanced centres in materials science and technology; these are quite effective in training students and conduct excellent research. The students trained in India have thus satisfied the manpower requirements of India's industry and research institutions; in addition, some students trained in India have been very successful in the advanced countries, going on to reach the top of their professions.

In addition to the teaching institutions, the government also established the Geological Survey, the Department of Mines and Steel, the Forest Research Institute, the Jute Research Institute and a series of national and regional research laboratories under the Council of Scientific and Industrial Research – including the National Metallurgical Laboratory, the Central Glass and Ceramics Research Institute, the Regional Research Laboratories at Bhubaneswar, Trivandrum, Bhopal, Jorhat, Jammu and Hyderabad, and the National Chemical Laboratory and the National Physical Laboratory. The Atomic Energy Commission and the Space Research Organization established very strong materials research and production facilities to meet their materials requirements. More recently, the Indian Ministry of Defence has established the Defence Metals Research Laboratory, which conducts materials-related R&D. In addition, the public sector corporations have set up research laboratories like the R&D Centre for Iron and Steel established by the Steel Authority of India.

In the private sector, one of the first materials laboratories was set up by the Tata Iron and Steel Company. Smaller research laboratories are maintained by other private materials industries like Hindustan Aluminium. However, the size and the excellence of private-sector laboratories does not, by and large, compare well with the laboratories in the private sector in developed countries. This is one of the important consequences of the import of technology, the fact that it is a sellers' market, and the relatively small size of the industry. Several industrial associations have set up sizeable laboratories – for example, the Textile Research Associations and Cement Research Institute – which have performed reasonably well; they do not, however, compare with similar organizations in the industrial nations.

In addition to the above, there are several government testing laboratories and standards institutions, as well as the departments of industry and the ministry of railways – all of which maintain research laboratories in materials.

India has, by and large, established an excellent foundation in its undergraduate education, good postgraduate education and research facilities, a few notable advanced centres for materials science research, competent testing and standards authorities, and possesses a country-wide infrastructure carrying out research, development and production. The Indian experience is a good example to other developing countries setting up these instititions, and the facilities in India can serve as the training ground for their materials personnel. In the past, for instance, India has assisted Latin American countries to set up national institutes of standardization.

In terms of testing, quality control and utilization of local materials including local minerals and forest products, the materials institutions in India have done quite well. India was one of the first developing countries to do technology forecasting and assessment in materials, and is considering setting up a National Materials Advisory Board. The institutions in India have done research into and begun production of some of the most modern materials. However, the development of new materials has been less than satisfactory. This, however, will change rapidly as the preoccupation with reproducing advanced materials developed in the industrialized nations gives way to a new-found pride and commitment to developing materials using local raw resources to cater for local basic human needs. Some regional research laboratories have already done significant pioneering work in materials science and technology of local materials: for example, developing composites using local natural fibres and forest products, and extending the life of local agriculture-based materials used for housing.

## 10. International Institutional Collaboration: A Strategy for Development

International cooperation in materials science and technology would go a long way toward building capability and increasing the availability of materials in developing countries. Indeed, it has already resulted in significant advances in, among others, India, China, and Korea. A survey of case studies from these countries should be compiled and recommendations for future collaboration drawn up. The following is a list of the kinds of changes that would constitite the improvement in international cooperation that is required:

- The developing countries, through international bodies, as well as bilaterally, should request greater international cooperation in such matters as setting materials standards, care of the environment, conservation of resources, a common approach to a wide range of materials research, exchange of materials scientists and engineers, and more extensive cooperation between private organizations at home and their counterparts abroad.
- The developing countries should seek cooperation from developed as well as sister developing countries to move from the position of being mainly raw materials suppliers to selling more value-added goods and finished products. This may even involve judicious and selective cooperation with multinationals who, in spite of all the drawbacks involved, are able to generate advanced production in developing countries reasonably quickly.
- The developing countries should request cooperation in evolving a world materials policy through which a more equitable distribution and availability of materials can be arrived at.
- A cooperative centre of technology forecasting, technology assessment and materials policy should be established in a developing country like India, which has experience in forecasting, assessment, research, development and production of fairly advanced materials. Such a centre should conduct materials-related studies from the developing countries' perspective, and thereby help towards the creation of a materials policy in different countries. The centre should stimulate education in materials policy at all levels in every developing country to encourage the young to view materials development as an important step on the road to national excellence. The centre should examine the materials policy work now underway in advanced countries such as the USA and the UK, and derive information and

**FIGURE 7.8** US net import reliance on selected minerals and metals as % of consumption, 1980

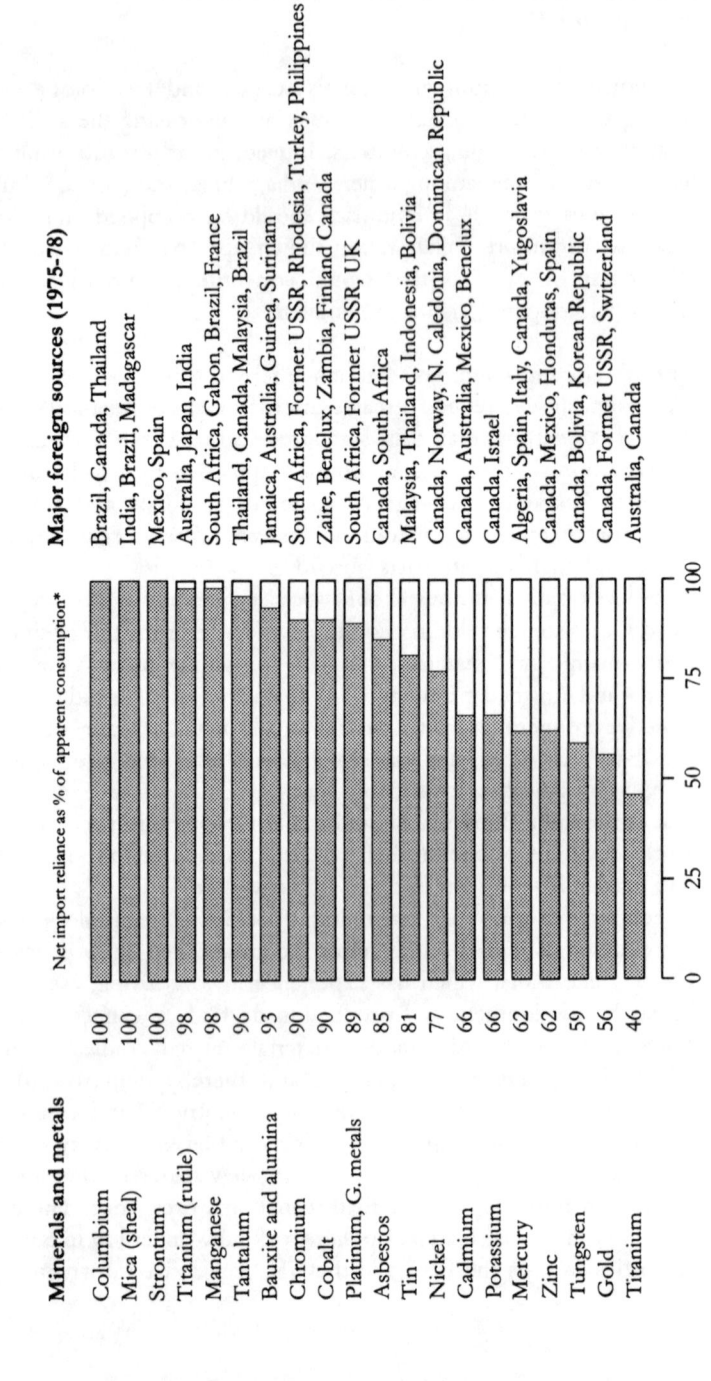

| Minerals and metals | Net import reliance as % of apparent consumption* | Major foreign sources (1975-78) |
|---|---|---|
| Columbium | 100 | Brazil, Canada, Thailand |
| Mica (sheal) | 100 | India, Brazil, Madagascar |
| Strontium | 100 | Mexico, Spain |
| Titanium (rutile) | 98 | Australia, Japan, India |
| Manganese | 98 | South Africa, Gabon, Brazil, France |
| Tantalum | 96 | Thailand, Canada, Malaysia, Brazil |
| Bauxite and alumina | 93 | Jamaica, Australia, Guinea, Surinam |
| Chromium | 90 | South Africa, Former USSR, Rhodesia, Turkey, Philippines |
| Cobalt | 90 | Zaire, Benelux, Zambia, Finland Canada |
| Platinum, G. metals | 89 | South Africa, Former USSR, UK |
| Asbestos | 85 | Canada, South Africa |
| Tin | 81 | Malaysia, Thailand, Indonesia, Bolivia |
| Nickel | 77 | Canada, Norway, N. Caledonia, Dominican Republic |
| Cadmium | 66 | Canada, Australia, Mexico, Benelux |
| Potassium | 66 | Canada, Israel |
| Mercury | 62 | Algeria, Spain, Italy, Canada, Yugoslavia |
| Zinc | 62 | Canada, Mexico, Honduras, Spain |
| Tungsten | 59 | Canada, Bolivia, Korean Republic |
| Gold | 56 | Canada, Former USSR, Switzerland |
| Titanium | 46 | Australia, Canada |

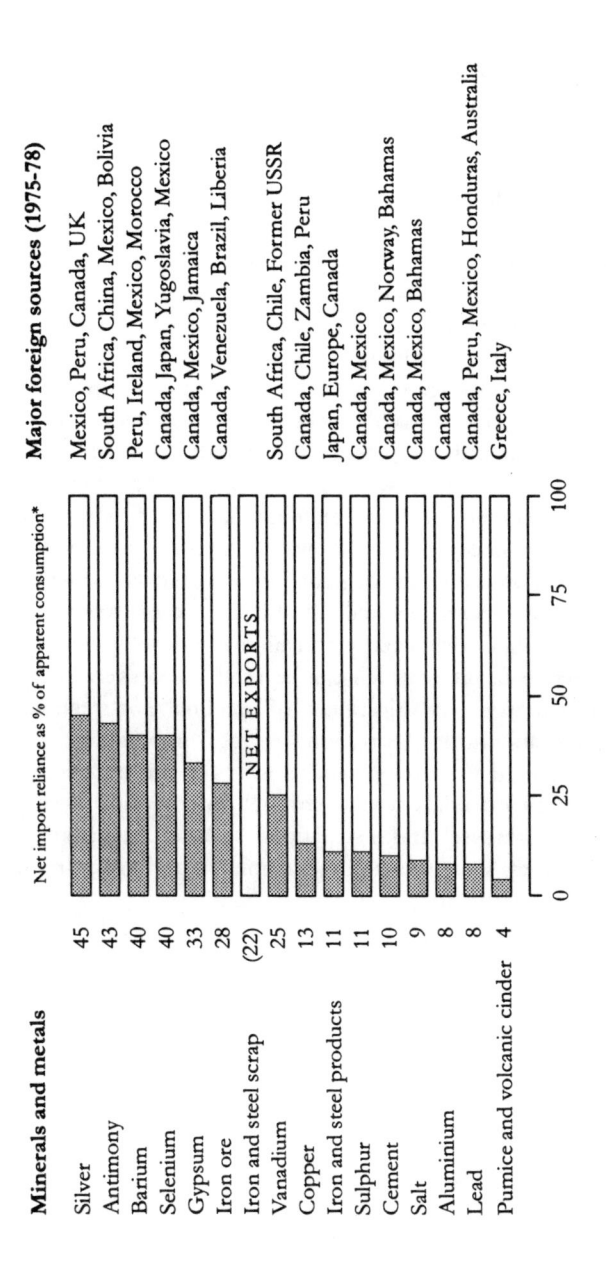

| Minerals and metals | Net import reliance as % of apparent consumption* | Major foreign sources (1975-78) |
|---|---|---|
| Silver | 45 | Mexico, Peru, Canada, UK |
| Antimony | 43 | South Africa, China, Mexico, Bolivia |
| Barium | 40 | Peru, Ireland, Mexico, Morocco |
| Selenium | 40 | Canada, Japan, Yugoslavia, Mexico |
| Gypsum | 33 | Canada, Mexico, Jamaica |
| Iron ore | 28 | Canada, Venezuela, Brazil, Liberia |
| Iron and steel scrap | (22) NET EXPORTS | |
| Vanadium | 25 | South Africa, Chile, Former USSR |
| Copper | 13 | Canada, Chile, Zambia, Peru |
| Iron and steel products | 11 | Japan, Europe, Canada |
| Sulphur | 11 | Canada, Mexico |
| Cement | 10 | Canada, Mexico, Norway, Bahamas |
| Salt | 9 | Canada, Mexico, Bahamas |
| Aluminium | 8 | Canada |
| Lead | 8 | Canada, Peru, Mexico, Honduras, Australia |
| Pumice and volcanic cinder | 4 | Greece, Italy |

NOTES: * Net import reliance = imports − exports; apparent consumption = US primary < secondary production > net import reliance.

SOURCE: E.D. Hondras, 'Materials, Year 2000', in Tom Forester, ed., *The Materials Revolution* (Cambridge, Mass. and London: MIT Press, 1988), p. 61.

draw conclusions relevant to the developing world. It should do normative exercises to draw up priorities in order to remove deficiency in materials as a constraint to development.
- Cooperation should be sought with developed as well as some relatively advanced developing countries for setting up interdisciplinary materials science, technology and processing research centres in academic institutions in developing countries, since this model has worked reasonably successfully. However, care should be taken to ensure that the activities of these countries remain strongly responsive to local materials and local needs.
- Cooperation should be sought for easier exchange of material scientists and technologists from academic, research and industrial organizations between developed and developing countries, and within developing countries. This is the best mechanism to facilitate transfer of information, know-how and attitudes.
- International cooperative research institutes should be set up under the United Nations and other world-level bodies to address common problems. These include: development of materials using advanced material and technology processes from renewable and abundant resources such as biomass, silicon, aluminium, oxygen and iron, specifically for basic needs like food, water, housing, health and energy; recycling and prevention of degradation of materials; reduction of pollution associated with materials industries; use of renewable resources of energy like solar energy and biomass; development of standards and testing facilities which can be established even in the least developed countries; establishment of national and regional materials research laboratories to respond to local imperatives.
- Cooperation is required to develop mechanisms for the transfer of open information on materials science through international access to computer-based databases and instructional services, exchange of educational videotapes, standards, testing equipment and procedures.

**TABLE 7.1** Per-capita consumption of metals, selected countries, 1985

| Country | Per-capita GDP (US$) | Alumin. | Copper | Lead | Zinc | Tin | Crude steel |
|---|---|---|---|---|---|---|---|
| Austria | 9 120 | 16.8 | 1.9 | 8.1 | 4.2 | 0.066 | 297.1 |
| Belgium/Luxembourg | 8 280 | 26.0 | 30.1 | 6.4 | 16.4 | 0.087 | 349.7 |
| Denmark | 11 200 | 4.3 | 0.3 | 2.6 | 2.4 | 0.020 | 353.5 |
| Finland | 10 890 | 3.4 | 14.7 | 4.9 | 5.3 | 0.020 | 360.8 |
| France | 9 540 | 10.6 | 7.2 | 3.8 | 4.5 | 0.125 | 267.3 |
| Germany (Federal Republic) | 10 940 | 19.0 | 12.4 | 5.7 | 6.7 | 0.272 | 504.7 |
| Greece | 3 550 | 8.9 | 3.9 | 2.3 | 1.5 | 0.040 | 156.9 |
| Italy | 6 520 | 8.2 | 6.3 | 4.0 | 3.8 | 0.088 | 381.7 |
| Netherlands | 9 290 | 6.1 | 1.2 | 3.1 | 3.5 | 0.352 | 290.4 |
| Norway | 14 370 | 30.8 | 3.1 | 3.3 | 5.0 | 0.095 | 340.2 |
| Portugal | 1 970 | 3.1 | 1.5 | 2.5 | 0.8 | 0.069 | 109.9 |
| Spain | 4 290 | 5.1 | 3.0 | 2.7 | 2.5 | 0.088 | 175.2 |
| Sweden | 11 890 | 11.1 | 13.1 | 3.2 | 3.8 | 0.048 | 392.0 |
| Switzerland | 16 370 | 22.0 | 1.4 | 1.6 | 4.0 | 0.123 | 350.8 |
| Turkey | 1 080 | 2.3 | 1.5 | 0.4 | 1.0 | 0.018 | 98.6 |
| United Kingdom | 8 460 | 6.2 | 6.1 | 4.9 | 3.4 | 0.166 | 257.2 |
| Yugoslavia | 2 070 | 9.1 | 6.4 | 5.1 | 4.6 | 0.061 | 220.0 |
| India | 270 | 0.4 | 0.1 | 0.1 | 0.2 | 0.003 | 18.1 |
| Japan | 11 300 | 15.1 | 10.2 | 3.3 | 6.5 | 0.262 | 607.4 |
| Republic of Korea | 2 150 | 3.5 | 5.0 | 2.0 | 2.9 | 0.063 | 243.2 |
| Taiwan | 3 690 | 7.7 | 4.8 | 2.1 | 2.6 | 0.063 | 238.2 |
| South Africa | 2 010 | 2.4 | 2.1 | 1.5 | 2.6 | 0.059 | 165.0 |
| Canada | 13 680 | 13.5 | 8.8 | 4.6 | 6.2 | 0.150 | 524.3 |
| USA | 16 690 | 18.1 | 9.0 | 4.7 | 4.0 | 0.155 | 439.8 |
| Argentina | 2 130 | 2.7 | 1.3 | 0.9 | 0.8 | 0.026 | 71.8 |
| Brazil | 1 640 | 2.6 | 1.4 | 0.5 | 1.1 | 0.032 | 88.1 |
| Chile | 1 430 | — | 2.1 | — | 0.5 | — | 47.4 |
| Mexico | 2 080 | 1.0 | 1.5 | 1.1 | 1.3 | 0.013 | 95.8 |
| Australia | 10 830 | 17.9 | 7.8 | 3.7 | 5.2 | 0.171 | 363.3 |
| New Zealand | 7 010 | 10.5 | 0.6 | 3.0 | 7.5 | 0.030 | 223.9 |

SOURCE: Federal Institute of Geosciences and Natural Resources, Hanover, Federal Republic of Germany.

**TABLE 7.2** Per-capita consumption of metals by region, 2000 (extrapolated values)

| Region | Per-capita consumption (kg) | | | |
|---|---|---|---|---|
| | *Steel* | *Aluminium* | *Copper* | *Zinc* |
| Western Europe | 710 | 20.24 | 10.50 | 6.83 |
| Japan | 1450 | 45.86 | 21.40 | 13.19 |
| Other developed countries | 680 | 22.32 | 11.98 | 8.58 |
| USSR | 850 | 17.47 | 7.84 | 3.92 |
| Eastern Europe | 610 | 13.87 | 5.41 | 5.92 |
| Africa | 20 | 0.24 | 0.16 | 0.07 |
| India | | | | |
|   Low growth | 26 | 0.51 | 0.20 | 0.32 |
|   High growth | 51 | 0.98 | 0.44 | 0.70 |
| Asia | 30 | 0.50 | 0.22 | 0.31 |
| Latin America | 100 | 1.72 | 0.91 | 0.95 |
| China | 60 | 0.79 | 0.63 | 0.54 |
| USA | 890 | 52.25 | 14.63 | 9.41 |
| World | 240 | 7.27 | 3.06 | 2.09 |

SOURCE: A. Lahiri, *Conservation of Mineral Resources in Commerce*, 1976, pp. 47–9.

**TABLE 7.3** Some important targets for materials technology for development

- Genetic engineering for plants with stronger timber and fibres, which can be pyrolized to form high-performance fibres, and carbon–carbon composites.
- Microbial processes to extract metals from ores and ocean nodules, and to remove sulphur and silica from coal, bauxite and other minerals.
- Microbial processes to extract fibres and ultrafine ceramic particles from agricultural products and wastes.
- Solar photovoltaic materials with increasing efficiencies and decreasing costs; solar furnaces for processing materials.
- Materials for fusion energy.
- Membranes made from polymers, ceramics and composites, producing decreasing costs and increasing performances in purification of water.
- Improved and inexpensive materials for housing from abundant and renewable resources like sand, clay, rock, stones, laterites, plant-based materials.
- Composites and ceramics with improved performances, based on abundant elements like aluminium, silicon, carbon, nitrogen and plant materials.
- Direct reduction of iron and aluminium using low-energy processes, employing solar and biomass energy.
- Recyclable materials with cascading down-graded application with longer life and resistance to corrosion, oxidation, wear and fatigue.
- Rapidly solidified materials for reducing energy losses.
- Surface and interface processed materials with tailored structures and properties to meet specific needs.
- Room-temperature superconductors.
- Parts consolidation and component miniaturization.
- Cement production at a much lower energy cost.
- Use of indigenous plant products to meet basic needs for shelter and infrastructure

**TABLE 7.4** Examples of new materials that will be important in the future

- Shape memory alloys
- Coal refuse fibres
- Slag cement
- Metastable and amorphous materials
- Composite materials (metal matrix, polymer matrix, cermets)
- Intermetallic compounds ordered alloys including aluminides
- Soft/hard magnetic materials
- Superconducting materials
- Structural and superplastic ceramics
- Electronic materials
- Photovoltaic materials with ultra-high efficiencies
- Laminates, claddings, surface coatings, surface modification
- High-temperature alloys
- Building materials (insulation, cinder blocks, glass)
- Wear-resistant materials, hard materials
- High-performance materials from renewable resources
- High-strength/low-density alloys (aluminium, titanium, etc.)
- High-technology cements/concretes
- Conducting polymers
- Glasses
- Low-temperature materials
- Battery materials
- Biomaterials
- Nanostructure materials
- Layered materials
- Fullerene carbon

**TABLE 7.5** Some new materials-processing technologies

- Molecular beam epitaxy
- Die-casting of high-melting-point alloys
- Rapid solidification of bulk components (amorphous sheets to bulk materials, Osprey-type spray deposition)
- Superplastic forming
- Plasma melting/spraying/synthesis
- Surface modification (ion implantation, laser, electron beam)
- Coating using melt spraying
- Directional crystal solidification/single crystal growth
- Near net shape processing

- Thin strip casting
- Dynamic powder compacting
- Direct steel making
- Explosive bonding/consolidation
- Laser production of ceramic powders
- Laser glazing
- Sol-gel processing
- Laser machining, electron beam processing
- Water jet machining
- CAD/CAM/Robotized process
- On-line sensors with feedback loops to expert systems and artificial intelligence for controls

**TABLE 7.6** Examples of materials applications/markets

- Energy (solar, nuclear, fossil, geothermal, battery, motors, transformers)
- Transportation (of all types)
- Construction (housing, bridges, highways)
- Environmental (chemical and radioactive waste disposal)
- Agriculture and food processing
- Water (purification systems)
- Medical (implants and health-care systems)
- Home appliances
- Communications
- Textiles and clothing
- Packaging
- Electronics (computers)
- Automation, robotics
- Shipbuilding
- Aerospace applications
- Mining, *in situ* cral gasification
- Manufacturing (machine tools, primary metals)
- Catalysts (chemical industry)

**TABLE 7.7** Examples of material problem areas

- Corrosion
- Wear
- Ductility (hot and cold)
- Formability
- Temperature Resistance
- Toughness
- Joining (welding, etc.)
- Weight savings (density, modulus)
- Heat transfer
- Cost (materials, fabrications, life-cycle cost)
- Earthquakes
- Pollution (related to processing)
- Producibility
- Availability (strategic materials)
- Appearance (colour, surface)
- Optical, magnetic, electrical
- Nuclear waste disposal
- Garbage (energy conversion, incineration)
- Catalysts
- Weight
- Recyclability

## Notes

1. World Resources Institute, *World Resources: A Guide to Global Environment 1994–95* (Oxford: Oxford University Press, 1994).
2. M. Yacaman, *Indicators of the Scientific and Technological Activities in Mexico* (Mexico City: National Council of Science and Technology, 1993); L. Martinez, Panel on Materials Science and Engineering in Mexico, Ixtapa State of Guerrero, January 1993; A. Flores, 'Good for Science, But Not for Technology', *Politicas*, vol. 23 (1993), no. 3; B. Bowonder and P.K. Rohatgi, 'Technology Forecasting Applicability, Relevance and Future Crisis Analysis in Materials', *Technological Forecasting and Social Change*, vol. 7 (1975), p. 233; P.K. Rohatgi and B. Bowonder, 'Potential of Technology Forecasting in Planning Materials Research', *National Metallurgical Laboratory Technical Journal*, vol. 14 (1972), p. 1; P.K. Rohatgi and C. Weiss, 'Technology Forecasting for Commodity Projections: A Case Study on the Effect of Substitution by Aluminum on the Futher Demand for Copper', *Technological Forecasting and Social Change*, vol. 11 (1977), p. 25; P.K. Rohatgi, B.J. Vyas and B. Bowonder, 'Foundry Industry of India Towards the Year 2000: Technology Forecasting', *Journal of Scientific and Industrial Research*, vol. 37 (1978), no. 1, p. 1; D. Suresh, S. Seshan and P.K. Rohatgi, 'Melting and Casting of Alloys in a Solar Furnace', *Solar Energy*, vol. 23 (1979), no. 6, p. 553; Bram Fond, ed., *Tomorrow's Revolution, Microbe Power* (London: Macdonald and Jane's Publishers, 1976).
3. Yacaman, *Indicators*.
4. M. Cohen, 'Materials Science and Engineering, Its Evolution, Practice and Prospects', *Materials Science and Engineering*, vol. 37 (1979), no. 1; National Research Council, *Materials Science and Engineering for the 1990s: Maintaining Competitiveness in the Age of Materials* (Washington DC: National Academy Press, 1989); G.O. Barney (Study Director), *The Global 2000 Report to the President: Entering the Twenty-First Century*, vol. 1, Council of Environmental Quality and the Department of State (Washington DC: US Govermnent Printing Office, 1980); Fond, ed., *Tomorrow's Revolution*; 'Biology and Materials Synthesis, Parts I and II', *Materials Research Society Bulletin* (October and November 1992); Federal Program in Materials Science and Technology, 'Advanced Materials and Processing'.
5. P.K. Rohatgi, K. Rohatgi and B. Bowonder, *Technology Forecasting: India Towards the 21st Century* (New Delhi: Tata McGraw-Hill, 1979); P.K. Rohatgi, S. Mohan and K.G. Satyanarayana, 'Materials Science and Technology in the Future of India', Regional Research Laboratory, Trivandrum, Kerala (CSIR, 1980); P.K. Rohatgi, B. Nagraj and K. Rohatgi, 'Future Technologies for Metals and Materials in India', *Journal of Scientific and Industrial Research*, vol. 41 (June 1982), pp. 351–60; P.K. Rohatgi, 'Materials Technology and Development', 5th Issue, Bulletin of the Advanced Technology Alert System, Center for Science and Technology, United Nations (New York: UN, 1988).
6. Rohatgi, Mohan and Satyanarayana, 'Materials Science and Technology in the Future of India'.
7. Ibid.
8. Fond, ed., *Tomorrow's Revolution*; 'Biology and Materials Synthesis, Parts I and II'.

9. National Research Council, *Materials Science and Engineering for the 1990s*.
10. Rohatgi, 'Materials Technology and Development'; Federal Program in Materials Science and Technology, 'Advanced Materials and Processing'.
11. Rohatgi, 'Materials Technology and Development'.
12. National Research Council, *Materials Science and Engineering for the 1990s*.
13. 'World Resources: A Guide to Global Environment 1994–95'.
14. Rohatgi, 'Materials Technology and Development'.

## References

'Biology and Materials Synthesis, Parts I and II', *Materials Research Society Bulletin* (October and November 1992).
Barney, G.O. (Study Director), *The Global 2000 Report to the President: Entering the Twenty-First Century*, vol. 1, Council of Environmental Quality and the Department of State (Washington DC: US Government Printing Office, 1980).
Bowonder, B. and P.K. Rohatgi, 'Technology Forecasting Applicability, Relevance and Future Crisis Analysis in Materials', *Technological Forecasting and Social Change*, vol. 7 (1975).
Cohen, M., 'Materials Science and Engineering, Its Evolution, Practice and Prospects', *Materials Science and Engineering*, vol. 37 (1979), no. 1.
Federal Program in Materials Science and Technology, 'Advanced Materials and Processing: The Fiscal Year 1993 Program', Office of Science and Technology Policy (Gaithersburg, Md.: National Institute of Standards and Technology, 1993).
Flores, A., 'Good for Science, But Not for Technology', *Politicas*, vol. 23 (1993), no. 3, pp. 15–17.
Fond, B., ed., *Tomorrow's Revolution, Microbe Power* (London: Macdonald and Jane's Publishers, 1976).
Hondras, E.D., 'Materials, Year 2000', in Tom Forester, ed., *The Materials Revolution* (Cambridge, Mass. and London: MIT Press, 1988).
Martinez, L., Panel on Materials Science and Engineering in Mexico, Ixtapa State of Guerrero, January 1993.
National Research Council, *Materials Science and Engineering for the 1990s: Maintaining Competitiveness in the Age of Materials* (Washington DC: National Academy Press, 1989).
Rohatgi, P.K., 'Materials Technology and Development', 5th Issue, Bulletin of the Advanced Technology Alert System, Center for Science and Technology, United Nations (New York: UN, 1988).
Rohatgi, P.K. and B. Bowonder, 'Potential of Technology Forecasting in Planning Materials Research', *National Metallurgical Laboratory Technical Journal*, vol. 14 (1972).
Rohatgi, P.K., S. Mohan and K.G. Satyanarayana, 'Materials Science and Technology in the Future of Kerala', Regional Research Laboratory, Trivandrum, Kerala (Council of Scientific and Industrial Research [CSIR], Government of India, New Delhi, 1980).

Rohatgi, P.K., B. Nagraj and K. Rohatgi, 'Future Technologies for Metals and Materials in India', *Journal of Scientific and Industrial Research*, vol. 41 (June 1982), no. 6, pp. 351–60.

Rohatgi, P.K., K. Rohatgi and B. Bowonder, *Technology Forecasting: India Towards the 21st Century* (New Delhi: Tata McGraw-Hill, 1979).

Rohatgi, P.K., B.J. Vyas and B. Bowonder, 'Foundry Industry of India Towards the Year 2000: Technology Forecasting', *Journal of Scientific and Industrial Research*, vol. 37 (1978), no. 1.

Rohatgi, P.K. and C. Weiss, 'Technology Forecasting for Commodity Projections: A Case Study on the Effect of Substitution by Aluminum on the Future Demand for Copper', *Technological Forecasting and Social Change*, vol. 11 (1977).

Suresh, D., S. Seshan and P.K. Rohatgi, 'Melting and Casting of Alloys in a Solar Furnace', *Solar Energy*, vol. 23 (1979), no. 6.

World Resources Institute, *World Resources: A Guide to Global Environment 1994– 95* (Oxford: Oxford University Press, 1994).

Yacaman, M., *Indicators of the Scientific and Technological Activities in Mexico* (Mexico City: National Council of Science and Technology, 1993).

# 8

# Generic Skills of Management and Organization: The Energy Sector in Africa

*Stephen Karekezi*

## 1. Energy and Development

Access to adequate levels of energy services is a critical prerequisite to development. Energy services are essential to virtually all productive and service sectors. Agriculture, mining, industry and transport are all critically dependent on adequate energy services. One of the most dramatic and current demonstrations of how critical energy is to national economies of developing countries is a recent statement by the president of the Philippines, Fidel Ramos, in which he specifically noted that the absence of reliable power was currently the single most important impediment to national economic growth.[1] Even service sectors such as trade, commerce, communications, education, health and entertainment all require reliable energy to function properly. Consequently, the development and dissemination of energy technologies is of interest to a wide range of sectors in developing countries.

For many countries in the South, energy investment accounts for a large and growing share of national investment. In 1988, USAID estimated that between 1988 and 2008, developing countries will require 1500 GW of generating capacity and concomitant transmission and distribution facilities.[2] This translates to a total of about US $2.5 trillion or in the region of US$125 billion per year – which is of the same order of magnitude as an estimate of the World Bank for the 1990s of US$100 billion per year.[3] It is estimated that in India, planned power-sector investments will account for between 30 and 40 per cent of the country's total projected national investment for the last decade of this century.[4] This pattern is repeated in virtually all developing countries. P. Kessler, a director of Shell International, estimates that LDCs (least developed countries) will require an additional capacity of 40 million barrels/day between now and 2020. At current production levels, this

**TABLE 8.1** Per-capita commercial energy consumption for selected developing countries, 1992

| Country | Kgoe | Country | Kgoe |
| --- | --- | --- | --- |
| Tunisia | 567 | Togo | 46 |
| Brazil | 681 | Burundi | 24 |
| Saudi Arabia | 4463 | Senegal | 111 |
| Malaysia | 1445 | Madagascar | 38 |
| Trinidad | 4910 | Tanzania | 30 |

SOURCE: World Bank, *World Development 1993* (Washington DC: World Bank, 1993); *World Development 1994* (Washington DC: World Bank, 1994).

is equivalent to five Saudi Arabias or eight North Seas of new production capacity.[5]

Although developing countries currently account for about 25 per cent of global energy consumption, their rapid economic, population and urbanization growth rates are expected to result in very rapid increases in energy consumption. An increasingly larger share of global energy investments will probably occur in developing countries as the energy industry begins to meet the pent-up massive demand for modern energy services. Some projections indicate that by the year 2020, energy consumption in developing countries would account for about 40 per cent of global energy demand.[6] In many respects, the technology choices that energy policy-makers in the South make are likely to have a fundamental influence on the world's energy industry.

One of the fundamental dilemmas that faces any attempt to analyse developing countries is the enormous diversity that is found in the energy sector of the South. This is illustrated by the examples listed in Table 8.1. Consequently, generalizations that treat developing countries as an undifferentiated entity should be treated with a great deal of caution.

## 2. Energy Technologies and Technological Change in Developing Countries

From time immemorial, energy technologies have and continue to be an important driving force of change in national economies. The open woodfire, the steam engine, the internal combustion engine, and finally the electric motor mark critical shifts in the technological development of human society.[7] A similar pattern in dominant energy fuels can also be discerned, with a gradual shift from wood through coal, oil, hydro, nuclear, natural gas, and finally to increased use of renewable sources of energy such as wind, biomass and solar energy. The bulk of future energy projections indicate a major role by renewables in meeting global energy demand.[8] A wide range of energy technologies have recently come on stream (or are near-commercial) in both the supply, transformation and demand sectors. The energy technologies that are likely to be of great interest to developing countries[9] include:

## Supply and transformation technologies

- *Advanced natural gas turbines.*
- *Geothermal power plants* for countries endowed with geothermal resources. Global installed capacity is estimated to be about 5 GW. Substantial geothermal potential exists in developing countries[10] such as China (160 GW), Indonesia (130 GW), Peru (86 GW), Mexico (77 GW), Ethiopia (50 GW) and Kenya (24 GW). New duel flash geothermal technologies can increase overall efficiency by 20 per cent.[11]
- *Advanced gas turbines* coupled to coal gasifiers that allow the cost-competitive development of relatively small (100 MW) power generation units suitable for developing countries with small grids.[12]
- *Fluidized-bed combustion* and integrated-gas combined cycle combustion. Fluidized-bed combustion technology is now in use in both India and China.[13]
- *Beneficiation technologies* for high-ash-content coal such as improved coal processing before combustion and enhanced treatment of waste gases.
- *Electronic systems for power sector* load management, dispatching, optimization or distribution network, metering, billing and detection of losses and thefts in transmission and distribution.[14]
- *Advanced diesel generators.*
- *Biomass integrated/gas turbines (BIG/GT)* technology based on the rapid progress made in the development of aero-derivative turbines and coal gasifiers.[15]
- *Co-generation technologies* for the simultaneous generation of electricity and heat.
- Large-scale and decentralized *briquetting of low-grade biomass* residues to produce high-grade multi-purpose fuel briquettes.
- *Wind turbines, photovoltaics and solar thermal technologies* that produce electricity and heat at prices that are competitive with conventional

technologies. Large-scale and well-managed wind farms can generate electricity at about 5 US cents per KWh, which is cost-competitive with conventional coal-fired power stations.[16] Solar photovoltaic technologies, the costs of which have dropped by an order of magnitude in the last 30 years and are now competitive with decentralized diesel-powered electricity generation. An important driving force behind the wide-scale use of PV technology in the region has been a dramatic drop in cost experienced over the last ten years. As explained by Derrick and Bokalders,[17] the price of PV in 1975 was estimated to be about US$30 per peak Watt. Current prices are estimated to be below US$5. The production of photovoltaic modules has increased over the past two decades, rising from about 1 MWp in 1976 to over 35 MWp by mid-1988.

Other energy technologies that have been perceived by some energy analysts as important to the world's energy future include *modular nuclear plants* that are passively safe and have diversion resistant fuel cycles. Their wide-scale use largely depends on the extent to which acceptable solutions are found for nuclear waste disposal and problems associated with proliferation. To date, no satisfactory response to the enormous uncertainties associated with nuclear energy is at hand.

## Demand-side, end-user technologies

- *Cookstoves with improved efficiency and reduced emissions.* The humble wood-fuelled cookstove is likely to be the most widely used energy device in the developing world. Substantial progress has been made in the development and dissemination of energy-efficient cookstoves. Using both grassroots approaches and centralized mechanisms, impressive numbers of improved cookstoves have been disseminated by some developing countries.[18] In certain cases, the numbers have increased by several orders of magnitude. For example, in Kenya an estimated 6000 stoves had been disseminated by 1981. By 1987, the figure was over 200 000 and by the early 1990s, over half a million improved cookstoves had been disseminated. India and China have disseminated an estimated 3 million and 60 million stoves, respectively.[19]
- Moderately sized and efficient *domestic refrigeration and ventilation units.*
- Efficient *compact fluorescent lamps* use 60 to 75 per cent less electricity unit of light output than equivalent incandescent lamps.[20]

- *Advanced lighting systems* such as multilevel switches, timers, photocell controls, occupancy sensors and daylight dimmers can realize energy savings of 15 to 30 per cent.[21] Studies in Brazil, India and Thailand have demonstrated that advanced lighting technologies and systems can reduce energy consumption for lighting by 22 per cent, 33 per cent and 70 per cent, respectively.[22]
- Energy-efficient climate-dependent *building design and construction techniques*.
- *Advanced windows technology* that controls optical and thermal energy flows.
- *Efficient vehicles* and mass transit systems.
- *Efficient motors* and variable speed drives can reduce electricity use by 15 to 40 per cent.[23] Proper system sizing and good housekeeping measures such as periodic testing of motors, proper lubrication, and appropriate rewinding temperatures can reduce losses by up to 80 per cent.[24]

In contrast to the North where energy technologies were developed and introduced over relatively long periods, which allowed for coherent transitional phases in which the productive and service sectors had sufficient lead time to reorganize and adapt to changes, energy technologies have been introduced in the South in a helter-skelter fashion in which the traditional woodfire coexists with massive multi-million-dollar coal, hydro and nuclear-energy complexes. Consequently, the requisite institutional, management and maintenance infrastructures are either non-existent or unable to cope with the wide range of needs that such a diverse energy system requires.

Another important feature of the energy sector in developing countries is the limited linkages that exist between the different sectors.[25] Biomass energy resources are largely utilized in the household sector,[26] while petroleum is mainly used by the transport sub-sector. Industry meets its energy needs through electricity which is often sourced from hydro resources or from thermal power stations. There are very few opportunities for substitution in either the main demand or the principal supply subsectors.

In many respects, the plethora of energy technologies that are now found in the developing countries, combined with the fact that there are very limited linkages between the various energy subsectors, places a high premium on effective management and organization of existing energy assets and technologies rather than on the development of new energy technologies.

The next section will discuss management and organization with reference to three energy technologies, namely: wind-turbine technology, biomass power generation, biomass-fuel production, and demand-side management.

## 3. The Role of Organization and Management in the Cost-effective Exploitation of Existing and New Energy Technologies in Developing Countries

Wind electricity generation technologies have registered dramatic improvements in the last 20 years. Most notable of all is that this development occurred at a time when the key source of major basic research and development (R&D) funding, the USA, had been significantly reduced. During the Reagan administration, the US budget for wind-turbine technology development experienced a sevenfold reduction (although it is important to note that the R&D budgets for wind technology in a number of European countries, notably Denmark and the Netherlands, did not experience such reductions).

In spite of significant cutbacks in R&D support, major reductions in costs and efficiencies have been realized. A recent global review of wind energy technology estimates that in areas with good wind regimes (450 watts per square metre of wind power density at hub height), current wind technology can now produce electricity at US$0.053 per KWh (excluding taxes and assuming 6 per cent interest on the investment capital borrowed), down from an estimated US$0.15 per KWh.[27] This is expected to be reduced to US $0.029 per KWh,[28] which would transform wind technology into a mainstream competitor with other conventional power generation technologies.

The bulk of improvements in the performance of wind energy technology were realized mainly through better organizational and management practices, such as:

- Factory-based improvements in turbine manufacturing technology which have significantly reduced capital costs.
- Improved design and manufacture of key components such as mechanical and tip brakes.
- Better siting of wind-turbines (at greater heights, winds are more stable and consistent).
- Scheduling of maintenance when there is no wind, thereby minimizing losses attributed to maintenance downtime.

- Enhanced integration in the grid to allow use of wind power to meet higher value peak loads.

One of the key advantages of wind-turbine technologies is the relative simplicity of its key components, which can be procured as off-the-shelf industrial parts – such as gearboxes, drive shafts and generators. The modularity of the technology facilitated early and rapid advance of the wind industry.[29] This contrasts sharply with conventional and centralized energy technologies, which require custom-made components and fabrication and assembly technologies.

Few of the above-listed improvements were the result of basic research and development, but were instead largely linked to gradual and incremental improvements in the management and organization of the technology. A number of large-scale wind projects are now underway in India and Africa. In Mauritania, for example, the Global Environment Facility is financing a US$2 million windpower project. Building on the successful experiences of developed nations such as the USA and Denmark, this project has adopted a modular-technology approach which has enabled the local manufacture of some 90 per cent of the parts used by the project.[30]

The value of centralized organization and management is also demonstrated in the case of bagasse power generation, one of the most established large-scale biomass energy systems with a sound economic track record. In comparison with fossil fuels such as coal, biomass-fuelled power plants are less polluting. Biomass contains limited amounts of sulphur and nitrogen (less than 0.05 per cent) and thus its contribution to the generation of pollutants that cause acid rain is relatively low.[31] In addition, biomass energy industries are not net contributors to the build-up of carbon dioxide (the major greenhouse gas) if the biomass used is regrown.

Co-generation using bagasse as feedstock to produce both process heat and electricity is an established and proven technology with a history going back almost 100 years. Experiences in a number of developing countries have been encouraging. Incremental improvements in the technology have led to the successful establishment of competitive power generation in the sugar industries of Brazil, Mauritius and Reunion. Over the past decade, exciting developments have taken place, particularly in increasing the efficiency of biomass-fuelled power generation systems. Of particular note is the development of biomass integrated/intercooled-steam injected turbines (BIG/ISTIG), which produce more than twice as much electricity per tonne of sugar cane as modern condensing/extraction steam turbines (CEST). It is estimated that if BIG/ISTIG technology was fully deployed in the 80 sugar-cane growing countries, the generated electricity would be equivalent to between 50 and 80 per cent of total electricity currently produced by utilities.[32]

One of the key organizational/institutional innovations in these countries has been the successful establishment of a durable alliance between the sugar industry and the power utilities. Further development of this alliance is perceived to be the key hurdle facing rapid expansion of bagasse power generation in developing countries.

In Mauritius, substantial progress has been made in co-generation. To ensure round-the-year production of electricity from bagasse-power plants, a number of plants are of a dual coal/bagasse nature. When no bagasse is available, these co-generation units can burn coal. Many of the constraints upon co-generation in the region are of an institutional and legal nature, such as the generation and distribution monopoly held by utilities. In an extensive review of the power sector in Mauritius, Baguant confirms that the country's sugar industry is self-sufficient in electricity and sells excess power to the national grid. In 1989, close to 10 per cent of the country's electricity was generated from bagasse, a by-product of the sugar industry.[33]

The major hurdles to wider deployment of large-scale biomass energy power generation are largely organizational. Questions such as the legality of institutions other than the national utility generating, distributing and selling power often hamper wider dissemination of this important technology. Even when the national or local utility agrees to purchase power generated by independent producers such as sugar firms, price setting becomes the sticking point. These questions are particularly acute in developing countries, where national utilities often exercise monopoly over the generation and distribution of power and have no institutional interest in fostering competing sources of electricity.

Baguant provides convincing evidence that the installed equipment in the sugar factories of Mauritius is sufficient to double the current level of electricity generated.[34] The main stumbling-block is the unattractive prices offered by the national utility to independent power producers. Modest capital investment combined with judicious equipment selection, modifications of sugar manufacturing processes and proper planning could yield a thirteenfold increase in the amount of electricity generated by sugar factories and sold to the national Mauritian power utility. A more recent review of the potential of biomass gasification technologies for electricity production in Mauritius estimates that by the year 2010 the combined production of power from bagasse and cane tops and leaves (including trash) was potentially equal to the country's total electricity requirement.[35]

The importance of good management is demonstrated by the fact that the Global Environment Facility (GEF) multi-million-dollar programme for bagasse-based power generation in Mauritius is placing the major emphasis on enhancing the efficiency and performance of the transport of sugar-cane rather than on increasing the efficiency of the generation technologies.

The production of ethanol as a replacement fuel for gasoline is another well-established and proven large-scale biomass energy technology. Biomass can also be used to produce methanol, but the production of ethanol is, at the moment, the most commercially attractive option.[36] Ethanol programmes that produce a blend of ethanol and gasoline (gasohol) for use in existing fleets of motor vehicles have been implemented in Brazil, Malawi, Zimbabwe and Kenya. Available evidence indicates that these programmes have generated substantial benefits. The Zimbabwe alcohol programme has reduced annual gasoline imports by 40 million litres.[37] Alcohol production is not, it should be noted, a fully environmentally benign technology: it produces acidic stillage (fermentation effluents) whose disposal can cause problems. In the Zimbabwe ethanol plant, the stillage is mixed with irrigation water and returned to the sugar-cane fields.

The ethanol programmes in both Brazil and Zimbabwe demonstrate the critical importance of organization and management. The Brazilian alcohol programme indicates that the single most important determinant of ethanol costs is the cost of delivered cane, which varies significantly within Brazil. Cane growing, harvesting and transportation costs also vary greatly among developing countries. The extent to which by-products such as bagasse, stillage and sugar-cane leaves are used also has an important bearing on the economic viability of an ethanol programme.

The potential for combining BIG/ISTIG technologies for power generation with sugar cane-based alcohol distilleries for vehicle fuel appears very attractive.[38] It is estimated that this combination could produce electricity at US $1,500/kw, which is cost-competitive with new hydro-power plants and alcohol at an equivalent oil price of US$20/barrel.[39]

The nature of most demand-side management technologies is embedded in management and organizational issues. Many of the innovative demand-side approaches are linked to new analytical methods rather than to new hardware. For example, integrated resource planning (IRP) is a relatively simple tool that facilitates comparative assessment of both demand-side and supply-side options. Many of the industrial efficiency technologies are largely linked to organization and management. For example, it is estimated that simple housekeeping measures in industry such as switching equipment off when not in use can significantly reduce electricity consumption. In their study of institutional and planning aspects of the Kenya industrial energy management programme, Okech and Nyoike demonstrate that simple housekeeping measures in the country's industrial concerns could result in energy savings of between 10 and 30 per cent.[40]

Substantial energy savings in industry can be obtained through low-cost interventions. Examples include improving the efficiency of generic energy conversion and end-use equipment such as motors and boilers. This is particularly the case for equipment such as motors that are found in industry in large numbers. A small saving on a single motor can yield significant energy-cost savings if hundreds of the same type of motor are in use. A special issue of *Scientific American* magazine on energy estimated that with low-cost retrofits, the efficiency of a typical motor-pump system can be raised to 72 per cent.[41] Low-cost measures include proper adjustment of boiler and furnace controls, appropriate temperature settings in process industries, and optimum loading of process machinery. Substantial energy savings can be realized as a by-product of efforts to improve other aspects of industry – for example quality, or productivity.[42]

A similar case is found in the transport energy subsector, where there exist substantial opportunities for energy conservation. One simple measure would be the rationalization of models to facilitate an improvement in spare-parts procurement and service; this would lead to lower fuel consumption and reduce emissions of environmentally harmful noxious fumes. According to a recent report of the Kenya Motor Industry Association, the country has over 300 different car models for a fleet (of just under 300,000 vehicles) that can support at most 40 models.[43] The rationalization of models would greatly increase fuel efficiency and enhance the effectiveness of motorized transport in the country.

Inspection and maintenance programmes for vehicles have been shown to increase fuel efficiency and lower emissions by almost 25 per cent.[44] A typical vehicle inspection and maintenance programme would include: (i) prohibiting the use of vehicles that fail the fuel-efficiency inspection test, and (ii) promotion and education programmes that encourage vehicle owners to adhere to high standards of maintenance. Equipment for effectively using waste heat can also result in substantial energy savings. More effective insulation and prevention of energy loss can also further assist energy efficiency. The technical know-how required to effect low-cost interventions in industrial energy conservation is of a rudimentary nature. Many of these technologies are within the bounds of systematic application of common sense. Those that require more specialized skills are often available within the public domain. What is often missing in developing countries is the institutional framework for the collation and dissemination of this information to appropriate energy end-use industries.

Recent experience in Zimbabwe illustrates the viability of low-cost interventions. One of Zimbabwe's leading footwear manufacturers, Superior Footwear Company, embarked on a major industrial energy conservation initiative. The combined effects of the low-cost interventions were expected to yield over 85 per cent of the total annual savings.[45]

Energy-efficient technologies are now widely available. Through a combination of technology breakthroughs and incremental improvements, new industrial plants require minimal energy to operate. The latest industrial machinery for the manufacture of, for example, steel, aluminium and cement are now highly energy efficient. For example, electric arc furnaces for steelmaking are now over 30 per cent more efficient than in 1965.[46] The major increase in the efficiency of energy use in the industrial sectors of developed countries is partly attributed to the wide-scale introduction of energy conservation measures. Even in the USA, the developed country that is said to be the least enthusiastic about energy conservation, the amount of fuel consumed by industry has declined by more than 50 per cent during the past 30 years.[47]

In contrast to the emerging market economies of Central Europe, many developing countries are not burdened with large-scale supply-side centralized energy systems. Consequently, judicious planning of energy investments would yield significant long-term benefits. This is a view shared by a somewhat dated but prescient report of the Economic Community of West Africa (ECOWAS) on technology and energy in the developing world:

> It is easier and far more cost-effective to influence the course of *future* energy consumption than it is to alter the *existing* pattern of energy use (although both activities are of extremely high value). Yet far less attention has been paid to a close examination of the energy implications of major *new* invest-

ments in agriculture, industry, and transport – the mainstream economic activities – than the potential energy/economic benefits would justify. It is virtually costless to introduce sound energy planning at the project development.[48]

## 4. Lessons for Development Aid Policy

Key lessons for development aid policy that can be drawn from the case examples of energy development in the South discussed above include:

- Better management and organization should be central to the debate on the development of a sustainable energy sector for developing countries. Management and organization should receive at least as much attention as that which is currently lavished on importing new technology from industrialized countries.
- Many of the new energy technologies and approaches that are coming on stream are management- and skill-intensive. Developing countries that focus on enhancing the management and organization capabilities will be well placed to enjoy the full benefits of new energy technologies and approaches.
- Management-intensive energy technologies and methodological tools tend to have substantial long-term spin-off benefits. Developing countries that are quick off the mark and begin to stress management and organization will be the first to reap the benefits of management- and organization-intensive technologies.
- Centralized and capital-intensive energy technologies constitute, in the absence of a critical level of management and organization capability, a major risk for developing countries. Preference should be given to decentralized and modular energy technology that would facilitate 'learning by doing' without placing an additional burden on the debt of developing countries.
- Organization, management and adaptation require a well-educated and skilled workforce; this would appear to place a premium on investment in research on the human-resource-development and capacity-building components of technology.

## Notes

1. 'Philippines Economy Hard Hit by Power Crisis', *Kenya Times*, Nairobi, 13 January 1993.

2. *Power Shortages in Developing Countries: Magnitude, Impacts, Solutions, and the Role of the Private Sector* (Washington DC: USAID, 1988).

3. D. Jhirad, *Implementing Power Sector Solutions in Developing Countries* (Stockholm: Stockholm Initiative in Energy, Environment and Sustainable Development [SEED], 1991); E. Moore and G. Smith, *Capital Expenditures for Electric Power in the Developing Countries*, World Bank Industry and Energy Department Working Paper, Energy Series Paper no. 21 (Washington DC: World Bank, 1990).

4. Jhirad, *Implementing Power Sector Solutions*.

5. P. Kessler, *Energy for Development* (London: Shell International Petroleum Company, 1994).

6. Office of Technology Assessment (OTA), US Congress, *Fueling Development: Energy Technologies for Developing Countries* (Washington DC: US Government Printing Office, 1992).

7. J. Debeir, J. Deleage and D. Hemery, *In the Servitude of Power – Energy and Civilization through the Ages* (London: Zed Books, 1991).

8. T. Johansson, H. Kelly, A. Reddy and R. Williams, eds, *Renewable Energy – Sources for Fuels and Electricity* (Washington DC: Island Press, 1993).

9. T. Willbanks, 'Institutional Issues in Capacity Building for Energy Technology Assessment', ATAS (Advance Technology Assessment System), Centre for Science and Technology for Development, *Bulletin on Energy Systems, Environment and Development – A Reader* (New York: United Nations, 1991); *Energy Technology for Developing Countries* (Berkeley: Lawrence Berkeley Laboratory, 1989); S. Karekezi and G.A. Mackenzie, eds, *Energy Options for Africa – Environmentally Sustainable Alternatives* (London: Zed Books, AFREPREN, FWD and UNEP Collaborating Centre on Energy and Environment, 1993).

10. OTA, *Fueling Development*.

11. R. DiPippo, *Geothermal Energy: Electricity Production and Environmental Impact – A World Wide Perspective*, Proceedings of the Energy and Environment in the 21st Century Conference, 26–28 March 1990, Cambridge, Mass.

12. OTA, *Fueling Development*.

13. S. Tavoulareas, Memorandum of 26 April, 1991 (Washington: USAID, 1991).

14. G. Schramm, *Issues and Problems in the Power Sector of Developing Countries* (Stockholm: Stockholm Initiative on Energy, Environment and Sustainable Development, 1991).

15. R.H. Williams and E.D. Larson, 'Advanced Gasification-Based Biomass Power Generation', in Johansson et al., eds, *Renewable Energy*.

16. A.J. Cavallo, S. Hock and D. Smith, 'Wind Energy: Technology and Economics', in Johansson et al., eds, *Renewable Energy*.

17. A. Derrick, C. Francis and V. Bokalders, *Solar Photovoltaic Products* (London: Intermediate Technology Publications, 1989).

18. K.K. Prasad, 'Whither Stoves?', paper presented to the 2nd FWD International Workshop on Woodstoves, 1987; and S. Karekezi and D. Walubengo, *Household Stoves in Kenya – the Case of the Kenya Ceramic Jiko* (Nairobi: KENGO Wood Energy Series, 1989).

19. F.M. Njoroge, 'An Overview of Improved Stove Dissemination Programmes in Kenya', paper presented at the 2nd International Workshop on Woodstoves held in Antigua, Guatemala (Nairobi: Foundation for Woodstove

Dissemination, 1987); S. Karekezi, *Improved Charcoal Production and Fuel Efficient Cookstoves* (Nairobi: FWD, 1993); B.M.L. Garg, 'Improved Chulhas (Cookstoves) Programme in India', paper presented at the 2nd International Workshop on Woodstoves in Antigua, Guatemala (Nairobi: Foundation for Woodstove Dissemination, 1987); W. Mengjie and H. Liang, 'Background and Measures for Promoting the Use of Fuelwood and Coal-Saving Stoves in China', paper presented at the 2nd International Workshop on Woodstoves in Antigua, Guatemala (Nairobi: Foundation for Woodstove Dissemination, 1987).

20. M.D. Levine et al., *Energy Efficiency, Developing Nations and Eastern Europe – A Report to the US Working Group on Global Energy Efficiency* (Washington DC: International Institute for Energy Conservation, 1991).

21. Ibid.

22. G. Jannuzzi, A. Gadgil, H. Geller and M.A. Sastry, 'Energy-Efficient Lighting in Brazil and India: Potential and Issues of Technology Diffusion', in E. Mills, ed., *Proceedings of the 1st European Conference on Energy-Efficient Lighting* (Stockholm: Swedish National Board for Industrial and Technical Development, Department of Energy Efficiency, 1991); S. Nadel, V. Kothari and S. Gopinath, *Opportunities for Improving End-Use Electricity Efficiency in India* (Washington DC: American Council for an Energy-Efficient Economy, 1991); J.P. Busch and S. Chirarattananon, 'Conserving Electricity for Lighting in Thai Commercial Buildings: A Review of Current Status, Potential and Policies', in Mills, ed., *Proceedings of the 1st European Conference on Energy-Efficient Lighting*.

23. Nadel, Kothari and Gopinath, *Opportunities for Improving End-Use Electricity Efficiency in India*.

24. P. Ibanez, 'Electromechanical Energy', in C.B. Smith et al., eds, *Efficient Electricity Use* (Elmsford: Pergamon Press, 1978); Nadel, Kothari and Gopinath, *Opportunities for Improving End-Use Electricity Efficiency in India*; Levine et al., *Energy Efficiency, Developing Nations and Eastern Europe*.

25. Karekezi, 'African Energy Research Networks'.

26. S. Karekezi, *Energy from Biomass and Associated Residues – A Global Review* (Nairobi: FWD and AFREPREN, 1991).

27. A.J. Cavallo, R.H. Williams and G. Terzian, *Baseload Wind Power from the Great Plains for Major Electricity Demand Centres* (Princeton, N.J.: Princeton University Press, 1994).

28. Cavallo, Hock and Smith, 'Wind Energy: Technology and Economics'.

29. Ibid.

30. D. Barricklow, 'Wind Power Brightens Mauritania's Future', *Choices – The Human Development Magazine* (New York: UNDP, March 1994).

31. National Wood Energy Association, *Biomass and Biofuels* (Washington DC: National Wood Energy Association, n.d.).

32. R.H. Williams, 'The Outlook for Renewable Energy', in N.I. Meyer and P.S. Nielsen, eds, *Global Collaboration on a Sustainable Energy Development* (Lyngby: Technical University of Denmark, 1991); Williams and Larson, 'Advanced Gasification-Based Biomass Power Generation'.

33. J. Baguant, 'The Case of Mauritius', in M.R. Bhagavan and S. Karekezi, eds, *Energy Management in Africa* (London: Zed Books and African Energy Policy Research Network [AFREPREN], 1992).

34. Ibid.
35. R.P. Beeharry and J. Baguant, *Bagasse Gasification Technologies for Electricity Production in Mauritius*, Report of the Task Force on Research and Development and Technology Transfer Aspects Relating to Power from Biomass (Reduit: University of Mauritius, 1994).
36. R.H.Williams, *Biomass Energy Conversion Technologies for Large-Scale Power Generation and Transport Fuels Applications* (Princeton, N.J.: Princeton University Press, 1994).
37. J.M.O. Scurlock, A. Rosenschein and D.O. Hall, *Fuelling the Future: Power Alcohol in Zimbabwe* (Nairobi: ACTS Press and Biomass Users Network [BUN], 1991).
38. R.H. Williams, 'Potential Roles for Bioenergy in an Energy Efficient World', *Ambio* 14 (1987).
39. Williams, 'The Outlook for Renewable Energy'.
40. B.A. Okech and P. Nyoike, 'The Case of Kenya', in M.R. Bhagavan and S. Karekezi, eds, *Energy Management in Africa* (London: Zed Books and African Energy Policy Research Network [AFREPREN], 1992).
41. P. Arnold Fickett, W.G. Clark and A.B. Lovins, *Efficient Use of Electricity*, *Scientific American* (September 1990).
42. M.H. Ross and D. Steinmeyer, 'Energy for Industry', *Scientific American* (September 1990).
43. Sammy Masara, 'Kenya Needs 40 Car Models, Says Report', *The Standard*, Nairobi, 7 August 1990.
44. M.P. Walsh, 'Managing the Transition to a Low-Emissions Future in Developing Countries', in special Transport and Environment edition of *UNEP Industry and Environment Journal* (Tour-Mirabeau: UNEP, January–June 1993).
45. G. Augusto, *Energy Conservation Unit in Harare Steams Ahead, SADCC Energy Bulletin*, vol. VII (1989), no. 19 (Luanda: SADCC Energy Sector Technical and Administrative Unit [TAU]), p. 9.
46. Ross and Steinmeyer, 'Energy for Industry'.
47. Ibid.
48. ECOWAS and USAID, *Energy Efficiency and Conservation in West Africa: Proceedings of a Seminar held in Lome, Togo, 30 March–8 April, 1983* (New York: Brookhaven National Laboratory, 1983).

# References

Augusto, G., 'Energy Conservation Unit in Harare Steams Ahead', *SADCC Energy Bulletin*, vol. VII (1989), no. 19 (Luanda: SADCC Energy Sector Technical and Administrative Unit [TAU]).

Baguant, J., 'The Case of Mauritius', in M.R. Bhagavan and S. Karekezi, eds, *Energy Management in Africa* (London: Zed Books and African Energy Policy Research Network [AFREPREN], 1992).

Barricklow, D., 'Wind Power Brightens Mauritania's Future', *Choices – The Human Development Magazine* (New York: UNDP, March 1994).

Beeharry, R.P. and J. Baguant, *Bagasse Gasification Technologies for Electricity Produc-*

*tion in Mauritius*, Report of the Task Force on Research and Development and Technology Transfer Aspects Relating to Power from Biomass (Reduit: University of Mauritius, 1994).
Busch, J.P. and S. Chirarattananon, 'Conserving Electricity for Lighting in Thai Commercial Buildings: A Review of Current Status, Potential and Policies', in E. Mills, ed., *Proceedings of the 1st European Conference on Energy-Efficient Lighting* (Stockholm: Swedish National Board for Industrial and Technical Development, Department of Energy Efficiency, 1991).
Cavallo, A.J., S. Hock and D. Smith, 'Wind Energy: Technology and Economics', in T. Johansson, H. Kelly, A. Reddy and R. Williams, eds, *Renewable Energy – Sources for Fuels and Electricity* (Washington DC: Island Press, 1993).
Cavallo, A.J., R.H. Williams and G. Terzian, *Baseload Wind Power from the Great Plains for Major Electricity Demand Centres* (Princeton, N.J.: Princeton University Press, 1994).
Debeir, J., J. Deleage and D. Hemery, *In the Servitude of Power – Energy and Civilization through the Ages* (London: Zed Books, 1991).
Derrick, A., C. Francis and V. Bokalders, *Solar Photovoltaic Products* (London: Intermediate Technology Publications, 1989).
DiPippo, R., *Geothermal Energy: Electricity Production and Environmental Impact – A World Wide Perspective*, Proceedings of the Energy and Environment in the 21st Century Conference, 26–28 March 1990, Cambridge, Mass.
ECOWAS and USAID, *Energy Efficiency and Conservation in West Africa: Proceedings of a Seminar held in Lome, Togo, 30 March–8 April, 1983* (New York: Brookhaven National Laboratory, 1983).
Fickett, P. Arnold, W.G. Clark and A.B. Lovins, *Efficient Use of Electricity, Scientific American* (September 1990).
Garg, B.M.L., 'Improved Chulhas (Cookstoves) Programme in India', paper presented at the 2nd International Workshop on Woodstoves in Antigua, Guatemala (Nairobi: Foundation for Woodstove Dissemination, 1987).
Ibanez, P., 'Electromechanical Energy', in C.B. Smith et al., eds, *Efficient Electricity Use* (Elmsford: Pergamon Press, 1978).
Jannuzzi, G., A. Gadgil, H. Geller and M.A. Sastry, 'Energy-Efficient Lighting in Brazil and India: Potential and Issues of Technology Diffusion', in E. Mills, ed., *Proceedings of the 1st European Conference on Energy-Efficient Lighting* (Stockholm: Swedish National Board for Industrial and Technical Development, Department of Energy Efficiency, 1991).
Jhirad, D., *Implementing Power Sector Solutions in Developing Countries* (Stockholm: Stockholm Initiative in Energy, Environment and Sustainable Development [SEED], 1991).
Johansson, T., H. Kelly, A. Reddy and R. Williams, eds, *Renewable Energy – Sources for Fuels and Electricity* (Washington DC: Island Press,1993).
Karekezi, S., *Energy from Biomass and Associated Residues – A Global Review* (Nairobi: FWD and AFREPREN, 1991).
Karekezi, S., *Improved Charcoal Production and Fuel Efficient Cookstoves* (Nairobi: FWD, 1993).
Karekezi, S., 'African Energy Research Networks: Impact on Policy Formulation and Implementation', in Anton Eberhard and Paul Theron, eds, *International*

Experience in Energy Policy Research and Planning (Cape Town: Energy for Development Research Centre, University of Cape Town, 1993).

Karekezi, S. and G.A. Mackenzie, eds, *Energy Options for Africa – Environmentally Sustainable Alternatives* (London: Zed Books, AFREPREN, FWD and UNEP Collaborating Centre on Energy and Environment, 1993).

Karekezi, S. and D. Walubengo, *Household Stoves in Kenya – The Case of the Kenya Ceramic Jiko* (Nairobi: KENGO Wood Energy series, 1989).

Kenya Times, 'Philippines Economy Hard Hit by Power Crisis', *Kenya Times*, Nairobi, 13 January 1993.

Kessler, P., *Energy for Development* (London: Shell International Petroleum Company, 1994).

Lawrence Berkeley Laboratory, *Energy Technology for Developing Countries* (Berkeley: Lawrence Berkeley Laboratory, 1989).

Levine, M.D., et al., *Energy Efficiency, Developing Nations and Eastern Europe – A Report to the US Working Group on Global Energy Efficiency* (Washington DC: International Institute for Energy Conservation, 1991)

Masara, Sammy, 'Kenya Needs 40 Car Models, Says Report', *The Standard*, Nairobi, 7 August 1990.

Mengjie, W. and H. Liang, 'Background and Measures for Promoting the Use of Fuelwood and Coal-Saving Stoves in China', paper presented at the 2nd International Workshop on Woodstoves in Antigua, Guatemala (Nairobi: Foundation for Woodstove Dissemination, 1987).

Moore, E. and G. Smith, *Capital Expenditures for Electric Power in the Developing Countries*. World Bank Industry and Energy Department Working Paper, Energy Series Paper no. 21 (Washington DC: World Bank, 1990).

Nadel, S., H. Geller, F. Davis and D. Goldstein, *Lamp Efficiency Standards for Massachusetts: Analysis and Recommendations*, Massachusetts Executive Office of Energy Resources (Washington DC: American Council for an Energy-Efficient Economy, 1989).

Nadel, S., V. Kothari and S. Gopinath, *Opportunities for Improving End-Use Electricity Efficiency in India* (Washington DC: American Council for an Energy-Efficient Economy, 1991).

National Wood Energy Association, *Biomass and Biofuels* (Washington DC: National Wood Energy Association, n.d.).

Njoroge, F.M., 'An Overview of Improved Stove Dissemination Programmes in Kenya', paper presented at the 2nd International Workshop on Woodstoves, Antigua, Guatemala (Nairobi: Foundation for Woodstove Dissemination, 1987).

Office of Technology Assessment (OTA), US Congress, *Fueling Development: Energy Technologies for Developing Countries* (Washington DC: US Government Printing Office, 1992).

Okech, B.A. and P. Nyoike, 'The Case of Kenya', in M.R. Bhagavan and S. Karekezi, eds, *Energy Management in Africa* (London: Zed Books and African Energy Policy Research Network [AFREPREN], 1992).

Prasad, K.K., 'Whither Stoves?', paper presented at the 2nd International Workshop on Woodstoves, Antigua, Guatemala, 1987.

Ross, M.H. and D. Steinmeyer, 'Energy for Industry', *Scientific American* (September 1990).
Schramm, G., *Issues and Problems in the Power Sector of Developing Countries* (Stockholm: Stockholm Initiative on Energy, Environment and Sustainable Development, 1991).
Scurlock, J.M.O., A. Rosenschein and D.O. Hall, *Fuelling the Future: Power Alcohol in Zimbabwe* (Nairobi: ACTS Press and Biomass Users Network [BUN], 1991).
Tavoulareas, S., Memorandum of 26 April, 1991 (Washington DC: USAID, 1991).
USAID, *Power Shortages in Developing Countries: Magnitude, Impacts, Solutions, and the Role of the Private Sector* (Washington DC: USAID, 1988).
Walsh, M.P., 'Managing the Transition to a Low-Emissions Future in Developing Countries', in special Transport and Environment edition of *UNEP Industry and Environment Journal* (Tour-Mirabeau: UNEP, January–June 1993).
Williams, R.H., 'Potential Roles for Bioenergy in an Energy Efficient World', *Ambio* 14 (1987), pp. 201–9.
Williams, R.H., 'The Outlook for Renewable Energy', in N.I. Meyer and P.S. Nielsen, eds, *Global Collaboration on a Sustainable Energy Development* (Lyngby: Technical University of Denmark, 1991).
Williams, R.H. and E.D. Larson, 'Advanced Gasification-Based Biomass Power Generation', in T. Johansson, H. Kelly, A. Reddy and R. Williams, eds, *Renewable Energy – Sources for Fuels and Electricity* (Washington DC: Island Press, 1993).
Williams, R.H., *Biomass Energy Conversion Technologies for Large-Scale Power Generation and Transport Fuels Applications* (Princeton, N.J.: Princeton University Press, 1994).
Willbanks, T., 'Institutional Issues in Capacity Building for Energy Technology Assessment', ATAS (Advance Technology Assessment System), Centre for Science and Technology for Development *Bulletin on Energy Systems, Environment and Development – A Reader* (New York: United Nations, 1991).
World Bank, *World Development 1993* (Washington DC: World Bank, 1993).
World Bank, *World Development 1994 – Infrastructure for Development* (Washington DC: World Bank, 1994).

*Part III*

Views from the Industrialized Countries

# 9

# Functional Markets and Indigenous Capacity for Sustainable Development: What Can Transnational Corporations Do through Technology Transfer?

*Richard Adams*

Technology and economic development have been inseparably linked throughout history. Their interdependence increased from the start of the industrial revolution and was further emphasized with the growth of the transnational corporation (TNC). Technology – applied arts and sciences – has been harnessed by all governments to achieve their objectives: peace and war, social change and social stability. But although governments can stimulate technological processes, they are increasingly in the hands of the TNC when it comes to application and implementation. It is hard to overestimate the degree to which the large corporation regards itself as the specialist, *par excellence*, in making new technology functional.

In the last 35 years the pace of technological evolution has accelerated, changing the structure and influence of a number of global industries and having massive impact on home–country industry and indigenous markets. In this context, technology transfer has been seen as an essential component to bring about added value, to help in the process of import substitution, and to create indigenous capacity in management, industrial and infrastructure development. The TNCs grasped the fact a long time ago that new materials technology, to take just one example, could not be applied or marketed by outdated management techniques. Those that did not grasp this failed to survive in an increasingly competitive and rapidly evolving international business world. Today's corporation places as much emphasis on its employees being graduates from Harvard Business School or INSEAD as from MIT.

Three factors drive technological development via innovation, new products or cost saving: the desire for profit; the need to exercise and maintain power, particularly through military technology; and the impetus of the research establishment to know what can be known. With varying degrees of speed, developing nations are coming to realize that new technology is not just a formula, a process or a production line. It is part of a complex web involving sophisticated operational management; it is a thing which upgrades and improves itself by the link through to the international market, and it seems to do this quickest and most efficiently through the system we know as post-industrial capitalism, a system which in the last ten years has swept all before it.

This is not a pain-free or value-free process; nor am I assuming that we are on the right course. There are many that argue that industrial consumerism has exacted an enormous price, eating away at both human society and the foundations of life on earth. There is a lot that we can call positive and progressive, yet efficiency-orientated consumer societies are increasingly entangled in contradictions. The level of demand is constantly stimulated to be above the level of supply, new needs are created as soon as the old are satisfied. The targets of our own standards of living are raised with every improvement in the supply situation. In the First World, the Second World, and amongst the middle classes in the Third World there is a dynamic expectation of increasing prosperity and a satisfying life.

But in the Western industrialized countries intangibles like happiness, fulfilment and social contentment have not noticeably increased with the increase in standards of living. The sense of community continues to be eroded. At the same time the impact of modern economies on our environment, local and global, is giving great cause for concern. Modern economic systems, surprisingly, have also emphasized the continually widening gap between rich and poor countries, which has doubled in the last 30 years.

In the majority of OECD countries, technology transfer has been a steadily incremental process spread over a period of one or two centuries. Even so, social and political change and corresponding shifts in power and influence have been dramatic. By contrast, in most of the developing world, technology has arrived in the form of ready-made products, processes and applications, evolved in other cultures. As different waves of technology developed – water, steam power, electricity, mass production, microelectronics[1] – they depended increasingly on the support of an integrated and developed infrastructure.

Not only has technology become dependent for its effectiveness on a complex web of support structures, it has also become subject to control and ownership of a type unknown to the inventor of the wheel or the watermill. In the recently concluded Uruguay Round of GATT negotiations, intellectual property rights became a significant bargaining element for some developed countries, to be set against tariff barrier reductions and other concessions to the developing world. In the United States, in the ten years between 1980 and 1990, patents granted to corporations rose in number from 46 800 to 73 300 per year. Increasingly the invention and application of new technology relies on a well-established research base. A productive alliance has developed between government-sponsored pure research and the applied research of the large corporation. TNCs are responsible for the greater part of all technological development. Beginning with Thomas Edison, who in 1870 established the first dedicated corporate research and development unit, R&D has assumed ever greater importance in company budgets. In 1990 the 20 TNCs with the highest spending on R&D allocated $48 billion in this area worldwide, an amount which exceeded the GNPs of the vast majority of developing countries.

The contribution of technology to economic development is graphically illustrated in figures for research and development-intensive production for export from the developing world. In 1989 Latin America exported $6.8 billion of such products, Asia $12.2 billion and Africa just $9 *million*.[2] Even though some suppression of data took place for Africa, the contrast is massive. In the same year US-owned TNCs operating in Africa spent one-tenth of the amount (as a proportion of sales) as their counterparts did in Asia and Latin America.[3] While Africa as a whole received 12.3 per cent of direct investment into the developing world in the period 1976–83, this fell to 3.9 per cent in the period 1984–89.[4] These figures illustrate the way that TNCs can strengthen the technological base of a nation where there already exists the capacity to apply, modify and disseminate, and where well-trained human resources are in place. By contrast even the most technologically open and innovative TNC is unlikely to create these conditions. 'The shift of [developing countries'] economies towards high-technology activities is impossible without a critical mass of domestic advanced materials design, synthesis, processing and application competences. Moreover, the in-house and domestic acquisition of materials technological capabilities is not only a prerequisite for, but, also requires, fast changing foreign scientific and technical knowledge.'[5]

Technology has also had an important effect on the nature of industrial policy in developing countries. As a result of innovative competition and the opening up of economies, industrialization is no longer

> **Case study 1** Guinness Nigeria PLC
>
> Guinness is one of the world's largest alcoholic-drinks manufacturing and marketing companies. In 1992 it had a turnover of $3.8 billion in distilling and $2.6 billion in brewing. Founded in 1759 in Dublin, Ireland, the company is now UK based. Sixty per cent of turnover is in Europe, 17.5 per cent in North America, 15 per cent in Asia/Pacific and 8 per cent in the rest of the world. Guinness Nigeria PLC, which, although in 1962 became home to the first Guinness brewery outside the British Isles, is now only 25.5 per cent owned by Guinness PLC – though links remain strong. The black stout known as Guinness is produced in 22 countries in Africa and sold in 37, and it accounts for 7 per cent of all beer sales in the continent, excluding South Africa.
>
> In the 1980s the Nigerian government introduced severe controls on imports. Although Guinness was being made in Nigeria, the barley used in brewing was imported. A programme to replace foreign grain totally with locally grown sorghum and maize was set in hand. Guinness established their own farm and encouraged local farmers to grow new varieties from seed strains specially developed in partnership with the original parent company. Considerable modification of the brewing and handling technology had to be undertaken with new machinery for these different grains being designed and manufactured. Work is still under way on the production of locally sourced and produced enzymes for fermentation. All this has required the development of an extensive R&D division in Nigeria, creating a significant indigenous capacity. In a further transfer, the Nigerian technology has been used in another Guinness brewery in Benin and could have extensive application in the developing world.

invariably a process of building up a sheltered national industry base. Increasingly the policy-makers in a developing country have to decide whether to stimulate an attempt to achieve entry into a world industry, a very different objective from that of import substitution within a protected market. Two examples illustrate this choice: that of Guinness Nigeria, and of Proton/Mitsubishi in Malaysia (see Case studies 1 and 2). In both cases the corporations involved had to acquire and retain significant technological capabilities.[6] In both cases the existence or nurturing of a strong local market was involved.

Such examples illustrate the perspective of the corporate world on one of the conditions necessary for the successful transfer of technology. The Business Council for Sustainable Development endorses the view put forward by its chairman, Stephan Schmidheiny, that technology cooperation will be most successful when it takes place between provider and recipient companies with strong motives of self-interest. 'Business enterprises exist to generate wealth by adding value. Their level of return on investment is a measure of efficiency. Business and industry can make technology available only on competitive, commercial terms.'[7] This position has been reinforced by the increasing complexity of technology management. Increasingly a company requires expertise in several technologies simultaneously in order to be viable. This encourages the formation of strategic alliances and creates clusters of cooperating companies.

All this takes place in the context of a market capitalism designed to satisfy the consumer, effectively the only economic option worldwide. We have to ask in what way this affects the TNC — essentially a business organization designed to be competitive in a demand economy. The first point to note is that in today's world 'demand' is defined not as what people want but what they can afford to pay for. Hence there is no 'demand' for air-conditioning in the villages of Uganda. The TNC responds effectively to demand but not to need. It is not an immoral organization, just one which has evolved appropriately to meet prevailing economic conditions. Just 30 years ago the UN Commission on Trade and Development was established as it became clear that developing nations were not sharing equally in accelerating prosperity. Originally the voice of the South in the trade debate, each agenda item was introduced by the developing countries. Today, most of the agenda is set by the industrialized world, who leave the South to determine how they will fit into the structures set by the North. Efficiency has replaced need, and the test of efficiency is the ability to get rich. So now we only listen to countries like China, India, Brazil and the newly industrializing countries like Malaysia because it may be profitable to deal with them. Africa and many parts of the former Soviet Union can be ignored because there are no profits to be made.

Recent years have seen the development of new non-Marxist critiques of consumer capitalism as a vehicle for development. One of the earliest was articulated in 1970 in the seminal book *Limits to Growth*. 'Faith in technology as the ultimate solution to all problems can divert attention from the most fundamental problem — the problem of growth in a finite system — and prevent us from taking effective action to solve it.'[8] An emerging school of 'Green' economists are arguing that the consumer society must become the conserver society and that the full social and environmental costs of business must be factored in to the pricing process. They have suggested that advances in technology coupled with the dominance of the market system have led to growth

> **Case study 2**   Proton and Mitsubishi – Malaysia
>
> Proton was conceived from the Malaysian government's desire to create a national car industry to develop its heavy industry base. A joint venture was established between HICOM (Heavy Industries Corporation of Malaysia; 100 per cent government-owned) and Mitsubishi Motor Corporation and Mitsubishi Corporation of Japan. Since its inception in 1985, Proton has produced over 500 000 units and has won 72 per cent of the domestic passenger car market, with assistance from protective measures. It is a successful exporter with 1 per cent of the UK market and 7 per cent of that in Singapore and is establishing a joint venture with Mitsubishi in Vietnam.
>
> Proton cars are 80 per cent locally sourced. From 17 vendors supplying 228 parts in 1985, there are now 112 vendors supplying 1200 parts. Mitsubishi continues to provide specialist training to Proton personnel in Japan, and Proton in turn helps to improve the technical and managerial capability and productivity of its local suppliers. It is currently developing its own R&D facility and building a casting plant. Although described by the *Financial Times* as 'an anomaly feeding a nation's greater need', the Proton plant is seen as a shining example by many governments in developing countries and has hosted numerous fact-finding delegations at its Kuala Lumpur plant.

of a type which tends to disregard negative long-term implications, particularly social and ecological aspects. 'Economic conservatives and the business establishment are already assuming that, thanks to technology, GNP growth can continue practically unimpeded, albeit in somewhat different directions, and that such growth is even necessary for environmental conservation.'[9] Sustainability is defined by such economists not as that level of continued growth which *can* be sustained by the economic process but as the reduction of environmental impact in developed countries through reducing consumption and redistribution, allowing economic growth in developing countries sufficient to provide a reasonable standard of living.

In this context the definition of 'better' practice when related to technology may be linked to the concept of 'appropriate' technologies (ATs). Advocates of ATs have emphasized their small-scale nature. It has been shown, for example, how profitable and efficient small-scale technologies for brick-making and sugar production have been passed over.[10] While AT might be thought to contrast with the technological *modus operandi* of the average TNC, the recognition of flexible specialization as the possible successor to mass production, and its role in the restructuring of sunset industries in the older industrialized nations, has brought it into the strategy portfolio of even the largest corporations. Technology, in its various forms, is all pervasive. Deciding whether it is, on the one hand, a passive instrument of business strategy or economic policy, or, on the other, a virulent agent of change, is a global question and not exclusive to developing countries. Nor is technology applied solely by the TNC; its use and effects are endemic in economic life at every level. But TNCs are at the cutting edge of technological development, they regard it as one of the key components in long-term strategic planning, and they spend billions of dollars annually in pushing out technological frontiers. Technology is also rarely socially neutral, often having a significant impact on social, cultural and political life.

In the most influential management book of the 1980s, *In Search of Excellence*, the authors quote several studies where response to the market, the end user, has set the seal on effective innovation. 'Successful innovators innovate in response to market needs, involve potential users in the development of the innovation and understand user needs better.'[11] In spite of the doubts increasingly raised about the 'excellence' approach, the appropriate transfer of technology into developing countries must take account of this. The following factors have been identified as those most often raised in the technology and development debate, but they should be explored with one important point in mind: the best companies are the best listeners – the lesson is obvious in relation to indigenous markets.

## Technology Transfer

Technology flows to developing countries grew rapidly in the 1960s and early 1970s, but since the beginning of the 1980s such transfers have stagnated.[12] The World Commission on Environment and Development placed disadvantageous terms of technology transfer at the head of the list of factors preventing the world's poorest countries from diversifying in ways that would both alleviate poverty and reduce ecological stress.[13] But, with technological input increasingly becoming a greater factor of production costs compared with labour, the sharing of production technology raises real issues of competitive advantage for most companies.

An owner or developer of technology considering exporting such technology can transfer it to either a fully owned subsidiary or to a joint venture; can license the process, equipment, material or technique, or can export it already embedded in a product or service. In addition, technology can be transferred through scientific and technical publications. The almost invariable requirement with technology transfer – through whatever method – is the need for a continuing provision of both managerial and technical knowledge.

It must also be recognized that some governments in developing countries have sought to regulate technology inflows, and numerous market restrictions remain in force. India, for example, through the 1980s, averaged payments for external technology of some $280 million; Thailand averaged $325 million in 1985–87; the Republic of Korea was exceeding $900 million by the mid-1980s; and Brazil averaged $2140 million annually during 1980–88.[14] The same study, looking particularly at technology transfer in India, concluded that government restrictions on royalty payments 'limited the quality and quantity of technology inflows and the accompanying technical support that may have been required. It is necessary to consider whether, in the future, the country will be able to acquire and absorb the wide range of technologies, particularly in microelectronics, biotechnology and new materials, for the level of payments that have been permitted in India, until recently.'[15]

A further barrier to effective technology transfer can arise when this is done on a government-to-government basis or in the context of an aid agency programme. The Business Council for Sustainable Development offers the following critique:

> It is not surprising that much of the developing world is littered with agricultural, medical, processing and industrial equipment that is idle for lack of spare parts, fuel, trained maintenance workers, supplies of raw materials or a market for products. In far too many cases aid has focused mainly on capital-intensive hardware that has prestige appeal for officials of the recipient government. A contractor procures the necessary technology under license from an operator in the industrial country, but the specialized management, maintenance, environmental, and operating know-how of the supplier – essential for the safe and efficient operation of the facility – is not included in the project. When the ribbons have been cut, both politicians and the contractor leave satisfied, but the facility often fails to operate at an acceptable level of productivity, and has adverse impacts on the environment and the surrounding population, whose traditional livelihoods may have been destroyed.[16]

While the unfortunate pattern described above remains all too common, many of the pitfalls have been sidestepped by the promotion of the joint venture. In a joint venture the parties have a mutual interest in success, and in continuing efficiency and profitability. It is hardly surprising that from the end of the 1970s many governments in the developing world began to review and liberalize joint venture legislation, often with dramatic results. Between 1979 (when joint-venture legislation was adopted) and 1988, China approved over 6000 joint ventures with a total committed investment of $7.9 billion, of which over $3.8 billion was invested in that period and in which over 3000 joint ventures actually became fully operational.[17]

The joint venture, however, needs careful examination, not least because it cannot be assumed that the partners have common objectives or equal bargaining power. The Nigerian government reported that in 1987, 40 per cent of joint-venture agreements they had examined contained no provision for training, 50 per cent were of excessively long duration, and 42 per cent had no provision for guarantees. Since the inception of Nigerian legislation, 30 per cent of agreements contained export restrictions, 10 per cent lacked provision for R&D, and 40 per cent were under foreign jurisdiction.[18] Nevertheless, pursuing a deliberate policy of 'technology inducement' has its rewards. South Korea did this through various incentives and waivers through the 1980s. In the twenty years up to 1981 South Korea obtained $1.87 billion of foreign investment; in the six years from 1982 to 1987 this rose to $2.82 billion.[19]

It remains the case that many developing countries have been slow to articulate a clear policy on technology. For example, in 1988 the Chief Legal Adviser to the Ghana Investment Centre stated that 'there is a need for a conscious and systematic effort to identify national needs and technological gaps in order to determine the role which foreign technology is expected to play in the economy.'[20] However technology is transferred, questions remain to be asked about the effect on local innovative capacity, whether it stimulates or depresses the creation of new domestic firms, and whether the often overwhelming strength and technological resources of an incoming TNC undermine the competitive ability of local enterprises.

## Technology: Appropriate for Whom?

A TNC entering a host country has an understandable tendency to replicate technology and associated systems that are currently serving it well elsewhere. Technological, financial and management systems are already bound together, and adaptation for new production outlets will involve additional costs. Particularly where output is for the international export market, there will be the further disincentive to adapt in case quality is affected. A series of studies have shown that, where TNCs do adapt processes in host countries, it is either to take advantage of lower labour costs or because of smaller output in the new location. In both cases adaptation is more likely to take place if production is designed for the local market – import substitution – than for export. One study showed that, of 24 export-oriented projects in developing countries, two adapted product design, four adapted production equipment and four adapted production techniques. By contrast, in 53 import-substituting projects 18 adapted product design, 18 adapted equipment, and 23 adapted operating methods.[21]

The previous example emphasizes the linkage between technology and the market. The same consumer market, worldwide, not just in developing countries, has a tendency to devalue traditional methods, skills and processes in favour of the new. What is also clear is that much 'appropriate' technology relevant to the economic needs of the poorest developing countries lies in a market which their enterprises have never entered or have long since vacated. An executive of a manufacturing company operating in a developing country has said, 'we respond to income growth by offering products to people ascending the consumer products "ladder".'[22] By contrast, it appears that the NGOs working in the field of 'appropriate' technology hold to the view that economic freedom for the poor in the developing world can be based on simple productive technologies. With an emphasis on small-scale producers, agencies like the UK Intermediate Technology encourage community control and participation and a re-evaluation of previously underrated skills, such as women's technical knowledge.[23] Practical know-how and low-level manufacturing systems for use by individuals or micro-enterprises are funded through charitable donations and grants.

There is a huge middle ground between large-scale or technically sophisticated consumer goods and the 'small is beautiful' technology. There is some evidence that TNCs with vertically integrated operations in developing countries recognize the value of appropriate technologies. Unilever, for example, gives illustrations of appropriate technology introduced to their plantations benefiting local farmers and smallholders. The introduction of a weevil which aided pollination to Malaysian oil palm estates (after due environmental considerations) and the development of a specialized buffalo cart for crop transport in Africa and Malaysia are examples.[24] But these are exceptions that prove the rule. 'At present, the most powerful interests emphasize labour productivity, profitability and military potential. Very different objectives are needed for a technological development appropriate for solving problems of unemployment, Third World poverty, and environmental destruction.'[25]

## Technology and Defence: The Development or Security Dilemma

Having technological superiority is held in as high esteem in the military establishments of the world as it is in the boardrooms. These two groups have a long history of linkage, generally known as 'the military-industrial complex', whether in market or centrally planned economies. Conflict or political instability is a significant cause of unsustainable development, the real cost of militarization being the loss to a non-productive sector of massive resources. It has been frequently pointed out that global expenditure on military equipment, if applied to non-defence economic programmes, could fuel massive development, and nowhere more so than in developing countries. It is certainly true that, as a proportion of GNP, developing countries devote a considerably larger element of national resources to military expenditure than does the industrialized world. In 1988 military spending in developing countries totalled $145 billion – a figure to be compared with less than $30 billion of inward foreign direct investment. Only a few of the NICs, such as South Korea and Taiwan, have been able to maintain significant economic growth simultaneously with high arms expenditure. The general pattern shows that, for nearly all developing countries, increased military spending has a negative impact on economic development.[26]

Companies engaged in military production, irrespective of their location, tend to employ a much higher proportion of technicians than the non-military sector.[27] In developing countries the loss of these specialists by the non-military economic sector can be ill-afforded, as can the loss of government resources which might otherwise have been channelled to general economic and infrastructure development, education or social welfare.

But the very real dilemma remains, for many developing nations, of ensuring security in an uncertain world. Since 1945 more than 21 million people have died in 127 wars in developing countries, many of these conflicts being, by proxy, an extension of East–West antagonisms.[28] The easing of Cold War tensions between NATO and former Warsaw Pact countries is already yielding a 'peace dividend' in the form of conversion of military productive capacity to non-military output. This has not been paralleled in developing nations. It is likely that the very withdrawal of client status in some countries like Vietnam, North and South Korea and Cuba has led to a greater feeling of insecurity and an increase in the proportion of internal resources allocated to defence. The weakness of international peace-keeping structures does little to reassure smaller nations that their territorial integrity is effectively guaranteed by the global community.

The end of the Cold War has lent an urgency to arms sales to smaller nations, and there is a need for codes of practice applying to the defence industry, covering decisions concerning new countries with which to engage, the type of military technology deployed, and the relationship between trade and home government political objectives.

## Intellectual Property Rights: The Corporatization of Knowledge?

It has already been pointed out that research-led technology is primarily the province of the TNC. This research is conducted overwhelmingly in the parent country; for example in 1989, 90 per cent of R&D by US corporations was conducted in the United States, with the vast majority of the balance taking place in other developed countries.[29] It has been estimated that in 1987 only 5.3 per cent of worldwide R&D expenditure took place outside North America, Western Europe and Japan.[30] As technology is a major element in the competitive advantage of a TNC, the protection of such technology through patents, licence agreements and other legal and regulatory mechanisms, is of increasing importance. The arguments for increased protection emphasize that a secured return on the high costs of innovation will encourage further investment in R&D and the development of new applied technology. The counter-argument suggests that diffusion of new technology to poorer nations will be slower, as it will be more costly than imitation-based technology.

The draft agreement on trade-related aspects of intellectual property rights (TRIPs), which was the subject of intense discussion at the GATT Uruguay Round, sets minimum standards of protection for copyright and related rights, trademarks, geographical indications, industrial designs, patents, layout designs of integrated circuits and undisclosed information (trade secrets). In addition it sets out the obligations of member governments for enforcement of those rights and establishes a multilateral procedure for the settlement of disputes between governments.[31]

> **Case study 3** The Endod plant
>
> The Endod plant has berries which have been traditionally used in many parts of Africa for laundry purposes. It was also noticed that it had a highly toxic effect on fish and snails, eventually leading to a collaboration between an Ethiopian biologist and Toledo University in the USA with a subsequent patent application for an Endod-based molluscicide, believed to have great commercial potential for the control of rampant snail infestation in the Great Lakes area of North America

Concerns have been expressed over the rights that this agreement may give to owners of 'traditional' knowledge. Case study 3 provides an effective illustration. This example is particularly applicable to the genetic resources of developing countries and has great relevance to the worldwide pharmaceutical industry. TNCs are frequently the sole possessors of the resources and technology to analyse, refine and develop a genetic resource. Should this mean that no financial recognition is given to the indigenous source of a subsequently successful commercial product? One attempt to tackle this dilemma has been made by the US corporation Merck, which, in exchange for exclusive rights to screen plants from Costa Rica's forests, will make payments for conservation.[32] Nevertheless, it has been pointed out that the payments so far revealed place an extremely low value on this potentially rich resource.

The international brand, and its protection, is also playing an increasingly significant role in stimulating technical application. Trademarks and brand names are seen as a form of international guarantee, with the technological content, whether electronics or footwear, playing a leading part. It is argued by industry that the 'brand' stimulates product renovation and development. Consequentially, market entry becomes increasingly difficult unless it is through a licensing or joint-venture arrangement.

## Nurturing Technology

Because of the increasing requirement for sophisticated support services for modern technology, there has been a tendency for industries using such processes to group together. This tendency has been encouraged by the adoption, in at least 27 developing countries, of the export processing zone (EPZ), an enclave of 10–300 hectares with the facilities of an industrial estate, specializing in 'manufacturing for export to more advanced market economies by offering exporters duty-free imports, a favourable business environment, few regulatory restrictions, and a

> **Case study 4**  Local industry upgrading in Singapore
>
> Singapore's Local Industry Upgrading Programme is a government-sponsored programme aimed at accumulating management and technical skills from TNCs operating in Singapore through their local suppliers. The first phase is to improve the operational efficiency of local companies supplying TNCs, the second to diversify products and processes with the aid of the TNC (import substitution and export) and finally to undertake joint product R&D. Today there are 32 TNCs and about 180 local companies involved in the programme.
>
> For instance, an early example of the programme involved Hewlett Packard advising FJ Industrial, a former producer of aluminium and plastic nameplates which now produces membrane switches for computer printer control panels. It hopes to capture 40 per cent of the S$12 million per annum local market for membrane switches and to export as well. Having developed the capability to produce membrane switches, FJ and Hewlett Packard worked jointly on keyboard membrane and calculator switches.

minimum of red tape.'[33] It is notable that the World Bank reports that, 'There is general agreement that the transfer of product and process technologies through EPZs is small, except perhaps in simple industries such as garments.'[34]

The EPZ also facilitates, but by no means exclusively, the establishment of a milieu where technology 'software' can thrive. This is an umbrella term for managerial culture and training and organizational technology; an environment in which company-wide systems like total quality management can be applied. It is not only in technological hardware that developing countries can find themselves at a disadvantage. The university, the business school, the management college form a symbiotic relationship with the corporate policy unit and human-resource division of the TNC, a relationship which flourishes in the developed world but is much rarer elsewhere.

### New Processes, Systems, Materials and Synthetic Substitutes

Although this chapter has been concerned with the application of technology within a developing country, historically the developing world has been seen as the source of raw materials for export, to which value is then added through processing in the industrialized world. Sugar, cotton, rubber, copper, and a dozen more commodity items fall into this category. One of the first tasks of many former colonies at independence was to consolidate or establish a domestic processing and conversion capacity for its commodity exports. But technological developments in recent years have undermined markets for some basic commodities in whatever form. High Fructose Corn Syrup (or isoglucose, which can be made from other grains such as rice or wheat as well as maize) and artificial sweeteners such as aspartame, are replacing sugar. Polymers and composites are replacing traditional minerals; and, thanks to the development of synthetic substitutes, there is a reduced demand for cotton, rubber and sisal. Fibre optics and plastic piping has had the same effect on demand for copper. In industries large and small technology and new processes introduced in the heavily industrialized countries have taken their toll in the developing world. The shellac industries of Southeast Asia were devastated by the switch to plastic insulators and vinyl gramophone records in the 1950s, just as indigo growers were ruined a century earlier with the introduction of chemical dyes.

It is also becoming apparent that the revolutionary transformation in Materials Science and Engineering (MSE) in the 1980s has major implications for many developing economies.[35] Leading-edge research and technology not only changes existing materials but, at the atomic and molecular level, creates new materials. Many developing countries, when considered in the light of their history as primary producers and their current economic weakness, are ill-equipped 'to build and acquire the necessary competences to assimilate and adapt the MSE revolution and the parallel educational, engineering, testing and standard, quality control, and institutional skills and structures that will enable them to use domestic resources to meet basic needs and participate more fully in a more dynamic and open world economy.'[36]

Technology has also introduced new techniques to the service industries that have opened up employment possibilities undreamed of a few years ago. Data processing, in various forms, is an example of what has become known as 'electronic social dumping'. Airline tickets and hard copies of bank transaction data are examples of where written data can be processed at a centre thousands of miles from source and electronically fed into a computer net.

## Technology as a Labour Substitute

The current new wave of technology, centred around microelectronics, generates process efficiency combined with innovation. It is largely developed in countries with relatively high labour costs, and tends towards labour reduction. This, combined with the introduction of integrated process systems reliant on computer-aided design and computer-aided manufacturing, militates against the traditional advantages of low-labour-cost economies. Technological innovation has not always had this result. The growth of the railways in the nineteenth century created vastly more jobs than were lost in other forms of the transport industry. Similarly, the last ten years have seen massive growth in the garment industry in Bangladesh, employing modest 'old wave' technology, an ironic contrast to the decimation of the muslin industry 200 years earlier in the same area.

Nevertheless, many countries in the developing world are caught in a pincer movement. The very complexity of new manufacturing methods creates an ever greater entry barrier for countries with a low-to-medium technological infrastructure, whilst applied technology in international markets reduces the need for the developing countries' primary commodities and labour resource. Even in the case of the clothing industry, the application of new technology is driving a move away from low-cost countries, back to Europe.

## Conclusion

There are an estimated 37 000 transnational corporations operating worldwide. They only employ 3 per cent of the world's workforce but account for one quarter of the combined gross national products of all the countries of the world, and their flow of foreign direct investment (FDI) exceeds $225 billion annually. In developing countries this investment now exceeds $36 billion annually and is immensely influential. The TNCs move vast sums of money around the world. Where they invest, it usually provides a massive economic boost for the local economy. In 1992 about 95 per cent of this investment went into Asia and Latin America. But Africa, with a population bigger than Central and South America put together, received the same amount of investment as did Portugal.

Two things are clear. The first is that we simply do not know whether the world's political, social and environmental systems can stand the strains that demand economies are generating – demand that is being both fed and encouraged by new technologies. Second, it is apparent that the world's poorest countries will be the last to benefit from modern economic and technological trends. Given the imperfections in technology markets and asymmetries in bargaining strength between buyers and sellers, developing-country governments need to formulate explicit objectives and specific measures to acquire and develop relevant and appropriate technology. Recipients must be allowed to participate in technological development to be able to modify or change the transferred technology in the future (the know-why). The absorption of high technology requires a special policy and institutional framework to continue developing and diffusing the technology acquired to suit different conditions. Without deliberate actions, the benefits of environmentally friendly and other technologies may not be diffused into the local economy.

Much of the business establishment continues to assume that, thanks to technology, GNP growth can continue practically unimpeded, albeit in somewhat different directions, and that such growth is even necessary for social improvement and environmental conservation. These views undermine the impetus for essential change by endorsing the most basic objective of 'business as usual'. However, it cannot be 'business as usual' next century; but this does *not* mean a no-growth scenario. We have to accept progressive bounding of economic activity by tight sustainability constraints and the explicit direction of that activity by and toward positive human values.[37] The market has a lot of life in it yet, and a lot of good things to commend it, and TNCs and new technology will take the leading role; but we have to confine greed and the desire to acquire and consume to those areas where individual and social self-interest really do coincide.

## Notes

1. Andrew Tylecote, *The Long Wave in the World Economy* (London: Routledge, 1991).
2. UNTNC, *World Investment Report – Transnational Corporations as Engines of Growth* (New York: United Nations, 1992), p. 146.
3. Ibid., p. 147.
4. OECD, *Declaration and Guidelines on International Investment and Multinational Enterprise: 1991 Review* (Paris: OECD, 1991).
5. Lakis Kaounides, 'International Business Strategies in Advanced Materials', *IDS Bulletin*, vol. 22 (April 1991), no. 2, p. 75.
6. United Nations Conference on Trade and Development (UNCTAD), *Report of the Ad Hoc Working Group on the Interrelationship between Investment and Technology Transfer on Its First Session* (Geneva: United Nations, 1993).
7. Stephan Schmidheiny, *Changing Course* (Cambridge, Mass.: MIT Press, 1992), pp. 121–2.
8. Dennis L. Meadows et al., *The Limits to Growth* (New York: Universe Books, 1972).

9. Paul Ekins et al., *Wealth Beyond Measure* (London: Gaia Books, 1992), p. 180.
10. Raphael Kaplinsky, *The Economics of Small Appropriate Technology in a Changing World* (London: IT Publications, 1990).
11. Thomas J. Peters and Robert H. Waterman, *In Search of Excellence* (New York: Harper & Row, 1982), pp. 193–9.
12. UNCTAD, *Joint Ventures as a Channel for the Transfer of Technology* (Geneva: United Nations, 1990), p. xi.
13. World Commission on Environment and Development, *Our Common Future* (Oxford: Oxford University Press, 1987), p. 29.
14. UNCTC, *Foreign Direct Investment and Technology Transfer in India* (New York: United Nations, 1992), p. 102.
15. Ibid., p. 103.
16. Schmidheiny, *Changing Course*, pp. 121–2.
17. UNCTAD, *Joint Ventures*, p. ix.
18. Ibid., p. 113–14.
19. Ibid., p. 83.
20. Ibid., p. 67.
21. G.L. Reuber et al., *Private Foreign Investment in Development* (Oxford: Clarendon Press, 1973), quoted in John H. Dunning, *Multinational Enterprises and the Global Economy* (Wokingham: Addison-Wesley, 1993).
22. I. Frank, *Foreign Enterprises in Developing Countries* (Baltimore, Md.: Johns Hopkins University Press, 1980), p. 74.
23. Intermediate Technology (Rugby), general literature and Annual Reports, 1988–93.
24. *Unilever's Plantations*, company literature (London: Unilever External Affairs Department, 1988) pp. 30–31.
25. Ekins, *Wealth Beyond Measure*, p. 174.
26. L. Taylor, *Military Economics in the Third World*, for The Independent Commission on Disarmament and Security Issues (1981).
27. Joyce Kolko, *Restructuring the World Economy* (New York: Pantheon Books, 1988).
28. World Development Movement, *Disarm to Develop* (London, 1990).
29. UNTNC, *World Investment Report*, p. 139.
30. Dunning, *Multinational Enterprises*, p. 300.
31. UNTNC, *World Investment Report*, p. 74.
32. Christopher Joyce, 'Prospectors for Tropical Medicine', *New Scientist*, vol. 132 (19 October 1991), no. 1791.
33. The World Bank, *Export Processing Zones* (Washington DC: World Bank, 1992), p. 1.
34. Ibid., p. 3.
35. Lakis Kaounides, 'The Materials Revolution and Economic Development, *IDS Bulletin*, vol. 21 (January 1990), no. 1, pp. 16–26.
36. Ibid., p. 25.
37. Ekins, *Wealth Beyond Measure*.

# References

Dunning, John H., *Multinational Enterprises and the Global Economy* (Wokingham: Addison-Wesley, 1993).
Ekins, Paul, et al., *Wealth Beyond Measure* (London: Gaia Books, 1992).
Frank, I., *Foreign Enterprises in Developing Countries* (Baltimore, Md.: Johns Hopkins University Press, 1980).
Joyce, Christopher, 'Prospectors for Tropical Medicine', *New Scientist*, vol. 132 (19 October 1991), no. 1791.
Kaplinsky, Raphael, *The Economics of Small Appropriate Technology in a Changing World* (London: IT Publications, 1990).
Kaounides, Lakis, 'The Materials Revolution and Economic Development, *IDS Bulletin*, vol. 21 (January 1990), no. 1.
Kaounides, Lakis, 'International Business Strategies in Advanced Materials', *IDS Bulletin*, vol. 22 (April 1991), no. 2.
Kolko, Joyce, *Restructuring the World Economy* (New York: Pantheon Books, 1988).
Meadows, Dennis L., et al., *The Limits to Growth* (New York: Universe Books, 1972).
OECD, *Declaration and Guidelines on International Investment and Multinational Enterprise: 1991 Review* (Paris: OECD, 1991).
Peters, Thomas J., and Robert H. Waterman, *In Search of Excellence* (New York: Harper & Row, 1982).
Reuber, G.L., et al., *Private Foreign Investment in Development* (Oxford: Clarendon Press, 1973).
Schmidheiny, Stephan, *Changing Course* (Cambridge, Mass.: MIT Press, 1992).
Taylor, L., *Military Economics in the Third World*, for The Independent Commission on Disarmament and Security Issues (1981).
Tylecote, Andrew, *The Long Wave in the World Economy* (London: Routledge, 1991).
UNCTAD, *Joint Ventures as a Channel for the Transfer of Technology* (Geneva: United Nations, 1990).
UNCTAD, *Report of the Ad Hoc Working Group on the Interrelationship between Investment and Technology Transfer on Its First Session* (Geneva: United Nations, 1993).
UNTNC, *World Investment Report – Transnational Corporations as Engines of Growth* (New York: United Nations, 1992).
World Bank, *Export Processing Zones* (Washington DC: World Bank, 1992).
World Development Movement, *Disarm to Develop* (London, 1990).

# 10
# Recycling Technologies and Engineering Challenges
## *Donald V. Roberts*

### Introduction

For an engineer, a sustainable system is one that is either in equilibrium, operating at a steady state, or one that changes slowly at a rate considered to be acceptable. This concept of sustainability is best illustrated by natural ecosystems. They function as semi-closed 'loops' that change slowly. For example, the hydrological cycle involves continuous evaporation from the oceans and other surface bodies of water up into the atmosphere. The vapour then moves over land where precipitation occurs as rain or snow. The water then returns back to the ocean through surface streams or groundwater, and the process is repeated over and over. The food cycle involving plants and animals represents another illustration. Plants grow and thrive in the presence of sunlight, moisture and nutrients. They are then consumed by herbivores and insects, which in turn are eaten by various classes of carnivores. The resulting waste products replenish the nutrients, which allows the process to be repeated again and again.

Slow changes occur in nature over centuries. Species of plants and animals evolve, and many ultimately disappear. In all natural ecosystems, however, change usually occurs at a rate that allows time for natural environmental adaptation. In contrast to nature, humans have used a linear approach to date (see Figure 10.1). Resources have been extracted as though they were inexhaustible. These resources have been modified or processed by industry in a manner limited only by economics and human ingenuity. The earth's natural resources and manufactured products are then transported to the consumer through engineered systems. As consumers, we have acted as though the world has an unlimited ability to produce goods to supply the needs of our

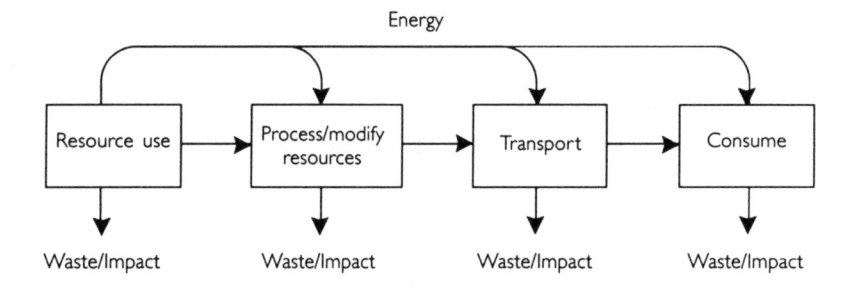

```
                              Energy
   ┌──────────────┬──────────────────┬──────────────┬──────────────┐
   │              ▼                  ▼              ▼              ▼
┌──────────┐  ┌──────────────┐  ┌──────────┐  ┌──────────┐
│ Resource │→ │Process/modify│→ │Transport │→ │ Consume  │
│   use    │  │  resources   │  │          │  │          │
└──────────┘  └──────────────┘  └──────────┘  └──────────┘
     ▼              ▼                ▼              ▼
Waste/Impact   Waste/Impact     Waste/Impact   Waste/Impact
```

**FIGURE 10.1**   Humans have used a linear approach to date

ever-growing population. And until recently, humans have assumed that our global surroundings could absorb any quantity of waste products.

## What Does Recycling Involve?

Sustainable development will require the adoption of an industrial ecology or artificial ecosystem patterned after natural processes. A proposed engineering model of such a system is illustrated in Figure 10.2. The extraction, processing, transportation and consumption of resources must flow continuously as a closed loop to the extent possible, rather than as a once-through system.

Renewable resources such as fish and trees must be harvested within the limits allowed by nature. The use of vital non-renewable resources, such as certain minerals and fuels, should be minimized. Industrial processes should be modified to use resources more efficiently and to minimize waste products. The manner in which we process, modify and transport resources must be conducted in harmony with the natural environment. Consumption patterns must be changed to ensure a more even distribution of goods.

Throughout this cycle, waste must be reduced, and the residual by-products from manufacturing and consumer use must be recycled over and over again as recovered resources. Some wastes are inevitable, but should be in forms that have minimal long-term impacts on the environment. The impacts from residual wastes should be offset by long-term

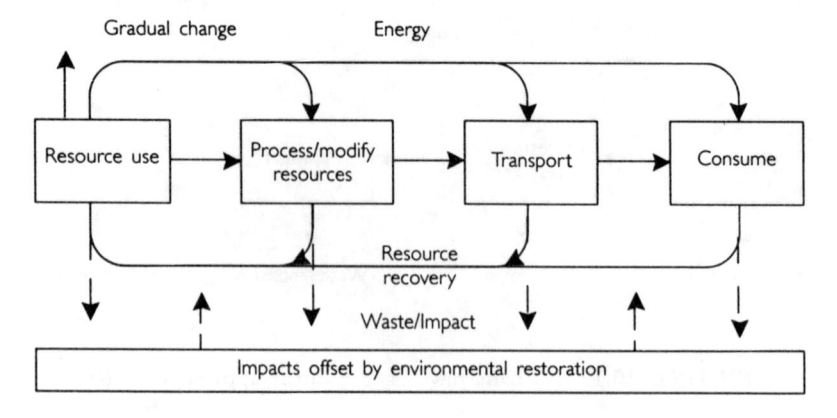

**FIGURE 10.2** Elements of a sustainable system for humans

programmes to clean up and reprocess old waste sites, along with other major environmental restoration programmes. The energy that drives the system should be minimized by engineered improvements that promote maximum efficiency.

Within man-made ecosystems some waste and loss of resources are to be expected, and long-term environmental changes are still inevitable. However, by slowing down and controlling the rate of change, it is hoped that humans and the natural environment could adapt to these changes in an acceptable manner.

### Engineering Challenges Ahead

Improvements are needed in mining operations and mineral processing to minimize land disturbance and to reduce pollution of surface and sub-surface water. Advanced mining methods are now being developed by some of the technologically leading countries. An example is the *in situ* leaching of minerals from underground ore bodies, with minimal disturbance to the ground surface.

The efficiency of gravity-flow irrigation systems can be improved. Trickle and drip irrigation systems can be used in arid regions to minimize evaporation and salination.

Industrial wastes often contain toxic metals and complex chemicals, which are not readily degradable under natural conditions or in

**LEVEL OF EFFORT**

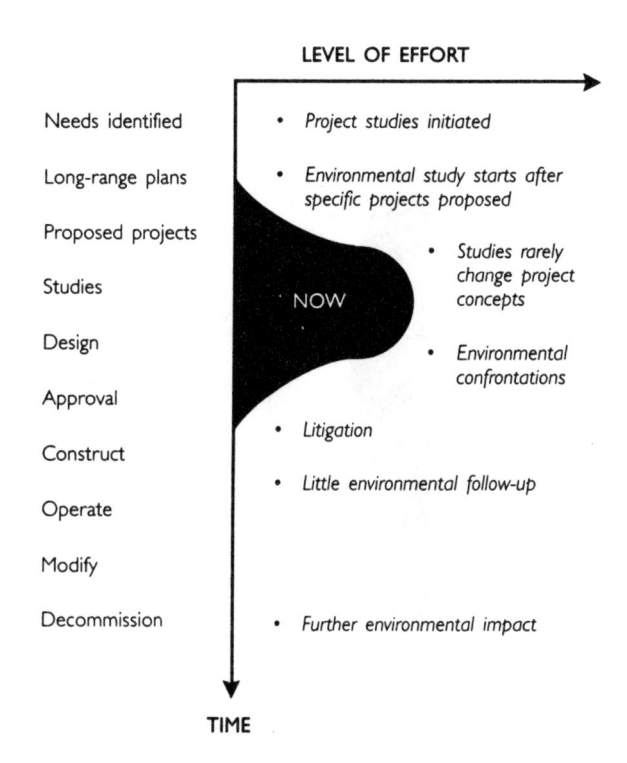

| | |
|---|---|
| Needs identified | • Project studies initiated |
| Long-range plans | • Environmental study starts after specific projects proposed |
| Proposed projects | |
| Studies | • Studies rarely change project concepts |
| | NOW |
| Design | • Environmental confrontations |
| Approval | |
| Construct | • Litigation |
| Operate | • Little environmental follow-up |
| Modify | |
| Decommission | • Further environmental impact |

TIME

**FIGURE 10.3** Environmental studies – now

conventional sewage treatment plants. Therefore, in the future, manufactured products should be so designed as to enable post-use recovery of materials and energy, and for their final waste forms to be biodegradable.

Developing new energy sources and improving existing ones represent significant challenges. Nuclear power may see a rebirth, assuming that engineering problems connected with nuclear safety and storage of nuclear wastes can be satisfactorily solved. Engineering improvements are needed to develop technically and economically viable systems of power generation through solar, geothermal, wind and biomass energy. It would be most rewarding for engineers to tackle the task of reducing energy use through higher energy efficiency in households, workplaces, mining and manufacturing industry, transportation and power generation. There are good grounds for believing that indirect increase in the

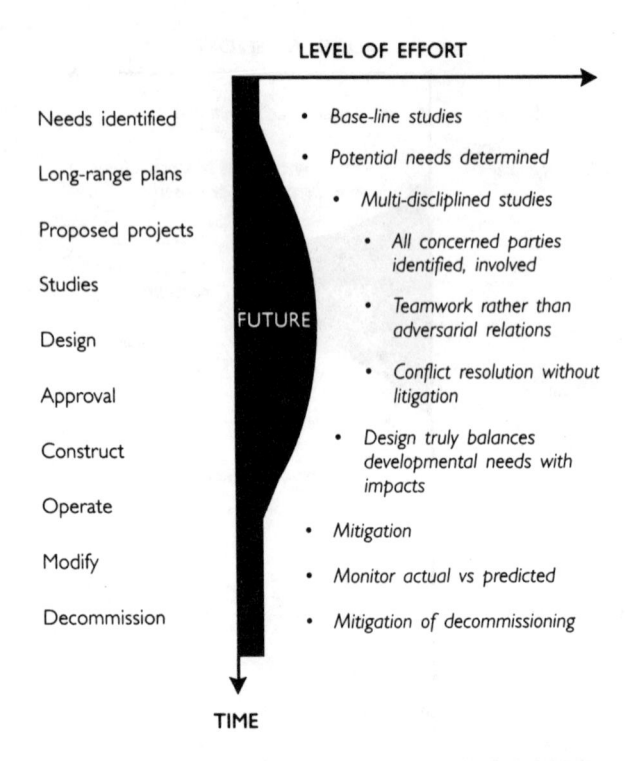

```
                          LEVEL OF EFFORT
                          ────────────────────▶
Needs identified      •  Base-line studies

Long-range plans      •  Potential needs determined

                      •  Multi-discliplined studies
Proposed projects
                         •  All concerned parties
                            identified, involved
Studies
                         •  Teamwork rather than
Design      FUTURE          adversarial relations

                            •  Conflict resolution without
Approval                       litigation

                         •  Design truly balances
Construct                   developmental needs with
                            impacts
Operate
                      •  Mitigation
Modify
                      •  Monitor actual vs predicted
Decommission
                      •  Mitigation of decommissioning
                          ▼
                        TIME
```

**FIGURE 10.4**  Environmental studies – the future

availability of energy through energy-efficiency is more economical than direct increase in energy production.

Over the past two centuries, engineers have continually made breakthroughs in the design and construction of canals, roads, railways, bridges, ports and airports. The ingenuity of future engineers will be taxed by, among other things, the challenge of creating vehicles and transportation systems with energy efficiencies much higher than at present. In terms of recycling, they would have to devise ways of salvaging and reusing a whole range of materials that go into civil engineering works.

Process technology should be improved to minimize the use of raw materials and reduce the use of water and energy in industrial processing.

Industrial effluents should be treated at industrial sites rather than being stored or moved or discharged into the surrounding environment.

Environmental restoration involves the rehabilitation of farm land; the recovery of sediments from reservoirs, harbours and deltas; and the cleaning up of polluted rivers, lakes and groundwater, as well as of sites where industrial and municipal wastes have been dumped. Approaches should be developed in which government, industry and the public at large work together on the task of cleaning up and rehabilitating areas where wastes have been dumped. This is best done in a spirit of cooperation, with neither faults nor risks assigned to any particular group. The engineering profession should be given incentives to undertake environmental rehabilitation tasks, and protected from the risk of expensive litigation.

## Improving environmental planning of projects

Environmental impact studies (EIS) as they are performed today are often wasteful and ineffective. Typically, they are performed in the sequence illustrated in Figure 10.3. A development project is proposed, with a specific site in mind. An EIS may be delayed until after the project has been evaluated and feasibility studies started or completed. By this time, the project may already have drawn the attention of, and provoked opposition from, environmentalists and government agencies. The EIS may become a battleground between those who wish the project to go ahead and those who wish to have it stopped. Both advocates and opponents may perform independent studies, generating massive amounts of data. But these may have little real influence on the outcome of the design. The studies may turn out to have been wasteful in terms of financial and personnel resources, often leading to confrontation with significant legal costs. Assuming that the project is finally approved, there may be little monitoring of the environmental consequences during the execution stage and after completion.

Some planners are now considering a better approach to EIS, as shown on Figure 10.4. This involves starting sooner and continuing longer in environmental planning and follow-through. It is based on a long-range perspective, beginning with the identification of the needs of a country and/or a community and ending with environmental mitigation, where necessary. Such a strategic approach should be able to strike a balance between a country's 'development needs' and a community's environmental imperatives.

EIS should take account of the concerns of all the parties who have legitimate interest in, and who are liable to be affected by, the outcome of the project. Further, they ought to include all the various types of costs, both direct and indirect, that will be incurred due to off-site pollution. Social and cultural issues should be integrated into the EIS. Provisions should be made to modify the design of the project, if it transpires during the course of executing the project that the environmental impact is significantly different than predicted. Environmental monitoring should continue throughout the life of the project, including the decommissioning stage, if there is one.

# 11

# Underutilized Capabilities in the Transfer of New Generic Technologies from Sweden to the Developing Countries

*Carl-Göran Hedén*

## Introduction

Eighty-five per cent of the world's population live in poor countries, but they only use between one-third and one-quarter of the energy and one-third of the metals. The Food and Agricultural Organization of the United Nations has repeatedly expressed deep concern about future food supply, because no less than 35 per cent of the area suitable for cultivation has been largely destroyed by soil erosion and salination. Even worse, it is likely to decline by 21 per cent over the next 20 years, and this will happen as fish catches drop by perhaps 10 per cent or so.

Since 97 per cent of the food we eat comes from the soil, the productivity that it supports is of course critical. Of equal importance to food productivity is our planet's biological diversity, which at present estimates is being eroded at the rate of some 150,000 species per year. It must therefore become a global priority to assist developing countries to develop and use selected generic technologies characterized by sustainable high productivity. This is a venture in which one can foresee a growing convergence of interests, as well as practices, on the part of the industrialized and the developing countries.

Within the rich world, knowledge-intensive technology is already being fostered that will permit it gradually to uncouple itself from the flywheel of heavy investments in scale-dependent equipment and in traditional systems of education. On the other hand, techniques of recycling materials and energy have been practised for generations in the rural economies of many developing countries. A convergence of such 'traditional' and modern scientific knowledge, resulting in a blending of technologies, will be mutually beneficial. This points towards the need for a new dimension in international cooperation based on the lessons that we have learnt from the fallacies and mistakes of over-industrialized economies and monocultural agriculture.

However, this will require a new level of global cooperation where the North will be obliged to forgo regional competition[1] in favour of a major effort designed specifically to assist the South, in particular to help it absorb those new generic technologies which promote environmentally sustainable development. For instance, the dynamic economies of East, Southeast and South Asia suffer from high levels of environmental pollution and degradation, caused for the most part by environmentally inappropriate technologies. But these countries, thanks to their dynamism, also have an impressive capacity to absorb new environmentally sound technologies and recycling processes. This process would in turn generate demonstration cases that could stimulate effective South–South cooperation.

In the past, Swedish scientists and engineers, in collaboration with Swedish industry, have demonstrated a quite remarkable ability and capacity for overcoming the problems caused by energy crises and environmental stresses. All the more reason therefore for them to contribute to the means of meeting this global challenge. There is also an increasing number of farsighted industrial and economic leaders in Sweden who are reacting to the folly of letting future markets 'go down the drain', and are therefore starting to regard themselves as development agents even when they operate under the pragmatic banner 'trade not aid'. In fact, this might not be a bad recipe for development cooperation, provided of course that fundamentals like equity, dignity and health are respected, and that distortions like protectionism and 'social dumping' are avoided.

## Swedish Development Assistance as Seen from the Sidelines

Official Swedish aid policy has recently been undergoing a series of painful and long overdue reviews. It is my hope that the argument for a gradual shift of focus from bilateral support for individual developing countries towards strategic cooperation in the field of generic technologies of agreed relevance will trigger a useful discussion. We should learn from our mistakes in the aid-field and concentrate on what we are really good at: for example, R&D and engineering in many areas of the natural sciences and medicine. Howevcer, we need to think very carefully before solutions suitable for our own society are recommended to the cultures in the South.

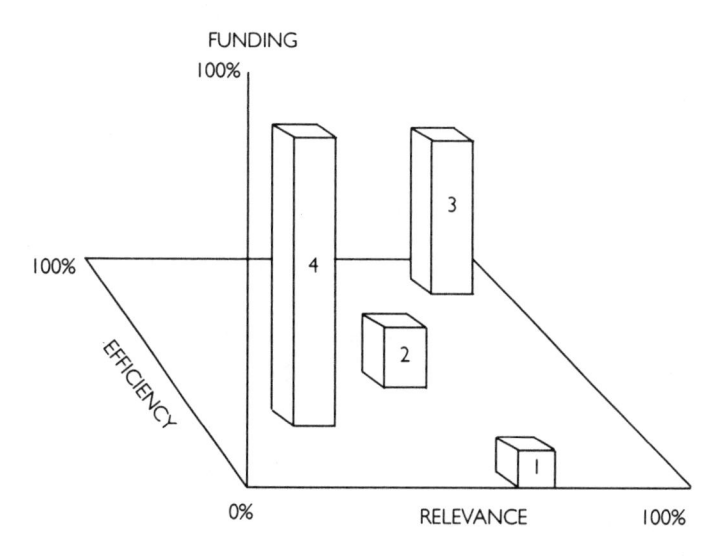

**FIGURE 11.1** An NGO's likely path in aid-management

In this chapter I want to argue for a strengthening of the entrepreneurial effort to 'reach the unreached', that is, landless farmers, unemployed youth and poor women,[2] and a stronger role for Swedish development assistance in new generic technologies through this approach. Sweden is well geared to this task because of certain advantages it currently enjoys: strong public support for the efficient provision of aid to disadvantaged people; an international reputation for non-biased aid efforts; an internationally oriented young generation; universities with considerable experience in research into problems of relevance to poor countries; and a high level of competence in generic technologies of particular significance to the development process.

'Reaching the unreached' may not always be a priority for the developing-country governments through which Sweden's official aid is now channelled, and perhaps not even for the international agencies where sovereign states meet. It is a task that is better practised by voluntary, non-governmental organizations (NGOs), which are much closer to the 'grassroots' than governments and aid agencies can ever be.

In Figure 11.1, I have sketched, in a highly schematic fashion, the dynamic element in aid-provision through the NGOs. It traces the path that an NGO is likely to take in a three-dimensional space defined by relevance, efficiency and donor-funding. NGOs start out with little money but from a position of high relevance (Pillar 1). The relevance declines somewhat, as they look for increased funding, and in the process are prepared to make compromises to accommodate the wishes of donors (Pillar 2). However, once assured of the commitment of donors and associated increases in funding, both relevance and efficiency can rise again to substantially high levels (Pillar 3). But, at a certain stage, bureaucracy may take the upper hand, leading to falls in both relevance and efficiency (Pillar 4).

## Entrepreneurship as an Aid to the Absorption of Technologies

When considering how Swedish expertise can assist developing countries in absorbing generic technologies of particular significance to their development process, the stimulation of indigenous creativity and potential for entrepreneurial activity in the aid-recipient countries must be given a very high priority. The reason is that poor countries badly need to have access to local experts who can help not only in the selection of technologies but also in adapting them to local conditions, so that they can be managed by local entrepreneurs. But the training of such experts cannot be done locally, partly because of inadequate equipment, but partly because their future usefulness depends on the network of contacts they can establish and keep active. This calls for good personal contacts with and regular visits to foreign centres of expertise.

In my field, which is bioengineering, the need for local expertise is striking, because the politicians in many poor countries are tempted to set their priorities after sales talks by big companies, perhaps supplemented with popular articles about the wonders of genetic engineering. At best they may have read some balanced reviews about its impact, but they easily forget the need to climb up a decision-making ladder like the one illustrated in Figure 11.2. The entrepreneurial spirit, and the incentives and resources available to entrepreneurs, form the base of the ladder.

## Controlled-environment Agriculture

Considering the turnover of organic matter, it is clear that developing countries are sitting on a treasure chest which contains the bulk of this

**FIGURE 11.2** A decision-making ladder for biotechnology-based entrepreneurship

**D  LEGAL AND ENVIRONMENTAL CONSTRAINTS**

1. Permission granted     NO →   Modify approach
   ↑
   YES

2. Warning signals from quality control    NO →   Improve process and
   and environmental monitoring                    management

**C  MARKETING**

1. Positive market analysis    NO →   Study alternative
   ↑
   YES

2. Responsive market    NO →   Accept slow growth
   infrastructure
   ↑
   YES

3. Good profitability    NO →   Study alternative

**B  NATURAL RESOURCE REQUIREMENTS**

1. Suitable site    NO →   Technical improvement
   ↑
   YES

2. Good climatic conditions    NO →   Protection
   ↑
   YES

3. Adequate quantity of water    NO →   Recycle
   ↑
   YES

4. Good water quality    NO →   Treatment

**A  HUMAN RESOURCE REQUIREMENTS**

1. Adequate basic schooling    NO →   Courses
   ↑
   YES

2. Biological knowledge and    NO →   Training/research
   infrastructure
   ↑
   YES

3. Technical knowledge and    NO →   Engineering support and
   infrastructure                       development work
   ↑
   YES

4. Entrepreneurial spirit    NO →   Incentives

planet's microbial diversity. Conservatively, this is represented by some 50 000 species of algae, 30 000 species of bacteria, 1.5 million species of fungi and 130 000 viruses. It is a sobering thought that less than 1 per cent of the microorganisms in the environment have been identified. As we march towards the bio-society of the next century we obviously have every reason to join forces to pry open this treasure chest, because it contains the genetic blueprints not only for many biodegradable materials and bioenergy systems but also for environmentally sound production methods for food, fodder and medicine.

Many areas of applied microbiology will eventually become as important to the industrialized world as they ought to be for poor countries right now.[3] Just as the utilization of bacteria, fungi and viruses involves the use of controlled environments, so it does in plant biology. An example of this is the rapid mass propagation of plants from tissue cultures (micro-propagation). This rapidly expanding area deserves special attention for a number of technical, economic and environmental reasons: it can stimulate entrepreneurship by generating valuable export products; being solar radiation intensive and labour intensive, it offers selective advantages to developing countries in the semi-tropical and tropical belts; it offers an avenue for training scientists in sterility culture and for building research capacity in plant genetics; and it provides experience with controlled-environment agriculture as a means to diversify agriculture, save water, utilize natural biological insecticides instead of chemical ones, and optimize growth. There are now many examples of joint North–South operations, where entrepreneurs in some developing countries dispatch, by air freight, plants produced through micro-propagation to several industrialized countries, where their cultivation is scaled up for sale to the horticultural, food, pharmaceutical, dye and perfume industries.

An important concept in promoting agricultural biotechnology among smallholder peasantry is that of 'Bio-villages' linked to 'Bio-centres', the latter providing scientific and technological inputs, as well as research results, for implementation by the former.[4] 'Bio-centres' could play substantial catalytic roles, not only by supplying knowledge-intensive inputs – for example, advanced seeds, plantlets and fishfry, microbial starter cultures and mushroom spawn – but also by disseminating methods for upgrading the primary products for sale in distant markets. Information technology can be harnessed to great effect in such ventures.[5] To achieve the desired dynamism in such an approach, the Biofocus Foundation (1993) has developed the concept of 'Devel-ease', which involves a combination of leasing in order to obtain the necessary hardware, and franchising to impart the necessary training and to ensure quality control. If successful in improving rural living conditions, such systems may help to reduce migration to the cities. In addition, they might also raise the range and productivity of 'urban agriculture'.[6]

Recent advances in glasshouse design in the North – for instance, Gunnarshaug's design for a 'zero-energy greenhouse'[7] – may be adaptable to arid and semi-arid conditions in the South, where the conservation of water and carbon dioxide would be a priority. The same is true of recent developments in 'cold water agriculture' by Craven and his co-workers.[8] This technique, which supports plant growth with water condensed on pipes buried in the sand along the shoreline, forms part of a 'cascade utilization' of deep ocean water. Earlier, the emphasis was primarily on using the difference in temperatures between deep and surface waters to run turbines to produce electricity (OTEC). However, since the water which leaves the OTEC device is still quite cold, and of course has its nutrient content intact, more attention is now being given to subsequent stages of an integrated system. Much electricity, and substantial quantities of freon, currently used for air conditioning, can be saved, at the same time as many types of aquaculture can be supported by the nutrients in the water. Since some 2 billion people live close to coastlines, deep ocean water can definitely be regarded as a neglected resource. Substantial research work along these lines is ongoing at a research institution in Keyhole Point in Hawaii.

## Northern Initiatives in Support of Generic Technologies: Examples that Sweden Could Emulate

Fairly early on, the Netherlands understood the need for inter-ministerial coordination in providing development assistance in biotechnology. Here an ideal mix of vision and pragmatism at ministerial level created a highly competent professional body for coordination, operating with a minimum of bureaucracy at the level of the Foreign Office. It also began publication of a very useful journal, *The Biotechnology Monitor*, which makes Dutch initiatives known all over the world, as well as reviewing various developments in biotechnology that might have policy implications for developing countries.

In Germany it is the very well-equipped Gesellschaft für Biotechnologische Forschung (GBF) in Braunschweig which takes care of the training in biotechnology for biologists and engineers from developing countries. They are given intensive short-term courses with a possibility for senior scientists to continue working at GBF laboratories for periods of up to two years. Also the agricultural university in Hohenheim is actively engaged in development-assistance-supported biotechnology projects in several countries in Africa, Asia and Latin America. These projects often have built-in components to promote entrepreneurial activities through the provision of soft loans and technical assistance.

In France, the government-funded ORSTOM research institutes (Institut Français de Recherche Scientifique pour le Développement en Coopération) have carried out important work on medical microbiology and fermentation technology in francophone Africa. In the UK, the Bioindustry Association maintains active contacts with many developing countries, and Rural Investment Overseas Ltd identifies and organizes agriculturally based projects and arranges finance for them. In the Far East, biotechnology training is centred on Japan, where large numbers of scientists from Asia have been trained at an excellent fermentation facility in Osaka.

In the USA it is the Appropriate Technology International, which receives its principal funding from AID (US Agency for International Development) and some of the major Foundations, that supports entrepreneurial activities with roots in biotechnology. Such activities are also promoted by the International Biotechnology Program set up by the Resources Development Foundation.

## Conclusion

Entrepreneurship has a lot to do with deliberate risk-taking, which is readily understood by all researchers, since science always has an interface with the unknown. However, highly differentiated decision-making structures, such as those created in Sweden, are designed to minimize risks. Consequently, administrators find it difficult to handle aid proposals that involve entrepreneurship, particularly when this is based on generic technologies, which of course have a confusing tendency to bridge the boundaries between established disciplines.

In this chapter, I have indicated why I think that 'business as usual' is no longer a good enough strategy for the future course of Swedish development assistance. I would argue instead for the establishment of a group of junior internationally oriented civil servants assigned principally to support the small-scale sector of the economy in poor countries, because this is a virtual fountainhead of entrepreneurship and creativity.[9] Needless to say, such a force of civilian 'blue berets' would need a thorough schooling in world affairs and cultural diversity. This I think ought to be given in joint courses with foreign students coming to Sweden. Given my mixed background in medicine, biology and engineering, I would not hesitate to give a human ecology profile to such training, for this discipline tells us why knowledge and responsibility should go together in the interaction between man and nature.[10]

I wish to conclude by repeating a story I once heard about a scientist who had been asked to look after his little daughter for a few hours. By being clever he thought he could buy himself an hour of peace and quiet, so he tore up a detailed map of the world, which he had seen in a newspaper. He then gave the small pieces to his daughter, thinking that putting them together would teach her some geography, while simultaneously he would get some time for his curves and tables. However, to his amazement and fatherly pride, she was back in ten minutes with the job done. When he asked how this was possible, she said: 'Well, it wasn't so difficult because there was a picture of a man on the back, and when the man came out right the world also came out right.'

## Notes

1. Gulbenkian Foundation, 'Limits to Competition', a study by the Lisbon Group, sponsored by the Gulbenkian Foundation (Lisbon: 1993).
2. M.S. Swaminathan, ed., *Biotechnology in Agriculture – A Dialogue* (Madras: Macmillan India, 1993).
3. E. da Silva, Y.R. Dommergues, E.J. Nyns and C. Rathledge, 'Microbial Technology in the Developing World' (Oxford: Oxford University Press, 1987).
4. Swaminathan, *Biotechnology in Agriculture*.
5. C.-G. Hedén, 'Bioinformatics for Development', in M.S. Swaminathan, ed., *Information Technology – A Dialogue* (Madras: Macmillan India, 1993).
6. International Development Research Centre (IDRC), 'Farming in the City: The Rise of Urban Agriculture', *IDRC Report*, vol. 21 (1993), no. 3.
7. J. Gunnarshaug, US Patent Number 4 733 506, 1993.
8. J. Craven, 'Coldwater Agriculture as Component of a Deep Ocean Water Recovery System for Self-sufficient Coastal Villages', Oceanography International 94, Proceedings of the Conference held in Brighton, March 1994.
9. H. de Soto, 'The Other Path' (New York: Harper & Row, 1989).
10. P.R. Ehrlich, A.H. Ehrlich and J.P. Holdren, *Human Ecology – Problems and Solutions* (San Francisco: W.H. Freeman, 1973).

## References

Biofocus Foundation, Update on Activities, 1993, Stockholm.

Craven, J., 'Coldwater Agriculture as Component of a Deep Ocean Water Recovery System for Self-sufficient Coastal Villages', Oceanography International 94, Proceedings of the Conference held in Brighton, March 1994.

da Silva, E., Y.R. Dommergues, E.J. Nyns and C. Rathledge, 'Microbial Technology in the Developing World' (Oxford: Oxford University Press, 1987).

de Soto, H., 'The Other Path' (New York: Harper & Row, 1989).

Ehrlich, P.R., A.H. Ehrlich, and J.P. Holdren, *Human Ecology – Problems and Solutions* (San Francisco: W.H. Freeman, 1973).

Gulbenkian Foundation, 'Limits to Competition', Study by the Lisbon Group, sponsored by the Gulbenkian Foundation (Lisbon: 1993).

Gunnarshaug, J., US Patent Number 4 733 506, 1993.

Hedén, C.-G., 'Bioinformatics for Development', in M.S. Swaminathan, ed., *Information Technology – A Dialogue* (Madras: Macmillan India, 1993).

International Development Research Centre (IDRC), 'Farming in the City: The Rise of Urban Agriculture', *IDRC Report*, vol. 21 (1993), no. 3.

Swaminathan, M.S., ed., *Biotechnology in Agriculture – A Dialogue* (Madras: Macmillan India, 1993).

*Part IV*

Donors' Experiences and Policy Approaches

## 12

# Information Technology Support to Developing Countries: The Canadian Experience

*Keith A. Bezanson*

## 1. Introduction

Today many of us live in a world where computers, telecommunications and other tools of the information technology age are becoming ever more pervasive in our work. But two decades ago, information technologies (IT) meant expensive mainframe computers tended by specialists, requiring arcane language skills and much care and tending. Remote sensing was a new science and the analysis was mainly visual, not digital. At that time, with telephone systems scarcely functioning in most developing countries, the idea of the developing world entering the IT age seemed an unrealistic notion to many. But this scepticism was not shared by the founders of the International Development Research Centre (IDRC), who, in framing the Act of the Canadian Parliament[1] which established IDRC in 1970, planted the first seeds with references to 'information and data centres and facilities for research' in developing countries. This early recognition of the importance of Information Sciences to research and development led to the establishment of programs in this field at IDRC.[2] IDRC and its partners in Canada and in developing countries have since accumulated valuable experience in the development of IT and its application to research.

Information technologies are primarily electronic-based technologies which can be used to collect, store, process, package and communicate information and provide access to knowledge. The emphasis is on technologies which are 'informatics'-compatible; that is, which can be integrated with computing capacity to provide for assisted information and knowledge processing. IDRC has concentrated on technologies that are at the 'leading edge' of their development, so that developing countries can have experience with their design, adaptation and use before the development/introduction cycle fully solidifies and excludes their interests. A key common element here is the computer, which is a prime example of a 'generic technology' if ever there was one. Through appropriate programming ('software') and interface devices ('hardware'), a computer can analyse data, control industrial processes, act as a telecommunications switch, provide 'expert' advice, or integrate multimedia presentations for entertainment or education.

As the world economy speeds up and becomes even more globalized, the issue of access and utilization of knowledge governs the survivability of all actors. The effective use of information technologies is an essential ingredient of success in this process. Without appropriate adoption, adaptation, transfer, development and use of information technologies, developing-country organizations and individuals will continue to participate in the global economy from an uncompetitive position: their disenfranchisement will grow. That is one of the reasons IDRC decided to provide support for information technologies in its programs.

## 2. IDRC's Approach to Information Technology and Development

Two key elements of IDRC's philosophy, as reflected in its Corporate Strategy, 'Empowerment through Knowledge',[3] are partnerships and capacity building. These are reflected in IDRC's approach to IT and development, and in its support to developing countries in this area. This approach can be summarized in the following principles:

1. *Work with the latest and best information technologies.* Keep on the 'leading edge', because that is where developing countries must also be for their survival. Try to anticipate technological trends and use imagination. Involve developing-country researchers in experimentation with new technologies and their application as early as is practical in their development.
2. *Support developing-country capacity building* through a mix of technology transfer, partnerships, and local technology research and development activities. Program activities should strengthen human and institutional resources in the area of IT research in developing countries. For collaborative projects, developing countries should be considered as full partners in research and development, and not simply as test-beds for technologies developed elsewhere.

3. *Look for research which is practical and results-oriented*, as opposed to theoretical or purely academic. It should lead to practical development applications and the solution of real-world problems. In many cases, this will require linkages to other programs and a multidisciplinary approach.
4. *Where appropriate, place emphasis on generic or generalizable outputs of global interest*, rather than only those specific to a particular country or institutional situation. Such projects should aim for a high multiplier effect and to maximize their impact. Where possible, link general global IT solutions to local/regional problems.
5. *Foster international collaboration in IT, through South–South, South–North, and East–West partnerships.* Networking (both human and electronic) is a key element, ensuring that developing-country scientists and researchers are connected to colleagues and sources of information and knowledge wherever they may be located.
6. *When appropriate, link the technologies and tools to past investments in the information content.* For example, over the last two decades IDRC's Information Sciences and Systems Division (ISSD) has supported the establishment of databases, and information and communications systems and services over two decades. Some of these can now provide extensive materials which can be restructured or repackaged using modern knowledge-representation methods and new information technologies for access. Many of the institutions with which the Division has been working are becoming very aware of the importance to development of information technologies, and are eager to begin work in the area.
7. *Encourage sensitization, evaluation and dissemination activities* which are vital to ensuring support for IT research and the effective utilization of research results (see also point 8 below).
8. *Support policy research which can provide an important framework for decision-making and allocation of resources.* For example, IDRC recently created an Information Policy Research Program within ISSD with the objective of influencing the information and information-technology policy environment, in order to promote and facilitate more effective utilization of information and related ITs in support of sustainable and equitable development. In order to effect change with respect to the adoption and sustainable use of information technologies, it is not enough simply to carry out technical research projects, pilot projects or feasibility studies. To make technologies workable, both broad and specific evaluations must be carried out, case studies must be analysed, linkages must be forged, project results must be

disseminated, awareness must be increased, and enabling policies must be nurtured.

9. *Establish specialized programs to support information-technology research and development (R&D) and applied research.* These programs should be in addition to support for 'straightforward' implementations or applications of information technologies in other program areas. Since the technologies are complex and fast-moving, organizations need expertise in-house, not only to provide good technical advice and assessments, but also to interpret and understand the more specialized technical advice and assessments that they will have to obtain outside. Within IDRC, there are two programs with such a specialization: Information and Communication Technologies (ICT) and Software Development and Applications (SDA).

The objective of the *Information and Communication Technologies (ICT) Program* is to enable developing countries to carry out and/or benefit from research on new ITs in order to permit them to utilize effectively such technologies for solving their information problems and for improving access to knowledge for development. The Program supports the assessment, development, adaptation, testing and transfer of technological innovations, as well as research on the social, economic and political issues associated with their introduction. It has supported research in a wide range of IT disciplines: *informatics* (including software development; expert systems and related knowledge engineering methods; computer-based training; information- and knowledge-presentation and knowledge-communicating technologies such as multimedia systems; computer-based modelling tools; and system design and integration methods); *telematics* (computer-based communications and networking; satellite-based communications); and *geomatics*: automated cartography, remote sensing, Geographic Information Systems (GIS), Global Positioning Systems (GPS).

*The Software Development and Applications Group* (SDA) aims to increase capacity within developing countries to manage the software development process, and to provide an effective mechanism for increasing the availability of software products of practical application in the developing world. A specific example is the MINISIS software package.[4]

## 3. Examples and Case Studies

To illustrate IDRC's approach and the principles stated above, a number of examples are presented below from the ICT and SDA Programs:

## Telematics: linking people and information for development

Access to information technologies is uneven throughout the world, being most difficult in the case of the developing countries. This has serious implications for the increasing disparity between the 'haves' and the 'have-nots'. A telling latter-day illustration of this disparity is in the access to and availability of computer-based communications technologies (e-mail, bulletin-board systems, computer conferencing, and more recently the Internet). Recognizing this growing disparity early on, in 1981 IDRC organized a week-long workshop, Computer Based Conferencing Systems for Developing Countries,[5] to explore the state of the art and receive advice on the role of donors. At that time, only a few developing-country institutions were involved in related research activities. The recommendations of that workshop led to the creation of a Telematics Program at IDRC.

Throughout the 1980s this Program supported and promoted initiatives to facilitate more informed decisions by developing-country institutions concerning the transfer, adaptation and utilization of data communications techniques. One of the first such activity was the organization in 1983 of an international computer conference involving developing-country institutions.[6] Another was the support provided in 1982 for a feasibility study and pilot project on the implementation of a data transfer network for the Consultative Group on International Agricultural Research (CGIAR) institutions. This eventually led to the creation and continuing growth of the CGNET,[7] which has facilitated the research of the CGIAR institutions while saving them millions of dollars. Research in Latin America into the appropriateness of different conferencing and messaging systems among the non-governmental organizations (NGOs) and trade information agencies has been supported, as has software development in North Africa involving communications protocols and Arabization, policy aspects related to Transborder Data Flows, and research concerning the use of low-earth-orbit (LEO) satellites to solve 'last-mile' communications problems.

The early years of the Telematics Program can be characterized as one of promotion and sensitization since there was, with a few exceptions, next to no recognition of the importance of information and communication technologies in the development process. Towards the end of the 1980s, emphasis was placed geographically on Africa, where the needs were greatest and progress was minimal. A number of interrelated networking projects were developed to achieve the objectives of the Program in this region. Examples of these include the NGONET

Africa, linking four nodes and many NGO end-users in Nairobi, Harare, Dakar and Tunis for information exchange; the ESANET, connecting five universities in Kenya, Zambia, Tanzania, Uganda and Zimbabwe to experiment with different data communication topologies; the HealthNet, linking health institutions in Africa including medical faculties, hospitals and medical researchers using packet radio and LEO-satellite technology; the ARSONET, connecting standardization offices in Nairobi, Addis Ababa, Dakar, Tunis and Cairo; and the Pan African Development Information System (PADIS) net, a regional pilot effort utilizing a variety of data communications technologies to link development institutions in Africa, designed as a networking of networks project and coordinated by the United Nations Economic Commission for Africa (ECA) in Addis Ababa.

In response to the lessons learned through the constellation of projects supported in Africa and the recommendations made by the external evaluation of the Telematics Program, a major networking capacity building project, Capacity Building in Electronic Communications for Development in Africa (CABECA), was developed. Among the major findings that went into the designing of CABECA were that it makes more economic sense to build networking capacity for a collection of user groups and networks at the national level, with interconnections between countries, rather than to look for unique networking solutions for each network of interest to its proponents or funder; the necessity of developing systems where users are able to purchase services with local currencies, and provide for ongoing training and troubleshooting regimes; the importance of indigenization of skills in network implementation and systems operation; and the need for involving user groups' representatives, as well as institutions with the appropriate mandates and capabilites, in formulating and implementing policies related to network operation.

CABECA is coordinated by the PADIS/ECA. It uses FIDO-like software technology and dial-up communications over the existing telephonic infrastructure to build national nodes in sub-Saharan African countries. These national nodes are accessible by all types of institutions and interconnected internationally by long-distance polls and through gateways to the Internet. There are close to 20 such systems at various stages of development at this time in Africa. When compared with what needs to be done, this IDRC-supported project actually represents only a very modest contribution. It is meant to demonstrate a model solution and act as a stimulus for others to build appropriate partnerships to address formidable challenges.

Since this project was begun as recently as 1994, it remains to be seen what impact it will have. If successful, it should empower many individuals and institutions, including many donor-supported institutions, through the provision of ongoing access to information and communications. Further, the demonstration of a *truly* sustainable system in the difficult African environment could influence future attitudes about IT's possibilities on the continent. However, confidence in the CADECA and the increasing recognition of the need for such systems has led IDRC to replicate this approach in South and Southeast Asia, based at its regional office in Singapore.

At the 1981 workshop which led to the creation of the Telematics Program, one of the developing-country participants proposed the concept of a developing-country led initiative to build an LEO satellite dedicated to text communication among developing countries. This proposal sparked considerable excitement in the workshop and later in other international fora such as UNISPACE '82. It stimulated the development of a prototype payload which was launched on the UOSAT-2 satellite, followed by more sophisticated payloads on subsequent satellites. IDRC promoted the development of this concept through presentations and demonstrations at conferences, followed by a research project with an African telecommunications authority, acquisition of capacity on UOSAT-3, and support for the HealthNet, which involves 20 ground stations in information exchange in the health sector. As of now, more than a dozen African countries have already set up 'operational' systems using this LEO technology. This is to be contrasted with Motorola's 77 satellite IRRIDIUM system, Microsoft's USD 9 billion TELEDESIC LEO system, and others, which are still in the design or construction stage. HealthNet is an inspiring example of a successful low-cost development-centred effort preceding global high-cost initiatives.

Some general conclusions can be drawn from these experiences. Problems related to the transfer and utilization of networking technologies involve *institutional and attitudinal issues* much more than technical, and in many cases financial, issues. If donors and different sectoral actors work more collaboratively on addressing basic underlying problems, greater progress can be achieved. Some of the most important and enduring investments will be in the development of appropriate human resources, perhaps in the form of a cadre of indigenous consultants able to support local institutions.

By 1993, IDRC felt that the original objectives of its Telematics Program had largely been realized. Awareness had increased dramatically.

Expertise had been developed in many developing-country institutions. Developing-country organizations and interests were 'at the table' during related global events. And successful models had been demonstrated. Computer-based networking was being identified routinely by players on the development stage as being necessary to carry out their endeavours.

While much research remains to be carried out, the primary concern has shifted to operationalization. Given the current aid environment, IDRC is now placing more emphasis on policy issues related to information technology, in recognition of the fact that a greater impact might be realized through this route in a time of limited resources but growing demands.

## *Informatics: providing software for development*

Building local capacity within developing countries to improve the management and use of information is an important objective for IDRC. Software can be a valuable vehicle for achieving this objective.

### MINISIS

MINISIS is a computer-based information management system used by institutions of varying size and purpose to automate the administration of bibliographic and textual collections. The software allows users the flexibility to manage text-based information. Interaction with the software and storage of data can be in a combination of multiple character sets and languages.

At an early stage of the development of its program in Information Sciences in the 1970s, IDRC recognized the need for a text-oriented computer system capable of running on minicomputers. There was a requirement for an information storage, management and retrieval system suitable for installation in developing countries which would enable them to participate effectively in international cooperative networks. MINISIS was the result, and has since found a comfortable spread of uses in many areas including inventories, registries, student records, legislative full text, press clippings, real estate listings, and so on. The software is generalized to allow users to adapt it to their local applications and requirements themselves, without the need to have a commercial organization reconfigure any modules. The ability to handle multiple languages and character sets permits the use of the software in the local language.

The MINISIS software is currently installed in over 350 organizations, including universities, government ministries, research institutes and international organizations, in more than 60 countries. It is used by organizations as diverse as museums and genetics research institutes.

Developing-country organizations have worked in partnership with IDRC to develop and enhance, as well as to disseminate and support, the MINISIS package. Together, IDRC and developing-country partners have developed alternate character set processing modules which are used by organizations in both the South and the North. This has enabled the Arab League, for example, to use the MINISIS software and the Arabic language modules, which they developed, to establish a regional information system, ARISNET. The management and dissemination of information in Arabic and Roman scripts is a key component of this initiative.

Trainers from the MINISIS Resource Centres located in Beijing, Bombay, Addis Ababa, Cairo and Mexico City have trained hundreds of developing-country information providers and technicians on how to use, manage and apply a sophisticated information management tool in the development of which they have played an integral part. Developing-country MINISIS users have also produced applications and utilities which have been distributed to the entire MINISIS community through a library of user-contributed software. Most widely used contributions in this Library are the Chinese and Arabic language tools. These were developed by the local MINISIS Resource Centres for dissemination to users in the region, and they enable organizations to use the MINISIS software and to manipulate information stored in their local language.

Users of the software have developed the capacity to apply it in a broad spectrum of applications for their individual requirements. This goes beyond specialized bibliographic and text-based information systems. It includes land bank databases registering information about individual plots of land, germplasm databases, meteorological information systems, geological survey inventories, and varieties of applications which have been developed by the users to meet their own needs.

IDRC is now actively supporting the move to self-sufficiency of the developing-country MINISIS Resource Centres by supporting their activities relating to the commercialization of MINISIS products and services. This collaboration with developing-country partners is one aspect of IDRC's objective to promote the development, enhancement, marketing and application of selected software of obvious benefit to developing countries.

## REDATAM

In an another important software development activity, IDRC supported the development of the Retrieval of Data for Small Areas by Microcomputers (REDATAM) software package by the Latin American Demographic Centre (CELADE) of the UN Economic Commission for Latin America and the Caribbean. REDATAM was also cofinanced by the United Nations Family Planning Agency (UNFPA) and the Canadian International Development Authority (CIDA).

The projected need for such software in developing countries was based on an analysis of the requirements of statistical, planning, and other agencies there during the 1980s. REDATAM was initially designed to provide access to selected subsets of sample data (for example, data from a national census) according to small geographic areas, and to provide a variety of analytic functions. For example, a planner intending to construct a hospital in a municipality might wish to know how many women of child-bearing age live in selected districts around the proposed site in order to estimate the number of maternity beds required. This data could be buried in a national census, and may be very difficult to extract. REDATAM provides support for just such a task using a microcomputer. In addition, REDATAM can store and work with the complete census, or other similar survey sets, data for small- or medium-size countries (or regions or cities within a country), and can be used to generate statistics such as frequencies, cross tabulations and averages. Recent work on interfacing REDATAM to Geographic Information System software has extended its functionality and ease of use.

The software has been installed in more than 30 countries where it serves as a platform for developing specific applications in various domains: economic planning, impact assessment of the tourist industry, poverty alleviation programs, social services planning, and so on. Government and non-government institutions in developing countries use the REDATAM software in their own areas of interest. For example, in an IDRC-supported project in Bangladesh, 'Monitoring Adjustment and Poverty', the Centre on Integrated Rural Development for Asia and the Pacific (CIRDAP) is using REDATAM to process the data on key economic and social indicators relevant to the project. REDATAM has been installed at the Municipality of Conchali (Chile), for the Pockets of Poverty study. It is being used by the National Statistics Office in Costa Rica to provide services to users across a large variety of applications. The General Statistics Office and Institute for Information Technology in Hanoi recently produced a Vietnamese version of

REDATAM and are working on integration with the POPMAP package. The integration will allow them to generate maps for geographical presentation of data processed by REDATAM. In a current project phase, CELADE, in cooperation with the University of Waterloo in Canada, will develop new generic applications in Chile, Costa Rica and St Lucia concerning the following problems: delivery of primary health care and other services; planning for urban growth with constraints on expansion; encroachment of urban areas on agricultural land; assessment of the effect of tourism development on the local environment and population, with special consideration for small island countries.

Research institutions in Rwanda, Ghana, Mali, Egypt, Burkina Faso and Cameroun became familiar with REDATAM at a workshop held in 1992 in Ouagadougou. IDRC is now working with them to disseminate REDATAM to Africa and to provide the required back-up support.

*Expert systems*

An expert system is 'a computer program that uses knowledge, facts and reasoning techniques to solve problems that normally require the abilities of a human expert.... The goals sought by expert system builders include helping human experts, assimilating the knowledge and experience of several human experts, and providing requisite expertise to the project that cannot afford scarce expertise on site.'[8] Expert systems are an example of knowledge-based information technologies.

Since 1986, IDRC has supported projects in expert systems and advanced database applications in developing countries such as Peru, Indonesia and the Philippines. Regional seminars on artificial intelligence applications have also been held in Africa and the Far East. In the view of researchers in developing countries, emerging knowledge-based technologies should not only be a subject of technology transfer but also a field of genuine collaboration and exchange of expertise.

An important goal of IDRC's expert systems projects has been capacity-building and eventual delivery of operational expert systems. A recent example is expert system development at the Southeast Asian Regional Centre for Tropical Biology (BIOTROP) in Bogor, Indonesia. As a part of a larger project, a group of researchers developed a prototype expert system for weed identification and management on rubber plantations. In February 1994, BIOTROP organized a regional workshop in which 23 researchers from Malaysia, Thailand, the Philippines, and Indonesia participated. The program included familiarization with expert systems technology and hands-on experience in applying the

Weed Identification and Weed Management expert systems. Researchers from the agricultural institutions that attended received copies of the prototype systems, which they will evaluate and use to identify their particular requirements regarding expert system technology. Delivery of a fully operational expert system for weed identification and management and its potential commercial value will also be discussed by them.

Among the lessons learned from IDRC projects in the area of knowledge-based systems development[9] are, first, the recognition that researchers in developing countries are interested in the application of expert systems and have already produced interesting results; second, that a long-term strategy is needed in capacity building and supporting access to advanced development tools; and third, that collaboration and sharing of information are very important.[10] The potential of advanced information technologies, particularly knowledge-based systems, in developing practical applications for agriculture, health services, environmental protection, and decision-support has been recognized by many researchers in developing countries.

IDRC will continue to support projects in this area and is developing its program strategy aimed at achieving practical results. It will concentrate on building interdisciplinary teams which can identify real problems and the domains of knowledge required to solve them; identifying research teams which have sufficient expertise in computer technology to adapt and use advanced software tools; identifying institutions which can make a long-term commitment to the development of knowledge-based systems; fostering collaboration between recipients of grants in developing countries and Canadian researchers; and identifying applications with potential commercial value and wide distribution.

## Geomatics: connecting geography and development[11]

With the continuing evolution of geomatics-based information technologies over the last two decades, spatial or geographic information is increasingly being used for strategic decision-making in all aspects of earth resources assessment, management and monitoring. Three sets of spatial information technologies have proved to be of particular importance to environmental and natural resource information management in developing countries. They are: remote sensing (RS), which, because of its ability to obtain data on large areas on the earth and surrounding atmosphere, is seen by Southern researchers as a suitable tool to acquire information often lacking about their territories; Geographic Information Systems (GIS), which can assist in the management, analysis and presentation of information in a form suitable for policy- and decision-making; and Global Positioning Systems (GPS), which allow for the rapid determination of geographic positions required in medium-scale cartography and cadastral mapping for land registration and titling purposes. Incidentally, all three of these are also areas of technology in which Canada has demonstrated technical leadership, from the perspectives of technology development, implementation and technology transfer.

Unfortunately, new geomatics technologies are often still perceived as the monopoly of the North, with the exception of a few developing countries such as India and China which have developed some of their own systems (such as satellites and software). Southern countries are often considered as potential consumers of data, or at best as test-beds for applications that often cannot be undertaken in the North (for example, those related to tropical rainforests, high-mountainous environments, drought and desertification, mangroves, and so on).

In the late 1970s, IDRC recognized the potential offered by geomatics technologies for meeting the needs of developing countries for mapped information, and began to support applied research involving their use in several sectors, including forestry, agriculture, fisheries, rural development and natural resources management. Projects have been funded in over 25 developing countries, including Bangladesh, Brazil, Burkina Faso, Bolivia, Colombia, Chile, China, Costa Rica, Côte d'Ivoire, Egypt, India, Jordan, Kenya, Morocco, Mali, Malaysia, Nepal, Nigeria, the Philippines, Senegal, Sudan, Tanzania, Thailand, Uganda and Vietnam. These projects represented different experiences in technology transfer and applications research, and have provided an opportunity to test the feasibility of applying satellite data, GIS and GPS for the production of thematic information of importance to development. Many of these projects dealt with the assessment, monitoring and conservation of land and water resources.

IDRC advocates three main approaches in spatial information technology research: validation of technologies and/or their adaptation to the local environment; investigating the potential of new technologies for developing countries; and technology research and development. The approach selected depends on the state of readiness of the technology and the researchers who are to work with it.

For example, when a new technology is already operational and has proved its utility to solve development problems in the North, IDRC's strategy consists of stimulating the introduction and adaptation of the technology in developing countries to help solve similar problems through heuristic research (learning by doing). The technology transfer approach involves helping recipient-country researchers and practitioners to acquire the technology and test it in 'solution-oriented' research projects. This is the case with most of the research projects supported by IDRC's Geomatics Program. For example, in Morocco, a two-phase project contributed strongly to the promotion and strengthening of remote sensing and GIS disciplines in that country and in neighbouring ones (through the development of similar projects on different applications). Phase I, 'Remote-Sensing Contribution to Soil Mapping, Morocco', dealt with medium-scale soil mapping inventories in semi-arid areas using remotely-sensed data. Phase II, 'GIS for Agricultural Development, Morocco', is still active. It deals with the integration of data derived from various sources, maps, remote-sensing imagery, and statistics in an operational information system for agricultural development of semi-arid lands.

When a promising technology is still under development and testing in the North, its usefulness and capacity to help solve development problems in the South can often be explored. IDRC believes that an efficient way to prepare people to use a technology is to encourage their participation in its early fine-tuning. IDRC's strategy in this case is to stimulate the participation of developing-country researchers in global applications development plans together with researchers from the North. This is the case with a series of Synthetic Aperture Radar (SAR) projects which IDRC has been supporting worldwide. For example, the 'Radar Remote Sensing, Costa Rica' project involved over 20 institutions and 40 research topics and was carried out in partnership with the Canada Centre for Remote Sensing. A larger 'GLOBESAR' project involves over 14 countries, with 10 developing countries in Africa, the Middle East and Asia, and over 60 institutions. GLOBESAR's primary objective goes much deeper than other projects, since developing countries researchers are actively associated with the development of applications made possible by the first Canadian earth resources satellite, RADARSAT. This kind of partnership between North and South should go a step further towards the participation of developing countries in R&D and in sharing the resulting technological wealth.

IDRC also supports geomatics R&D in developing countries, in both the software and hardware areas. This is done when a number of developing-country institutions and researchers have the required capacity to undertake R&D in areas of interest to them and to IDRC, and when it is believed that the research and any resulting product has potential use in sustainable development and can be exploited commercially to generate income. IDRC's strategy is then to support the development of the product and assist developers in patenting and distributing it for the benefit of development. For example, IDRC is supporting the development of an active GPS Point System which will be used for cadastral mapping in Argentina (hardware and software). It is also working with the Canada Centre for Remote Sensing and a number of international partners to produce an Electronic Atlas of Agenda 21, with its first chapter on Biodiversity (software and databases).

## 4. Conclusions

The informatics revolution is bringing about many changes in the work we do and in the way we do it. It provides tremendous opportunities for progress and transformation (in the global economy, regions, countries and country blocs, societies, cultures, communities, institutions and individuals), but it also poses tremendous challenges (in issues like equity, information policies, social impacts, employment, privacy, access, and so on). This is especially true for developing regions that can benefit from effective use of new information technologies but that also run the risk of dramatically increasing the information and technology gap within their own societies. Faced with an ever-changing environment (about which the only thing that is constant is change), the development community generally, and organizations like IDRC in particular, need to examine carefully these opportunities, challenges and risks, and to establish strategies for the effective utilization of IT in their operations, programs, and relationships with partners. IT is also a key ingredient in organizational change, enabling new modalities of operation and relationships, with instant access to information. IT is *transformative* technology.

IDRC has invested a relatively modest amount of money in the IT area over its 25-year history, but believes that this has been a wise investment in development. For example, the ICT Program mentioned above has supported around 100 projects for a total investment of about $Can 15 million over the last ten years.

The development assistance community is becoming more conscious of the value of information and related information technologies. This is not necessarily a new phenomenon.[12] Other agencies are investing heavily in IT, although mainly as a component of other projects. A recent study by the World Bank[13] indicates that, although it does not have an information program *per se*, it is lending hundreds of millions of dollars a year for information technology. At an informal consultation on IT and Development organized by IDRC and the Commonwealth Secretariat in January 1993, selected donors and developing-country participants met to examine ways of working more closely together in this field. It was discovered that a significant stumbling block was the lack of identified programs or focal points within some of the agencies involved.

Nonetheless, donor agencies are considering IT much more actively in their support and work. Of course, developing countries have also recognized this, and the demand for projects that would bring the benefits of these technologies and improve access to information and knowledge has increased in recent years.[14] It is recognized that collaboration in the field of IT can lead to win–win situations among partners. Partners include not only development assistance agencies and developing-country recipients but also the private sector, the information industry in the North, and the information industry now starting up in the South.

Collaboration can also be facilitated by the effective use of IT. At a meeting of donors at the Bellagio Study and Conference Center in November 1993, it was decided to create a flexible and effective mechanism to enhance information sharing and collaboration among donors. To support this initiative, a communications and information system, BellaNet, has been established. A number of donor agencies, including IDRC, are involved in this venture. A key feature of the BellaNet model is that it will build on the existing resources and strengths of donors in information management and communications where these are located.

## Notes

1. International Development Research Centre Act, Revised Statutes of Canada 1970, 1st Supplement, Chapter 21.
2. See, for example, R. Valantin, *A Conceptual Model for Support for Information Technologies Research for Developing Countries* (Ottawa: IDRC, 1985).
3. Ottawa: IDRC, November 1991.
4. See the discussion on MINISIS below.
5. D. Balson, R. Drysdale and B. Stanley, *Computer-Based Conferencing Systems for Developing Countries, Report of a Workshop held at Ottawa, Canada, 26–30 October 1981* (Ottawa: IDRC, 1981).
6. D. Balson, *International Computer-Based Conference on Biotechnology: A Case*

*Study* (Ottawa: IDRC, 1985).
7. G. Lindsey, K. Novak, S. Ozgediz, and D. Balson, *The CGNET Story: A Case Study of International Computer Networking* (Ottawa: IDRC, 1993).
8. C.K. Mann and S.R. Ruth, *Expert Systems in Developing Countries: Practice and Promise* (Boulder, Colo.: Westview Press, 1992).
9. Z. Mikolajuk, *Expert Systems in Developing Countries: Learning through IDRC's Experience*, Proceedings of the 1993 DND Workshop on Advanced Technologies in Knowledge-based Systems and Robotics (Ottawa: November 1993).
10. For example, IDRC recently agreed to sponsor the establishment of a new interest group on the Internet dedicated to expert systems technology in developing countries.
11. Geomatics refers to the science and technology of spatial information management. It includes technologies such as automated mapping, remote sensing, Geographic Information Systems (GIS), and Global Positioning Systems, and covers the acquisition, storage, processing and analysis of data and the production of geographically referenced information products.
12. R. Valantin and M. Arkin, 'Informatics and Development – Selected Programmes and Considerations', in K. Haq, ed., *The Informatics Revolution and Developing Countries: Papers prepared for the North South Roundtable Consultative Meeting in Scheveningen, The Netherlands, 13–15 September 1985* (Islamabad: North–South Roundtable, 1986).
13. N. Hanna and S. Boyson, *Information Technology in World Bank Landing: Increasing the Development Impact* (Washington, DC: World Bank, 1993).
14. R. Valantin, 'Information Tools and their Transfer', in M.S. Swaminathan, ed., *Information Technology, A Dialogue* (Madras: Macmillan India, 1993).

## References

Balson, D., R. Drysdale and B. Stanley, *Computer-Based Conferencing Systems for Developing Countries, Report of a Workshop held at Ottawa, Canada, 26–30 October 1981* (Ottawa: IDRC, 1981).
Balson, D., *International Computer-Based Conference on Biotechnology: A Case Study* (Ottawa: IDRC, 1985).
Hanna, N. and S. Boyson, *Information Technology in World Bank Landing: Increasing the Development Impact* (Washington, DC: World Bank, 1993).
International Development Research Centre Act, Revised Statutes of Canada 1970, 1st Supplement.
Lindsey, G., K. Novak, S. Ozgediz, and D. Balson, *The CGNET Story: A Case Study of International Computer Networking* (Ottawa: IDRC, 1993).
Mann, C.K. and S.R. Ruth, *Expert Systems in Developing Countries: Practice and Promise* (Boulder, Colo.: Westview Press, 1992).
Mikolajuk, Z., *Expert Systems in Developing Countries: Learning through IDRC's Experience*, Proceedings of the 1993 DND Workshop on Advanced Technologies in Knowledge-based Systems and Robotics (Ottawa: IDRC, November 1993).

Valantin, R., *A Conceptual Model for Support for Information Technologies Research for Developing Countries* (Ottawa: IDRC, 1985).

Valantin, R., 'Information Tools and their Transfer', in M.S. Swaminathan, ed., *Information Technology, A Dialogue* (Madras: Macmillan India, 1993).

Valantin, R. and M. Arkin, 'Informatics and Development – Selected Programmes and Considerations', in K. Haq, ed., *The Informatics Revolution and Developing Countries: Papers prepared for the North South Roundtable Consultative Meeting in Scheveningen, The Netherlands, 13–15 September 1985* (Islamabad: North–South Roundtable, 1986).

# 13
# Support for Biotechnology in Developing Countries: The Dutch Experience
## T.J. Wessels

## 1. Introduction

Several developments during the 1980s have influenced thinking on the appropriateness and challenge of applying advanced biotechnology in the developing world. Biotechnology is considered dualistic in its impact: apart from the potential positive contributions it can make to the development process, there might be negative consequences as well.

Using biotechnology, it has become possible to produce substitutes for some traditional primary commodities. As commodities become increasingly interchangeable, the processing companies (which are mainly located in the industrialized countries) have a wide choice of materials for their products. Almost half of all primary commodities originate in developing countries, which, in contrast to other commodity-producing countries, depend largely on the export of a limited number of such products. Because of their economically weak position and narrow export base, developing countries will be unable in the short term to compensate for the loss of export markets for one or more commodities, and this could have serious consequences for their development.

A large part of current biotechnological plant research is carried out by multinationals and is aimed mainly at large-scale, commercial agriculture. The technology developed is adapted to the needs of this type of agriculture. More than 75 per cent of agricultural production in developing countries, however, is on a small scale. The currently available biotechnology is not adapted to the needs of small producers and their limited ability to invest, and is thus unsuitable for them.

In many developing countries there is well-organized, large-scale commercial agriculture as well as small-scale agriculture. Like their counterparts in industrialized countries, the large commercial producers will be able to make use of the technology, thus increasing their production, while that of the small farmers remains the same. Local, non-commercial and largely self-sufficient agriculture is, however, essential for local food security in rural areas. Another problem associated with the industrialization of agriculture is that more and more crops are being grown with the same genetic base. The narrowing of the genetic base of the main commercial and food crops increases the risk from disease and pests because the whole plant population becomes equally susceptible to disease and environmental factors. This genetic erosion threatens the world supply of food and plant improvement research, which is based on genetic diversity.

In industry, innovative knowledge is patented to protect the invention and to recover the development costs. Internationally a discussion takes place in order to secure intellectual property protection on living materials. The idea, however, of patents on products and processes in microorganisms, plants and animals is contrary to the principle of the free availability of natural genetic resources. At both national and international levels, there is ongoing debate over the application of patent law and plantbreeders' rights.

The desirability of patent protection in developing countries, where no infrastructure for research or for monitoring compliance exists, is questionable. The consequences for Third World countries have not as yet been investigated. In general it can be expected that these developments will form a threat to the export markets and small-scale food production of these countries (in particular, the price of seed is expected to increase).

The effects of biotechnology on people and the environment are not specific to developing countries, but, because of their weak economic position and the absence of legislation, they will feel some effects particularly severely. In many cases the application of biotechnology will involve the introduction of new or modified organisms to areas where they were not found in the same form or to the same extent before.

Relatively little is known as yet about the possible risks of introducing genetically modified organisms into the environment. Living organisms reproduce, and once released they are hard to control. The effects may be irreversible and there is little hard information about the mechanisms involved, so in many industrialized countries the introduction of genetically modified organisms is assessed in advance case by case and subsequently step by step. This assessment is based on risk analysis. Meanwhile, in the industrialized world, consistent legislation is being prepared at the national and international levels dealing with the introduction of genetically modified organisms. It is important that this legislation should take into account the consequences of introduction in developing countries. Given that many of them have no laws or regulations covering this area, there is a danger that they will be used for experimental studies. This has already occurred in sporadic cases. It is also important to realize that modified organisms which are relatively safe in a temperate climate could behave differently in a hot climate and pose a risk to people and the environment.

## 2. Potential Positive Developments

To make use of the potential of biotechnology as a means of combating poverty structurally, it should be made possible for developing countries to build up their own research capability, and development-oriented research in the industrialized countries should be encouraged. However, since the present economic balance of power makes it unlikely that biotechnological research will move in this direction, there is a need to find out how developing countries can be supported in applying biotechnology in an ecologically sound way, how biotechnological techniques can be used specifically to solve problems in developing countries; and how the negative consequences of biotechnology for developing countries can be minimized.

To arrive at a coordinated approach for a user- and problem-oriented use of biotechnology for and with the Third World, the Netherlands Ministry of Development Cooperation created the Stimulation Programme for Biotechnology and Development Cooperation. It consists of three elements: (1) Integration of the development dimension in Dutch biotechnology policy; (2) Demand-oriented technical cooperation; (3) International collaboration and coordination.

(1) *Integration of the development dimension in Dutch biotechnology policy*
Dutch biotechnology policy is based at present on the scientific concerns and economic interests of the industrialized world. To ensure that a wider range of interests are considered and to prevent biotechnological developments in the Netherlands from increasing the gap between rich and poor countries, it is necessary to take the position of the developing countries into account when discussing and taking decisions on biotechnology. On the agenda are harmonization of policy, the Netherlands' position at international fora with regard to intellectual property rights (GATT, UNCTAD, the EC), and biosafety.

Furthermore the integration of development expertise in relevant biotechnology commissions is sought. The objective is to promote research relevant to developing countries and to assess possible consequences of research projects for developing countries.

*(2) Demand-oriented technical cooperation*
Through technical collaboration projects, efforts will be made to achieve problem-oriented use of biotechnology in selected programme countries and regions. This will be done on the basis of the priorities set by these countries. These projects will make use of the potential of biotechnology as a means of structurally combating poverty. For this reason they will be concerned not only with the technological aspects but with the socioeconomic, legal, cultural and ecological aspects of the use of biotechnology. Special attention will be given to issues of particular importance to small-scale farmers and women in developing countries.

As part of technical cooperation the following activities will be undertaken, where possible in an integrated fashion: biotechnological research projects by and for developing countries; building of biotechnological research capability in developing countries; technology assessment, socioeconomic studies and risk analyses; support for biotechnology policy in developing countries; laws and regulations; provision of information; and free exchange and preservation of genetic resources.

*(3) International collaboration and coordination*
If biotechnology is to be used effectively for the benefit of the Third World, it is essential to have an internationally coordinated approach together with other donors both bilaterally and within international organizations such as those of the UN, the Consultative Group on International Agricultural Research (CGIAR), the EC and the World Bank. The international organizations are the fora where international policy should be formulated as regards biotechnology and related issues such as intellectual property and biosafety. The latter point is especially important in view of the lack of regulations in these fields in most developing countries.

Since several bilateral donors are preparing a biotechnology policy, the objective is to have collaboration and coordination at an early stage. This could be achieved through information supply, workshops, networks, and so on. Possible fields of common interest are laws and regulations, genetic and biological diversity, orphan commodities/diseases and orphan technologies (for example, biofertilizers, biopesticides).

## Approach, organization and funding of the Special Programme

The Special Programme focuses thematically on agriculture, human health care and environmental management. Technical cooperation, directed at local capacity-building and the formulation and implementation of research programmes, will take place in four developing countries. These were selected on the basis of local interest in the formulation of a national biotechnology policy, the existence of a certain local research capacity on which this can be built, a collaborative relation with the Netherlands' bilateral development programme, and the regional role the country could play. On the basis of these criteria and a certain geographical dispersion, Zimbabwe, Kenya, Colombia and India were selected.

Characteristic of the Special Programme is that the activities are identified on the basis of a participative bottom-up process. It means that needs and priorities are assessed locally, with the involvement also of policy-makers and researchers as NGOs and end-users (or their representatives).

The Biotechnology Programme has been set up for an initial period of five years and is implemented through a 'special project' within the ministry. Outside experts have been brought in for the implementation of the tasks in the priority areas. A special Advisory Group of private individuals representing the relevant specializations, experience and social sectors was installed by the minister. The advisory group meets at least four times a year and advises on the criteria, methodology and relevance of the activities.

For the implementation of the Special Programme an amount of Dfl. 10 million per year is available (a total of Dfl. 50 million over five years), excluding the operational costs of the earlier mentioned 'special project'.

## 3. Achievements of the Special Programme

### Integration of the development dimension in the Netherlands' biotechnology policy.

The Programme has been actively involved in the discussions regarding intellectual property rights (the EC, GATT, UNCTAD) and biosafety (UNCED). In interdepartmental consultations the interests of the developing countries were brought into the discussions. Furthermore, interdepartmental task forces participated in drawing up legislation on genetically modified organisms, and discussion on the ethical aspects of biotechnology, public acceptance, and the policy document Biotechnology of the Ministry of Agriculture, Nature Management and Fisheries.

The contribution with regard to intellectual property protection on biotechnological inventions has concentrated mainly on the necessity to differentiate between developing countries in terms of development levels (local infrastructure and research capacity), questioning the common view in most industrialized countries that adoption of patent law in developing countries would automatically lead to innovations. There has also been recognition of the sovereign rights of developing countries to genetic resources on their territory, and in that respect a need for a more balanced distribution of the income generated by the industrial processing of genetic material by foreign companies.

With regard to the biosafety of biotechnological inventions, the Biotechnology Programme advocated the importance of uniform regulations and the necessity of capacity-building in developing countries. In October 1993, together with the Zimbabwean Research Council and The Netherlands' Ministry for the Environment, a regional biosafety conference was organized in Harare. A similar conference in Latin America was organized in collaboration with IICA, CIAT and USDA in June 1994 in Cartagena.

Presentations to NGOs, universities and research institutions, and active input in technical and societal fora, where biotechnology and the Third World are discussed, have been important.

The Special Programme has become a member of the Programme Commission responsible for the organization in Amsterdam of the 4th congress of the International Society for Plant Molecular Biology (ISPMB), and is as such responsible for the coordination of the programme directed at developing countries. The Special Programme regularly makes funds available for the partipation of (young) scientists of developing countries at international scientific meetings.

### Technical cooperation

In practice, the Special Programme's objective is to apply biotechnology in the solution of constraints in marginal production systems, with special attention to women and small-scale producers. The starting point is the supply of biotechnology to those groups that normally do not have access to it. The focus is on the identification of – and feedback of generated technology to – a specific geographical region (model system), whereby the research topics selected will be of cross-regional importance.

Characteristic of the Special Programme is that activities are identified via a participatory process that starts with the end-users and ultimate (potential) beneficiaries of the relevant technology. It means that projects and programmes are formulated on the basis of a local-needs assessment and priority setting, in which the end-users, researchers, policy-makers and NGOs are involved. In terms of method the 'interactive bottom-up approach'[1] is used, employing the following procedure.

A fact-finding mission consults representatives of relevant ministries, universities, research organizations, extension service, international institutes, farmers' and women's organizations and NGOs. In coordination with local institutions, an attempt is made to realize a first geographical/agro-ecological demarcation of possible intervention regions. In the next phase a (biotechnology-neutral) local partner organization is identified, and a process of consultations is initiated with small-scale producers, NGOs, researchers and government. In the farmers' workshops small-scale producers are consulted on production and marketing characteristics and on the constraints encountered. Problems are prioritized and recommendations made for their solution. With the help of the local partner a number of inventory studies (agronomic, socioeconomic, marketing, and so on) are carried out in order to narrow down the intervention region and relevant issues.

The next step is a national workshop with the representatives of the different interests, in which a further delimitation of issues (commodities, technologies, diseases and pests, and so on) and priority setting take place. In this phase, either on the initiative of the developing country or of the Special Programme, a National Biotechnology Forum (or Steering Committee) of around ten members is set up, in which representatives of policy, research, farmers and NGOs participate in a personal capacity. The Forum functions as a platform for problem-oriented application of biotechnology to small-scale producers. It bridges the gap between biotechnology research and policy, promotes the integration of biotechnology in the national research institutions, and monitors and advises on the relevance of the activities generated in the framework of the Special Programme.

The Special Programme for Biotechnology supplies the Forum with a small secretariat and (limited) funds to enable them to meet, produce a newsletter, commission additional studies, and organize seminars and 'feedback' consultations with the end-users. To that end a workplan is drawn up, that will be discussed in a local workshop with the end-users. This process began in 1992, and between 1993 and 1995 workshops were organized, in which first rounds of priority-setting were conducted in Zimbabwe, Kenya, India and Colombia.

Currently, the phase of institutionalization is becoming more and more important in the Special Programme approach. Although very important, this phase does not receive much attention in the IBU approach. It is the aim of the programme to develop an extension of the methodology to provide a framework for implementation.

The transfer of ownership of the programme – including decision-making authority over resource allocation – to the steering committees will ensure that important decisions are taken at the appropriate level. However, questions of how to involve end-users in the process of project formulation and implementation, and how to ensure that funding for priority research projects will remain available in the future, still need to be answered.

## Results so far

### Zimbabwe

From the inventory studies and the 1993 workshop, the following priorities were drawn up: implementation of national biotechnology policy, which clearly indicates how biotechnology in the future will be applied for the improvement of the standard of living; introduction of laws and regulations regarding biosafety and intellectual property rights; priority crops for biotechnology designated as maize, small grains, legumes, vegetables, cotton and root and tubers – selected for their importance as staples, nutritional value in the small farmers' diet, and importance to food security; minimization of post-harvest losses through storage pest; introduction of better mechanisms for technology transfer; pilot projects in micro-propagated disease-free plant material (sweet potato, Irish potato) for immediate impact on small-scale farmers.

Maize, legumes, sorghum, millet, sweet potato and cassava were confirmed as being the crops for immediate intervention. Preceding project formulation, a number of socioeconomic studies were carried out in close cooperation with the Zimbabwe Biotechnology Advisory Committee (ZIMBAC). These supplied detailed information on the importance of each commodity and an estimation of the expected impact of the proposed biotechnological intervention. Target groups chosen were the smallholder farming communities in the Wedza and Buhera areas.

The following project proposals were discussed: micropropagation for root and tuber crops; marker-assisted breeding for drought and insect tolerance in maize; a continuation of the M.Sc. course in biotechnology at the University of Zimbabwe in Harare; virus resistance in cowpea; biofertilizer in beans; fermentation improvement in food. In the area of animal husbandry, priorities are not yet defined. In addition, a Regional Focal Point for Biosafety has been set up in Harare.

*Kenya*

The two intervention districts are the higher altitude region *Kakamega*, where the available arable land and erosion are limiting factors, and the semi-arid region *Machakos*, where climate and soil fertility are major constraints. Preceding the national workshop (in September 1993), consultations took place with farmers from the intervention regions, as well as workshops with researchers and policy-makers regarding constraints and possible solutions in the areas of crops, livestock and agroforestry. In addition, a workshop was convened to discuss the necessary policy framework for an adequate biotechnology programme.

The national workshop resulted in clear priority being given to biotechnologies that would deliver results in the short term and simultaneously could link up with existing technical capacity. Great importance was placed on the application of biotechnology in the combat of diseases and plagues in crops (via resistance breeding using genetic markers and/or genetic modification), although it was recognized that a longer period was needed to achieve results.

To advise on the programme and to assist in the formulation of projects in the priority areas, the Kenyan Agricultural Biotechnology Platform (KABP) was founded. Framework-setting socioeconomic studies were carried out and one-day thematic and project-formulation workshops were organized. Feedback to end-users is a continuous concern.

The following projects have been identified and are in the process of formulation: marker-assisted breeding in maize for drought and insect tolerance (joint project with Zimbabwe); diagnostic tools for livestock diseases; biological nitrogen fixation; biopesticides; tissue culture for banana, root and tuber crops and agroforestry species. A local capacity for the development and monitoring of biosafety guidelines has been set up.

On the national level, the feasibility of a biotechnology programme in the field of human health care was discusssed. A first consultation was held in Nairobi in 1994. Representatives of the most important participants in the field of primary health care, including traditional healers, met to discuss the need for technology (tools). The first step was to investigate the potential contribution to the sector of improved diagnostics, especially with regard to the diseases malaria, tuberculosis, schistosomiasis, leishmaniasis and venereal diseases. Possible projects will be assessed on their importance for the end-user patient or primary health-care worker and not for their research value. The possibilities for improved diagnostics within the framework of the Kenya National Tuberculosis Programme were discussed.

*Colombia*

In Colombia a biotechnological intervention is directed at improving a farming system based on maize, rain-fed rice, cassava and plantain. Many rural people depend for their daily food on these low- to non-external input systems, while the production systems, especially the growing of cassava and plantain, receive very limited attention in agricultural research. Since these systems operate in large areas of Latin America, an improvement could have a positive effect on the food situation of millions of people in the region. Following exploratory visits to several regions in Colombia, an area in the Atlantic coast (Montería) was selected in close cooperation with the National Steering Committee. The process of bottom-up needs assessment and the setting of priorities was continued in September 1993 through a baseline study followed by a consultative workshop with farmers, agricultural extension personnel and researchers. (In addition to the National Steering Committee, there is a regional farmers' committee – Comité Campesino Regional – on which 12 farmers from the intervention region sit. This committee is represented by two members in the National Steering Committee.)

The preliminary conclusion of this workshop was that the non-availability of sufficient disease-free plant material of cassava, yam and plantain was a major constraint. Biotechnology – especially *in vitro* propagation – could therefore make an important contribution. On the basis of the consultation and existing statistical data, it was decided to carry out an additional socioeconomic study. The study gave better information on basic data such as areas planted, production volumes, yields and market outlets for the crops. Furthermore, the study assessed the possible impact of project activities like the supply of disease-free plant material of plantain and other crops.

Projects will be developed in the areas of tissue culture of plantain, cassava and yam, and molecular breeding for disease/pest tolerance in plantain (black sigatoka) and yam. In addition, possibilities for marker-assisted breeding for disease resistance in papaya are be evaluated.

## India

The first fact-finding mission to India took place in November 1993. Several problems with the semi-arid agriculture and of possible interest to the Special Programme were presented by NGOs and research institutes. There is much concern for the environmental problems of the so-called agricultural wastelands – often eroded communal lands – where production of annual crops seems no longer feasible and the solution to which must be found in agroforestry systems. It was decided to concentrate on the state of Andhra Pradesh and semi-arid agriculture.

In early 1994 a local institute was commissioned to carry out an inventory of NGOs in Andhra Pradesh operating in agricultural and rural development, with the objective of identifying a local partner. Two organizations were selected: Youth for Action (YFA), active in Mahbubnagar district, and Sri Aurobindo Institute for Rural Development (SAIRD), in Nalgonda district.

It was decided that three themes warranted further preparatory work: resource management, including agroforestry, and tree crop development in which biotechnology could contribute via *in vitro* propagation of agroforestry and fruit species; improved diagnostics and vaccine development for cattle, sheep and poultry; and biotic and abiotic stresses in small grains, pulses and oil seeds – especially drought tolerance, due to its importance in Zimbabwe and Kenya.

A multidisciplinary team, consisting of researchers and NGO representatives, was formed to conduct a detailed study of selected farming systems in Andhra Pradesh and the problems and constraints faced by farmers. The team visited five villages in two districts. Based on the findings, a report was produced that highlighted the major problems in the farming systems as experienced by the farmers. Options for biotechnological solutions were also indicated. The report served as a major input to the priority-setting workshop. Supply and demand sides were discussed in the light of feasible biotechnological solutions.

Project formulation has taken place in the following areas: tissue culture in agroforestry and other tree species; resistance to diseases and pests in food and oil crops (sorghum, pigeon pea, castor and groundnut); improved diagnostics for livestock and storage and the processing of agricultural production.

A local steering committee, the Biotechnology Advisory Committee, is currently being set up. Future activities will be the formulation and implementation of specific projects according to established priorities and the transfer of ownership to the local steering committee.

## International collaboration and coordination

International activities can be divided into policy aspects of biotechnology (intellectual property, biosafety and biodiversity) and network formation. In 1988 the Netherlands was requested to chair the Taskforce on Biotechnology (BIOTASK) of the Consultative Group on International Agricultural Research (CGIAR). BIOTASK was created to further the integration of biotechnology into the research conducted under the aegis of CGIAR, paying special attention to the socioeconomic aspects of the introduction of biotechnology. The Taskforce, which functioned until December 1992, was formed by representatives of donor agencies, research institutes and developing countries. It operated as a platform for biotechnology and related issues. Some examples of the activities it conducted are: preparing an inventory of information needs in agricultural research institutions in developing countries, and the promotion of suitable information systems; fostering the availability, at reduced rates, of subscriptions to biotechnology journals for research institutions in developing countries; organizing the workshop Biotechnology Policy and the CGIAR, directed at the issues of biosafety and intellectual property; stimulating discussion within the CGIAR community regarding intellectual property rights on plants and animals, and the drafting of the CGIAR Statement on Intellectual Property, adopted at a meeting in Istanbul in 1992; fostering the foundation of the Intermediary Biotechnology Service (IBS), a mechanism for the transfer of knowledge and technology to developing countries; organizing a training workshop for researchers from developing countries on the science and technology of Restriction Fragment Length Polymorphism.

The Biotechnology Programme, as the chair of an international group of donors (UK, Norway, USA, Australia, Switzerland, the Netherlands and the World Bank) was instrumental in establishing the Intermediary Biotechnology Service (IBS) at the International Service for National Agricultural Research (ISNAR) in the Hague (ISNAR is one of the many CGIAR research institutions spread around the globe). The IBS is a demand-driven, problem-oriented neutral advisory service with the objectives of advising developing countries on policy aspects of biotechnology, assisting in identifying priority problems amenable to solution through biotechnology, and on integrating biotechnology into the countries' research programmes. Furthermore, the IBS aims at identifying biotechnological expertise and enhancing its availability to national programmes in developing countries. Moreover, it tries to bridge agricultural development constraints and utilize the available expertise in industrialized countries.

As a result of a workshop on Cassava and Biotechnology, organized by the Dutch Special Programme in Amsterdam in 1990, the Cassava Biotechnology Network (CBN) was set up at the Centro Internacional de Agricultura Tropical (CIAT) in Colombia (CIAT is another of the CGIAR institutions). This network aims at being a forum for cassava biotechnology – in consumption terms, the fourth largest food crop in the world, providing food for 500 million people – at integrating the priorities of small-scale cassava producers into biotechnology research on cassava, and at stimulating cassava biotechnology research in areas of established priority.

During the formulation of the Special Programme, high priority was given to the question of information supply in the biotechnology and related policy areas. An important activity was the initiation in 1989 of the quarterly journal *Biotechnology and Development Monitor*, a joint publication with the University of Amsterdam. The *Monitor* is directed at researchers and policy-makers in both developing and industrialized countries and is sent free to around 5000 individuals and institutions worldwide. From the beginning of the *Monitor* project high priority was given to increasing the participation of developing countries, and ultimately transfer of publication of the *Monitor* to either an organization in a developing country or an international organization.

Developing countries have a pressing need for information on recent scientific developments in the field of biotechnology. However, local libraries generally have no funds for subscribing to specialist journals. In line with the recommendation of BIOTASK, the Special Programme has made available, for a period of three years, to the libraries of research institutions in developing countries, 150 free subscriptions of the important biotechnology journal *AgBiotech* – a publication of the Commonwealth Agricultural Bureau international.

### Note

1. J.F.G. Bunder, ed., *Biotechnology for Small-scale Farmers in Developing Countries* (Amsterdam: VU University Press, 1990).

# 14

# Generic Technologies: New Factors for Swedish Aid for Technology Development

*Jon Sigurdson*

## Background[1]

Science and technology constitute important activities in all advanced industrialized countries and the technology component is generally regarded as the driving force for economic development. Thus government and agencies pay great attention to policies and programmes that promote science and technology for the well-being and security of the citizens of the nation. In a gradual shift from military security to economic security, nations are giving increasing emphasis to the promotion of national technological development, and it has in the past been common to consider the nation as a national system of innovation (NSI).

The last two decades have seen a burgeoning interest in generic technologies. This interest has centred on microelectronics, new materials and biotechnology and has been fuelled by the concern in the mature industrialized countries to enter into new promising fields of technological development that would provide the basis for future industries. Naturally it is desirable to make generic technologies available to support industrial research projects that are spontaneously emerging from the economic sectors. Thus, targeted research projects often have the objective to establish close links between researchers and users in order to speed up the adoption of generic technologies.

The concern for generic technologies has mainly arisen within governments and their agencies responsible for research and development (R&D). They assume that their development efforts will provide technological results that are reusable and will permit the matching of different user needs while maintaining universal access. Thus they try to identify generic research with multi-sectoral applications. However, generic R&D programmes have often been formulated with goals that are too broad and ambitious and have generally been influenced by fashions prevailing in the global R&D community. Such shortcomings, although serious in mature industrialized countries, have been even more grave in developing countries.

Another important concern is the globalization of R&D. Today, systems of innovations no longer have a national character in the way that most politicians and bureaucrats would like them to have. Industries rely on networks – for marketing, for manufacture, and for research and development. Many if not most of these networks long ago shed their national character, for most products when it comes to marketing, and also increasingly for manufacture. Today we see that the networks for R&D are also shedding their national character with momentous consequences for national science and technology policy. Today we see aerospace emerging as the world's first truly borderless industry, to be followed by the car industry, substantial parts of the electronics sector, the heavy equipment industry for electricity generation and distribution, and so on.

The past decade has seen numerous attempts to establish, within singular nations, civilian programmes in the advanced industrialized countries to foster emerging high-technology sectors, partly influenced by the state–partnership consortia which have long existed in Japan. In Europe there has been an obsession with support for national champions, a concept which was adopted during the 1980s in principle, though not in name, by the R&D Framework Program within the European Union, and Eureka. Many of these attempts have only been marginally successful. A number of the large globalizing corporations were always one step ahead of the national or supra-national bureaucracies in identifying needs and forging new alliances.

## The Role of Generic Knowledge[2]

The support for generic technologies has become fashionable in the industrialized countries. It draws its strength from the assumed potential that the support for certain general technologies will create in society. The basic assumption is that knowledge that can be applied to a large group of problems will eventually provide products and services that are highly relevant either from a social or from an economic viewpoint. In order to understand this situation it may be useful briefly to consider the changing role of the university system and the shift to other sectors of the society when it comes to generating new knowledge.

A revision of traditional research activities is taking place in most industrialized countries to take account of the changing environment. At the same time there is also the introduction of 'priority technology projects' which are more directly linked to key *generic* technologies. They are considered to provide support for industrial competitiveness. Greater cooperation between producers and users is usually sought, from the outset, within the projects themselves.

The often quoted trio of generic technologies comprises information technologies, biotechnology and new materials. However, technologies that are generally considered to exist at a high level of generic properties include genetic engineering, superconductor technology and nanotechnology. The basic assumption is that it is possible to identify and nurture a core area of fundamental research which will generate results that are more or less relevant to the design of a wide range of technical systems or processes.

Any discussion of generic technologies will soon turn to issues of definitions and the identification of fields and how they should be organized. It is obvious that the support for generic technologies will be different from the support for the disciplinary sciences. One important problem is the relation between scientific-problem solving and the generation of new disciplinary knowledge. It is necessary to organize knowledge generation that is conducive to understanding various processes and how to control them.

The present universities in industrialized countries, and also in many less developed countries, are generally good. In fact they are too good, considering that much of the knowledge produced is never utilized. They are producing new knowledge all the time, but in the way the system is currently organized society cannot consume it all. The universities employ highly educated people; seen from the broader interests of society, they are full of experts who are often underutilized. This bridging problem, combined with the increasing costs of funding the universities, has led to a proliferation of science parks and experts who specialize in knowledge transfer.

Such recent developments have tended to obscure the recent changes within the knowledge-generating system itself. There has in fact been an important shift in the condition of knowledge production. Traditionally, knowledge has been generated within a disciplinary, primarily cognitive, context.[3] Today knowledge is more and more generated in a broader, transdisciplinary context which has social and economic ramifications.

In the earlier mode, mode 1, problems were identified and solved in a context established and controlled, by and large, through academic interests of a specific community. Most of this work is disciplinary. However, there is an expanding mode of transdisciplinary and knowledge generation, mode 2, that is carried out in the context of application. This mode is generally transient, and usually not hierarchical. Research carried out in the context of application can be found in a number of the disciplines within the university system – most prominently in the engineering and medical fields, but also in other areas.

In terms of problem-solving behaviour, activities within mode 2 do more than assemble specialists from various disciplines. In fact they create a transdisciplinary approach which is driven by non-codified research approaches. Transdisciplinarity has four major features, according to Gibbons et al. First, it develops a distinct but evolving framework to guide problem-solving efforts. Second, it contributes to knowledge generation, but this knowledge can only rarely be characterized in easily identifiable disciplines. Third, communication and transfer of results are often integral parts of the knowledge-generation process. Ensuing diffusion occurs primarily through the movement of members into new problem-solving groups. Fourth, transdisciplinarity is dynamic, and discoveries often lie outside the boundaries of any particular discipline, and practitioners need not return to it for validation.

The Gibbons group noted that the composition of a problem-solving team changes over time as requirements evolve. They also note that the system for generating knowledge has the following characteristics. First, there is an increase in the number of potential sites where knowledge can be created. The universities no longer control the major knowledge domains; instead knowledge is increasingly generated in industrial laboratories, think-tanks, government agencies, and various types of research institutes not belonging to the university system. Second, the sites for generating knowledge are linked with various communications networks that are social and organizational, and increasingly supported by technological links. Third, the sites for knowledge generation differentiate into finer and finer specialities. The Gibbons group stresses that the recombination and reconfiguration of such subfields form the bases for new forms of useful knowledge.

The new structure for generating new knowledge can be described simplistically by means of a figure (see Figure 14.1). When a problem needs to be solved a problem-solving team is put together that draws

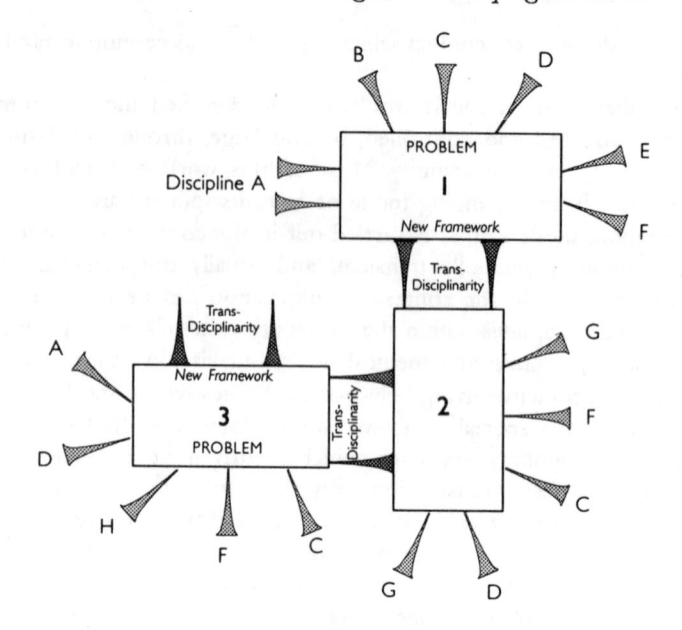

**FIGURE 14.1** Mode 2 networks

on expertise from many sources, involving people with very different backgrounds and (disciplinary) specializations. However, the team members are not selected primarily because of their disciplinary background but because of their assumed ability to contribute to solving a specific problem. When the problem has been solved the members dissipate, and most return to their original sites, although some may participate immediately in new teams where a newly acquired transdisciplinarity approach is considered highly valuable.

The sites for such teams are corporate laboratories, think-tanks, all types of government organizations, most non-university research institutes, and increasingly within the university system itself. The governments in most industrialized countries have a desire to support high-quality research that is relevant to, and supportive of, economic and social development. Thus it is not surprising that they look for mechanisms that support the transdisciplinary structures rather than to the traditional disciplines within the universities. The emphasis given to so-called generic technologies within biotechnology, microelectronics and new materials is a clear expression of new attitudes. It is not the desire

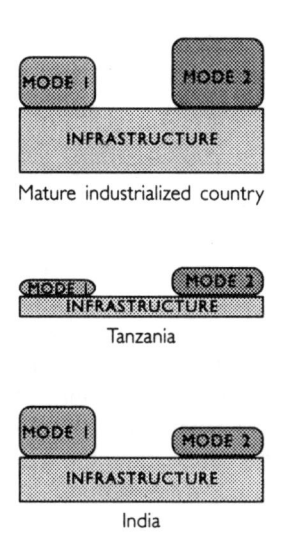

**FIGURE 14.2** Conditions affecting the support of generic technologies in selected countries

of governments and other funding bodies to support, for example, molecular biology or solid-state physics as such.

A number of conditions must be fulfilled if such support for generic technologies is to be successful. First, a fairly large number of specialists must be available or accessible – and some are to be found within the university system . Second, such people must be linked through social and organizational contacts. Third, they must, in most technological fields, be able to draw on the resources of an infrastructure that provides necessary instrumentation and various types of communication means. This is not always the case in the industrialized countries, and the situation is much less favourable in the developing countries. Figure 14.2 provides an illustration of the major problems.

In most mature industrialized countries there exists a good infrastructure. The disciplinary mode 1 of knowledge production, mainly based in the university system, is well developed. The problem-oriented mode 2 is well developed and usually bigger than the university system. These conditions do not guarantee that the system functions well but they are necessary, although not sufficient, conditions. In a large developing country like India the infrastructure is not so well developed. Mode 1 dominates with a highly developed university system. In a small and poorly developed country like Tanzania the situation is much worse. The infrastructure is very weak as is the university system. Thus the existing mode 2 depends to a considerable degree on specialists that come from outside the country and usually return to their original sites once a specific problem has been solved.

Thus the support for generic technologies that has become increasingly prevalent in the advanced industrialized countries cannot be easily transferred to a developing country. Certain requirements in infrastructure and in the quantity and availability of specialists operating in mode 1 and problem-solvers operating in mode 2 must be met. In addition, social and organizational communications must also be present. Therefore, the support for generic technologies must be carefully considered. There is a need to orchestrate the knowledge resources in the university system – both on a national and a global scale.

## Objectives of Swedish Development Assistance

The primary goal of Swedish development assistance continues to be to improve the standard of living of poor peoples. The sub-objectives for development assistance, established by parliament are: (1) growth of resources; (2) economic and social equity; (3) economic and political independence; (4) democratic development; and (5) long-term conservation of natural resources and care for the environment.[4] It is stated that the orientation of development assistance involves a 'greater emphasis on various forms of assistance which improve the ability of developing countries to provide good governance'.[5] It is further stated that the provision of development assistance must take into account how democracy is developing, how human rights are respected, how economic policy is working, and how effectively development resources are employed in recipient countries. These criteria translate into support for market economy, democracy, efficiency, a balance between military and social expenses, and good governance.

Over the last few years a shift of emphasis has occurred in the provision of development assistance. Disaster relief has increased as a result of the growing number of increasingly complicated disaster situations that have occurred in various parts of the world. Thus the linking of humanitarian initiatives, reconstruction and long-term assistance measures, and improved coordination in the disaster relief area is seen as a new challenge for development assistance. Furthermore, development assistance provided through non-governmental organizations continues to increase. The Swedish government view is that it is very important to utilize, as much as possible, the unique competence and resources that exist within non-governmental organizations. This applies both to the situation in recipient countries and in Sweden. These two items – disaster relief and support through NGOs – in fact constitute SIDA's largest budget items, amounting to SEK 1300 and 1000 million respectively, which amounts to almost 40 per cent of the SIDA budget proposed for 1994/95.

In July 1995, following a decision by the Swedish government,[6] the five government agencies dealing with development assistance – SIDA, SAREC, BITS, Swedecorp and the Sandö Training School – were integrated into one new agency entitled the Swedish International Development Cooperation Agency (Sida). The mandates, responsibilities and activities of the earlier agencies, relevant parts of which are described in the next section, have been transferred to the newly constituted Departments in the new Sida.

Multilateral procurement is another area that has also been reviewed. In a report released in mid-March it was indicated that the development banks and the UN agencies would handle procurement worth more than US$400 billion during the remainder of the decade, which indicates that the level has more than doubled during the past ten years. The review suggests that Sweden should initiate and implement more efficient and coordinated measures for multilateral procurement. One concrete proposal is that the Ministry for Foreign Affairs should establish a secretariat for project exports in order to facilitate more active Swedish participation in multilateral procurement.

Another significant analytical effort was a comparative review of development assistance provided by the European Union and by Sweden, conducted in preparation for the latter's membership of the EU. The objectives were fivefold: (1) to describe the development assistance of the EU in various areas; (2) to identify overlapping areas in the provision of development assistance by Sweden and by the EU; (3) to analyse the areas in which the EU may have comparative advantages or disadvantages with regard to current Swedish development assistance, whether of bilateral, Nordic or multilateral character; (4) to make relevant proposals for the distribution of resources for development assistance within the EU in order to complement Sweden's contribution; and (5) to prepare a basis for making priorities and facilitating an active Swedish involvement in influencing and implementing the development policy within the EU.

## Major Swedish Actors in Development Assistance

Until the reorganization in July 1995, the principal actors in Sweden providing support for technological development in developing countries were SAREC, BITS, SwedeCorp and SIDA, although the mandate for the latter has been drastically modified. The first three are briefly described below, and illustrated by a diagram (Figure 14.3) that shows their operational characteristics and gives an indication of their overlapping functions.

*Swedish Agency for Research Cooperation with Developing Countries (SAREC)*
Research cooperation with developing countries has always been an integral part of Swedish development cooperation. This task was originally administered by SIDA and the Ministry of Foreign Affairs. SAREC was established in 1975 to develop research cooperation with Third World countries. SAREC's policy is to concentrate its resources on a smaller number of countries with which it maintains long-term cooperation. SAREC focuses research cooperation in four broad fields: public health, natural resources and the environment, natural sciences and technology, and social sciences.

*Swedish Agency for International Economic and Technical Cooperation (BITS)*
BITS was established in 1979 as a government agency for promoting economic and social development in developing countries and for strengthening the relationship between such countries and Sweden. In addition to Africa, Asia and Latin America, BITS operates in Central and Eastern Europe. BITS does not enter into long-term agreements with the countries it assists, but instead acts on requests from recipient countries and takes decisions on the basis of the merit of each proposal. There is no predetermined allocation of funds between countries or sectors. BITS cooperates through the following instruments: technical cooperation, concessionary credits, and international training programmes.

*Swedish International Enterprise Development Corporation (SwedeCorp)*
In 1991, SWEDFUND, IMPOD and SIDA's industrial division were merged into SwedeCorp, with the aim of bringing together into one organization the major elements of Sweden's development assistance to the commercial and industrial sectors of developing countries and Eastern Europe. SwedeCorp operates in the promotion of business

environment and imports, while investments (joint-venture companies) are handled by Swedefund International. Projects are assessed from a commercial perspective and in a manner which ensures that the requirements of local enterprises determine the forms of cooperation. The investment portfolio is held by Swedefund International AB.

Management of and coordination between different agencies take place through channels such as the budget process, at board level in the agencies concerned, on the basis of specific government decisions and motions in parliament. There are two main criteria for the division of assignments among different agencies. First, SIDA and BITS differ in their focus of recipient countries and methods of operation. SIDA is mainly geared to administer long-term assistance to the poorest countries through their governments, to channel emergency relief, and to operate also through NGOs. BITS deals with assistance of a more short-term nature, in which the recipient countries bear the responsibility for procurement and management of the projects. BITS itself never enters into contracts with Swedish companies or other organizations. Hence, BITS primarily operates in more advanced developing countries that have the ability to negotiate with and control the suppliers.

The second criterion is the sector of operation. Both SIDA and BITS may, in principle, support activities in all sectors of the society. The Organizational Committee[7] considered that Sweden, as a modern industrialized country, has two main areas of know-how – industry and research – and that these require specialized agencies. Development assistance in industry and research areas is channelled through SwedeCorp and SAREC respectively. However, both SIDA and BITS may finance activities in these areas in special circumstances.

## Development Objectives and Scenarios for Generic Technologies

Swedish development objectives range from industrial support in middle-income countries to human development and infrastructural technical support in low-income recipient countries. Given this wide span, both in terms of contents and in terms of character of recipient countries, it is natural that other specific approaches are required if Sweden wants to include support for generic technologies in a major way in its future development aid. However, a clear distinction must be made among the different stages from basic research, to development, to manufacture,

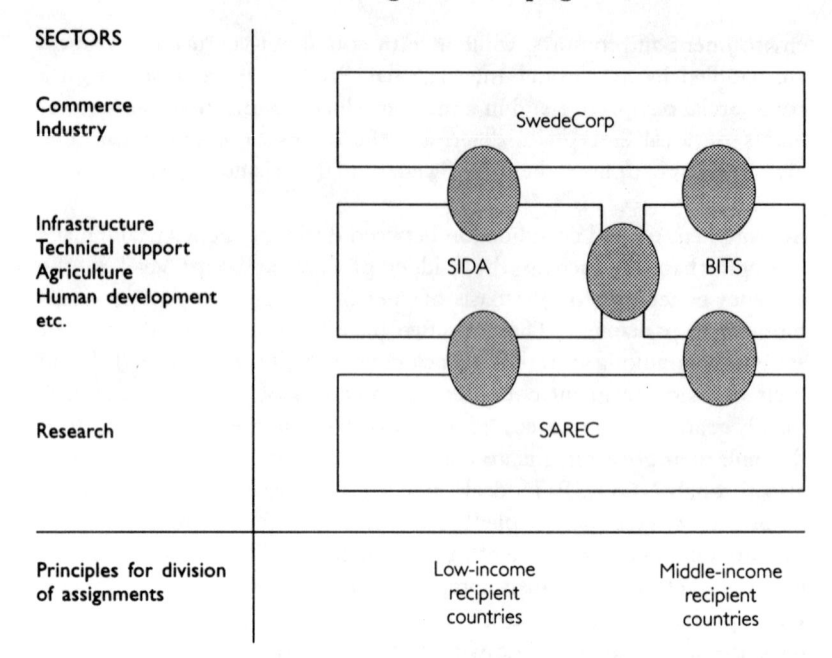

| SECTORS | |
|---|---|
| Commerce<br>Industry | SwedeCorp |
| Infrastructure<br>Technical support<br>Agriculture<br>Human development<br>etc. | SIDA — BITS |
| Research | SAREC |
| Principles for division<br>of assignments | Low-income recipient countries     Middle-income recipient countries |

**Figure 14.3** Areas of operation and potentially overlapping functions of Swedish overseas development agencies

SOURCE: *Styrnings- och samarbetsformer i biståndet* (Guiding Methods and Forms of Cooperation in Development Assistance), SOU 1 (Stockholm: Ministry for Foreign Affairs, 1993).

to servicing, and the use of generic technologies (or the product in which generic technologies are embedded). Three different approaches, or scenarios, can immediately be identified.

*Scenario one: utilizing the technological strength of Sweden*
Sweden has a high level of competence in telecommunications, and in certain areas of biotechnology related to medicine and agriculture. These can be used for developing the infrastructure in technologically weaker developing countries, as well as for obtaining results in product and process development in more advanced developing countries.

*Scenario two: utilizing the management capability of Sweden*
Sweden has a high level of management competence in areas like public health and agricultural research. This can be used as a platform for initiating programmes in generic technologies that draw on resources – scientific, financial, and so on – from elsewhere.

*Scenario three: 'portfolio investment in promising countries'*
Investing in recipient countries that have a high level of scientific receptivity and also economic receptivity. Several of the countries in East and Southeast Asia had these characteristics in the past, and some still have them, although this phenomenon is not limited to this part of the world.

It appears from past experience that resource-poor countries with limited vested interests in competing sectors of the economy are more likely to benefit from programmes that support technological change. However, the success of support programmes for generic technologies, as with many other programmes, is dependent on a number of conditions. First, minuscule projects are not likely to have much impact. Second, technology-support programmes are critically dependent on management and these skills are not always available in scientific research institutions. Third, programmes must be critically evaluated and should not be continued if they have been started on the wrong premiss or are not able to deliver on their promises – even if scientists themselves wish to continue a particular line of joint research. Finally, given the considerable differences in salaries that exist in the global scientific community, it is in many cases desirable for a low-income development country to draw on the scientific pool in, say, India or Pakistan rather than using manpower from Sweden.

## A New Technology Landscape?

A number of fundamental changes are affecting the technology markets, which are undergoing a secular shift towards globalization. These can be summarized in five specific changes. First, both absolute and relative levels of defence expenditure are going down; this forces the creation of technological alliances across borders in order to reduce military R&D costs. Second, government R&D expenditure is, in many countries, decreasing in relative terms as the corporate sector takes on a larger share of 'national' R&D expenditure. Third, defence R&D is in many fields losing its role in defining and advancing the technological front as the functional and quantitative demand shifts from the military to the civilian sector; thus civilian products have become the driving force in many areas. This has, for example, occurred in large segments of the microelectronics sector. Fourth, the corporate sector is modifying the global technology system through merger and acquisitions; this results in control of R&D being in countries other than that of corporate origin, and in systematic sourcing of R&D inputs in universities and in national research institutes throughout the world. Fifth, national governments are redefining the role of R&D away from direct support for 'national' companies and more toward the welfare of the nation – a change that can be interpreted as a shift from military security to economic security.

These changes greatly modify the security concerns of nation-states, as national control of both military and civilian technology is being simultaneously eroded. In a sense such a shift takes military and civilian technology systems closer to the fundamental research system of universities and basic research institutes, which have always been international with only limited restrictions imposed by national borders. The force of this change, as indicated above, is a reduced role for the nation-state as funder of R&D and an increasing role for corporate R&D. Naturally, the nation-state looks for new ways of retaining control over 'national' technology through military technology partnerships – the development of certain technologies and systems with friendly allies. The former approach is constrained by the process of civilizing military technology, already mentioned; while the latter approach, exemplified by the European Community and its R&D Framework Programme, is limited by the global character of a large number of big corporations. There can be no doubt that the globalization of big business, greatly facilitated by modern information technologies utilized both inside and outside the corporations, is a major destabilizing factor as seen from the perspective of the nation-state. This situation is in no way limited to the USA and Pacific Asia, but affects the whole global economic system – both the advanced industrialized and the less developed countries.

Strategic alliances, which have received much attention in recent years, are only one element in the long-term and fundamental shift towards a more globalized integrated economy. Equally if not more important are mergers and acquisitions. Furthermore, joint ventures to acquire R&D capacity and independent sourcing of R&D inputs are also key factors. In many if not most countries, there is still a lingering belief among politicians and the public at large – this is particularly true in the big countries – that policy measures can be instituted which retain a competitive advantage within the nation-state. Thus the institution of patenting and other means for protecting intellectual property rights have been strengthened in the USA and elsewhere. However, it is likely that such changes provide greater benefits to corporations than to nation-states.

There can be no doubt that R&D and the capability of controlling technology change are of fundamental importance for the well-being of a nation – including the military security dimension. Thus, the question is constantly posed as to what policy measures in the arena of technology will ensure national well-being. Since technology tends to diffuse between several actors – through the need to share development costs and through the behaviour of global corporations – it appears that a major policy direction must be taken toward the formation of an environment that is conducive to high value-added in the utilization of R&D and technology.

A new technology landscape is emerging with strong global characteristics, which result from corporate mergers and acquisitions, global sourcing of technology inputs, state–industry partnerships across borders, and corporate strategic alliances. The latter have become the focus of national concern in many countries as they are perceived to erode national security – both from a military and an economic perspective. Simplistically, it can be stated that the multinational companies – some of which are becoming global – are strong advocates of techno-globalism, while national and supra-national bureaucracies still appear to be striving for techno-nationalism.

The new partnerships among corporations are strongly influenced by the need to use more efficiently the two important assets of technology and market prowess. First, corporations are today sourcing inputs for science and technology in an international setting in order to achieve complementarity and cost reduction. Second, they are directing their efforts for expansion into large and rapidly emerging markets which are mainly located in Asia. The upper income crusts of China and India and other parts of Asia – outside Japan and the 'four tigers' – offer a combined market that is approaching the size of the EU. Furthermore, these sub-economies also offer a sophisticated science and technology structure, particularly if Russia is included, which potentially could provide great complementary and cost reductions for corporate R&D.

The driving forces behind these consequential changes are fourfold. First, market access has become increasingly costly for those products in which industrialized countries excel but that require large-scale production – which in turn requires constant technological change in order to adjust to changing market conditions, and thus technological sourcing becomes global. Second, the technological basis has become increasingly research-intensive and thereby augments the component of science, which by definition is international. Third, much of the *technology for technological change* has become international as telecommunications and computer networks are no longer national in character. Finally, the pooling of resources and sharing of costs may enable companies to make substantial savings.

Policy responses are sought in many countries. However, not only are the dynamic phenomena within the emerging technology landscape still poorly understood but the national policy-makers also have to respond to political constituencies with widely differing interests. The interests of depressed regions are different from those in regions where high-technology activities are clustered. Similar bifurcation also exists between global companies of a 'national' character which are expanding, and companies which are taken over by foreign global companies, where in the latter instance control of military technology may be compromised. The dynamism of technology changes would indicate that the best policy response is to create an environment where infrastructure and creativity enables 'national' industries to be actively involved in advancing the technological front.

## The Interests of the Industrialized Countries

Support for generic technologies in development aid policies has to take into account the fundamental changes that have taken place in the R&D system. The global character of R&D and the global interlinkage of technologies require inputs from many different sources. It is seldom realistic, either from the point of view of a donor country or from that of a recipient country, to have programmes that are primarily based on bilateral relations.

Research and development requires huge investment in personnel and equipment over long periods. The level of R&D investment has in several sectors surpassed that in plant and equipment. Unless this R&D investment, and in particular the very costly development phase, can be protected, it will cease to take place in market-oriented economies. What is at stake here may be best exemplified by the situation in the pharmaceutical sector where an original discovery of an active substance is heavily protected by patents and brand names. Similarly, the efforts to develop and utilize generic technologies are generally protected by intellectual property rights. Where this is not the case, the extension of such protection is being sought; this is evidenced in the field of genetically modified species, software programs and designs for integrated circuits. Thus, it is not surprising that industrialized countries are seeking to strengthen the IPR system while many developing countries maintain an opposite view. This conflict of IPR rights should also be understood in the light of the structural economic change that is taking place in the mature industrialized countries.

At the same time, it is important for governments to offer tax and financial incentives for the promotion of investment in R&D. Research on basic – as opposed to applied – technologies at universities and state institutes should also be enhanced. And new industrial infrastructures such as information technology should be further developed to help boost the sophistication of the industry in the mature industrialized countries.

Systems of innovation are no longer national in character, with the consequence that the centres of control and influence are increasingly located outside any specific country. Naturally this is even more pronounced for a developing country than for a mature industrialized country. Furthermore, the division of labour among industrialized countries, and in particular between industrialized and developing countries, is undergoing rapid change. This has already led to the migration of labour-intensive manufacture to developing countries, while the more sophisticated manufacture, including R&D, continues to be carried out at home. This situation is well illustrated by the structural change that is currently being implemented in Japan.

One important consequence is that industrialized countries are more eagerly protecting their investment in R&D, with the result that the system of intellectual property rights is gradually being modified – a change that in all likelihood will not benefit the countries at the lower levels of the technological ladder.

## The Disappearance of National Systems of Innovation

The disappearance of national systems of innovation implies a de-nationalizing of science that has three characteristics. One is the inherent transnational character of science, which is growing all the time. The other two are the growth of transnational science and technology, enabled by the replacement of public with private – mainly corporate – funding; and the regionalization of research, as exemplified by the R&D Framework within the European Union. However, the statistical and other evidence needed to analyse the phenomenon is still in short supply.

The present denationalization of science has important precursors in the early study of many aspects of nature. The transnational character has always been evident in field sciences, such as the study of geological structures, including the tectonic plates, flora and fauna, the ocean waters with their contents and behaviour, not to speak of meteorology. Interest in the study of these objects and related phenomena were in the early stages shared by many nations. The field sciences started to occupy what Crawford et al. refer to as the international space for science.[8] This space has been distinguished by its inertia, which means that it continues in its existing state unless that state is altered by an external force. Today we see more and more entrants and activity in the arena of the international space for science.

However, the authors emphasise that the sharing of a common interest among nations was a necessary condition, but not reason enough, for the field sciences to enter into the international space of science. They state that 'they would not have become common objects of research had it not been for socio-economic conditions favouring international organizing activities.'[9] Today, the scientific community is affected by similar changes, where the focus on international science and technology is set by social and economic considerations, mainly in a national context.

In parallel with the invasion of the international space by field sciences, the global community of science was also affected by the express need of national scientists to communicate across national boundaries. First, there was a necessity to 'accomplish the goals of standardization of nomenclatures, methods and units in both the laboratory and field sciences'.[10] Second, given the nature of the organizations needed to handle these tasks, the new scientific establishment occupied more and more of the international space of science. Among the many matters they regulated were the frequency, themes, venues, conditions of participation, and reporting of international meetings.

The attention paid to the standardization of scientific instruments and scientific measurement was also, to a considerable extent, fuelled by industrial and technological incentives that supported the process of internationalization. However, the two world wars and the drawn out Cold War stifled the internationalization of science and created conditions for strong national systems of innovations. During the First World War it became manifestly evident that science could make great contributions to the military power of a nation. This belief was not lessened by the discovery of radar and the development of the atomic bomb during the Second World War. Thus, academic science and state-controlled R&D became increasingly intertwined – within national systems of innovation – and the scientists became integral parts of the national security system. Crawford et al. state that, 'For the overwhelming majority of scientists who worked in defence installations, or on defence contracts in the universities, this meant that open communication with foreign scientists or even with conationals was blocked by security rules and closed sites.'[11]

The totalitarian character of regimes in countries like Italy and Germany, and later on in the Soviet Union and China, led to the mobilization of scientists for national goals, the authors point out. However, the nationalization of science was pursued in most industrialized countries in the West, drawing its strength from the earlier involvement and technological success of science and technology pursued for military purposes. As a consequence many scientists willingly accepted that it was the state's responsibility to set scientific objectives, even when it came to basic research.

The internationalization of science did not, of course, completely get off the ground during this period. The spontaneous process of organization that started before the First World War was reconstituted between the wars and then further expanded after the Second World War. One of its most important manifestations is ICSU – the International Council of Scientific Unions. ICSU today comprises the majority of international scientific unions and a large number of national members as well as many non-governmental professional organizations.

This spontaneous way of organizing international science has been overtaken by the bureaucratic mode of operating international organizations, such as UNESCO and many other UN bodies that are involved in scientific research. One of the most striking features in the internationalization of science since the Second World War has been the emergence of international consortia for funding and operating large and complex instruments, of which the European Centre for Nuclear Research (CERN) may be the most well known. Crawford et al. refer to these as Locally Grounded Transnational Research Sites (LGTRS). Another and partly related phenomenon is the desire on the part of scientists to gain access to certain experimental sites or to meet colleagues with a shared approach and understanding of a scientific problem. This has fostered scientific travel on a global scale, which consumes a large proportion of available research funds. In spite of this, most scientific endeavours retain a national character because of educational and career possibilities; and universities and labour markets for scientists still remain national, or even nationalistic, in many countries.

Furthermore, the state has, in almost all industrialized countries, attempted to mobilize major chunks of scientific and technological resources for its national economic and social development.[12] However, these efforts are increasingly at odds with the globalization of R&D that is pursued by the corporate sector. The expanding transnational links within companies, and between companies, dilute the national character of the national science and technology system. There are few countries where R&D funded from foreign sources exceeds 10 per cent, although in some small industrialized countries like Sweden, more than 25 per cent of corporate industrial R&D is done outside the country. The denationalization of science is also served by the increasing share of corporate R&D being pursued in alliances or consortia, often across national boundaries.

It may be useful to list the reasons why a company should pursue R&D in an international setting. All these factors have grown substantially in importance during the past decades and will continue to do so.

1. To support large production or service activities in other countries.
2. To search for research talent.
3. To obtain research results at lower costs.
4. To obtain scientific intelligence.
5. To speed up product adaptation in local markets.
6. To avoid legal or fiscal constraints.
7. To maintain R&D assets arising from mergers and acquisitions.

The above indicates clearly that there are strong forces propelling both nationalization and denationalization of science, although the latter is apparently gaining the upper hand in this period – following the end of the Cold War and as a result of the globalization of financial markets and production facilities. The denationalization of science is not only an expansion of transnational science; it also means the disappearance of national systems of innovation. This process is triggered by the gradual decline in the share of R&D being funded by the state and the increasing involvement of universities in industrial research that serves foreign companies. Another important element is the increasing role of regional organizations like the European Union in funding R&D.

## Notes

1. The information and views expressed in this chapter are partly drawn from the following documents: 'Technological Change and the New Concept of Security – Non-military Risks in a Geo-economic Environment', prepared for the National Board for Civil Emergency Preparedness (Överstyrelsen för Civil Beredskap – ÖCB); 'Is Sweden's R&D Controlled from Outside? Technology, Innovation and the Economy (TIE) – A New Understanding of National Systems of Innovations (NSI)', prepared for the conference 'Sveriges Innovationssystem', organized by the Swedish Academy of Engineering Sciences (IVA), 21 April 1994; 'Swedish Bilateral Cooperation Aid', prepared for Centro Studi di Politica Internazionale (CeSPI), as part of EC project coordinating the development aid policies of European countries, sponsored by DG VIII.
2. This section is inspired by a recent book, *The New Production of Knowledge*, by Michael Gibbons et al. (London: Sage, 1994), and draws on many of the insights presented by the authors.
3. 'Cognitive' means the action or process of knowing; this can be empirically verifiable or else analytic.
4. International Development Cooperation, *Summary of the Budget Bill 1994/ 95* (Stockholm: The Swedish Ministry of Foreign Affairs, 1994).
5. Ibid.
6. *Rena roller i biståndet – Styrning och arbetsfördelning i en effektiv biståndsförvaltning* (Pure Roles in Assistance – Control and Division of Labour in an Efficient Assistance Administration), SOU 19 (Stockholm: The Swedish Ministry of Foreign Affairs, 1994).
7. *Organisation och arbetsformer inom bilateralt utvecklings bistånd* (Organization and Working Methods in Bilateral Development Assistance), SOU 17 (Stockholm: The Swedish Ministry of Foreign Affairs, 1994).
8. Elizabeth Crawford, Terry Shinn and Sverker Sörlin, 'The Nationalization and Denationalization of the Sciences: An Introductory Essay', in E. Crawford, T. Shinn and S. Sörlin, eds, *Denationalizing Science – The Contexts of International Scientific Practice* (London: Kluwer Academic Publishers, 1993).
9. Ibid., p. 36.
10. Ibid., p. 16.
11. Ibid., p. 21.
12. Paul Krugman, *Peddling Prosperity: Economic Sense and Nonsense in the Age of Diminished Expectations* (New York: W.W. Norton, 1995).

## References

Crawford, Elizabeth, Terry Shinn and Sverker Sörlin, 'The Nationalization and Denationalization of the Sciences: An Introductory Essay', in E. Crawford, T. Shinn and S. Sörlin, eds, *Denationalizing Science – The Contexts of International Scientific Practice* (London: Kluwer Academic Publishers, 1993).
Gibbons, Michael, et al., *The New Production of Knowledge* (London: Sage, 1994).

International Development Cooperation, *Summary of the Budget Bill 1994/95* (Stockholm: The Swedish Ministry of Foreign Affairs, 1994).

Krugman, Paul, *Peddling Prosperity: Economic Sense and Nonsense in the Age of Diminished Expectations* (New York: W.W. Norton, 1995).

Swedish Ministry of Foreign Affairs, *Styrnings- och samarbetsformer i biståndet* (Guiding Methods and Forms of Cooperation in Development Assistance), SOU 1 (Stockholm: Swedish Ministry for Foreign Affairs, 1993).

Swedish Ministry of Foreign Affairs, *Rena roller i biståndet – Styrning och arbetsfördelning i en effektiv biståndsförvaltning* (Pure Roles in Assistance – Control and Division of Labour in an Efficient Assistance Administration), SOU 19 (Stockholm: The Swedish Ministry of Foreign Affairs, 1994).

Swedish Ministry of Foreign Affairs, *Organisation och arbetsformer inom bilateralt utvecklings bistånd* (Organization and Working Methods in Bilateral Development Assistance), SOU 17 (Stockholm: The Swedish Ministry of Foreign Affairs, 1994).

# 15
# Development Aid Policy Options for the North
*Charles Cooper*

There is a plausible argument that recent developments in technology and in the organization of the international economy have made the issue of technology in Third World industrialization more important than at any time since the 1960s. This is the context for discussing the implications for Northern aid policies. It is discussed very briefly in the first section of this chapter. In the second section, there is a discussion of the implications of new technologies for trade, investment and development, based on research ideas at the United Nations University Institute for New Technologies in Maastricht, the Netherlands. This is followed by a third section which draws conclusions for aid policy. The emphasis is on the industrial sector, but the arguments can be generalized.

## 1. The New Importance of Technology in Relation to Development Policy

Two major changes in the international context account for the accentuated importance of technology for developing countries. These are: (1) the acceleration in technological change and the appearance of generic technologies; and (2) the shift to open economy approaches to industrial development. The acceleration in the rates at which new technologies appear and are incorporated into production has been researched empirically and reported in the innovation literature. There is, of course, argument about why the acceleration and the accompanying shift to more science-based lines of production have not had more impact on the productivity data in the industrial countries, but by and large this has not greatly affected the general acceptance of an accelerated rate of change of production technology, accompanied by important changes in organizational techniques – the so-called Japanese management techniques.

As important as the *acceleration* in technological change has been a change in its character, especially the advent of *generic technologies*. These are families of technology which can be applied to industrial processes in many different sectors of the economy, including the so-called traditional sectors, like textiles, garments, leather and food processing. The importance of the generic technologies for developing countries is immediately apparent in the fact that they have resulted in renewed technological change in such traditional sectors, which were once considered technologically stagnant – and therefore particularly well suited to developing countries with scarce supplies of technological skills. We return to this important characteristic of technological change associated with new technologies later.

The changes in the organization of the international economy of the last ten years or so are too well documented to need much description. From the present point of view the key change has been the shift from protected industrialization in developing countries to various forms of open-economy policy in the industrial sectors. This has some critically important implications:

- increasingly entry into industrial production by firms in developing countries implies entry into a world industry ... a much more formidable undertaking than entry into a protected industry serving a domestic market;
- even firms which are primarily seeking to establish themselves in the domestic market are subject to international competition, so that in the absence of special conditions like high transport costs, entry into local markets is tantamount to entry into the world market.

Once firms have achieved entry, they have to survive in sectors where there is a high and increasing rate of technological change internationally. The opening of markets obviously accentuates the implications of acceleration in the rate of technological change for developing-country industry. Moreover, the spread of generic technological change to the once traditional sectors means that problems of maintaining competitiveness are present even in these sectors.

This sets the international economic context in which we need to consider the impact of technological factors on the position of developing (and other) countries in the international economy.

## 2. Current Thinking about Technology in the International Economy

Technological innovation is an intrinsic part of competition in many industrial sectors. The way competition is described in standard microeconomics texts fails to do justice to this. The standard picture presents firms that compete with one another by seeking to produce at the lowest possible costs and prices, whilst all using a technology which is assumed to be equally available to all producers. In reality firms compete in many sectors, as much by seeking technology which is different to that of their competitors as by minimizing costs of production on a given technology. At the extremes we can distinguish this *innovative competition* from *price competition*, though neither exists in a pure form. Dosi, Pavitt and Soete[1] suggest that the sectors of the industrial economy could be set out along a spectrum, with predominantly innovative competition at one end and predominantly price competition at the other. For example, the development of new personal computer technologies would be characterized by innovative competition, whilst most standard textile production would be characterized by price competition. Dosi et al. argue that trade experience in innovative sectors can be best explained by the concepts of innovative competition and that in less innovative sectors, where price competition rules, the more traditional concepts of comparative advantage may apply. This is further discussed below.

Starting out from this distinction, it is possible to make a series of points that will help to delineate the changes which technological factors have brought about in the international economy. The first group of points relate to sectoral differences in the patterns of technological change, whilst the second group concern national differences.

### *New forms of competition; sectoral patterns of technological change:*

- Innovative competition may be associated with product innovations, such as new consumer goods or new capital goods, or with process innovations – that is, new ways of making established products. In either case, the innovator firm attempts to privatize the technology to its unique use. The resulting monopolistic advantages are the commercial objective of innovation, and there would be no innovation in a market economy if firms were not reasonably sure they could achieve them.

- Product innovations by capital goods firms, machine-makers for example, become process innovations for user firms, though in the usual case it is only the product-innovating capital goods firm which can privatize the technology. The purchasing firms simply use the new technology and compete with one another in their established product markets.
- There is a distinction between these two cases. The product-innovating firm, supplying a new capital good – a machine perhaps – is involved in innovative competition with other firms in its sector. The firm that purchases the innovative equipment is usually involved in price competition with its competitors, all of whom will have incentives to buy the new equipment so as to keep their cost levels down to those of the original purchaser. In an industrial economy most sectors can probably be classified as either producing innovations, in which case they are predominantly engaged in innovative forms of competition, or as using innovations, in which case they are mainly involved in price competition in *their* final product markets. Pavitt calls the first type of sector 'user dominated' in regard to the sources of innovation it uses, and the second 'supplier dominated'.
- The importance of these distinctions from the point of view of the international economy is as follows. In sectors characterized by innovative competition – 'user dominated' sectors in Pavitt's terminology – the usual assumptions on which our approaches to international trade have been worked out do not work. It is inappropriate here to regard trade as determined by patterns of comparative advantage; it is in fact determined by the *temporary absolute advantages* of the innovators and the uses to which the they choose to put those advantages. In 'supplier dominated sectors', on the other hand, price competition prevails and the determinants of trade are likely to be those which we normally associate with comparative advantage theory.[2]

## *Summary*

This incorporation of the technological factor into the set of factors that determine advantages and disadvantages in international trade suggests that the different sectors entering trade relations can be classified along the following lines:

- *Product-innovating sectors* where there is innovative competition, and where innovative temporary monopolies are whittled down in time by imitation, often through licensing – essentially under the control of the innovating firms.

- A subgroup of *process-innovating sectors* in which innovating firms use the process innovation in their own production and keep it away from their competitors. This is not much different from the first group described above.
- *Process-innovating sectors* where the innovating firms are 'supplier dominated' – that is, obtain their technology from innovative suppliers. Here price competition reigns in the final product market, and the rules of comparative advantage will determine the pattern of trade (in a rough and ready way).
- *Non-innovative sectors* where there is neither process nor product innovation. This is the category to which the erstwhile 'traditional sectors' belonged. A very important implication of generic technological change is that these sectors are diminishing in number.

## *National patterns of competitive advantage*

It is fairly easy to see how the sectoral differences may map into different national patterns of competitive advantage. Plainly some countries, like the USA, Japan and Germany have firms with innovative leadership in international markets in a number of sectors. Others may have more limited and specialized patterns of technological leadership, and may rely far more on skilful imitation; still others may have firms which imitate without much mastery of technology and remain for that reason dependent on licensors. It would be easy to continue with descriptive categories.

However, more interesting than the classification of countries in this type of technological hierarchy are the dynamic processes through which countries have changed – in other words, have moved from one category to another – in the process of industrialization. The changes and struggles for technological leadership at the top of the system are well known. For present purposes, however, it is more interesting to consider the experiences of countries lower down in the list.

One way of doing this is to categorize the different types of experience that different countries have had in facing the opening of their economies to international trade, distinguishing the different approaches they have taken to technology. I think it is possible to distinguish three groups of countries at least:

- Some countries started industrialization in technologically undemanding sectors, and then, after accumulating a wider range of capabilities, moved up to technologically more and more advanced sectors characterized by increasing intensity of innovative competition. The clearest example today is Korea. Korea moved through a pattern of supplier-dominated technological change, through imitation and on to innovative leadership in some sectors. Japan went through a similar cycle. China, or at least parts of China, may be starting such a pattern. Korea, as is well known, illustrates the potential of this stepwise process: over the 20 years from 1969, exports by volume grew at an annual rate of 15 per cent. *Real wages* grew at 7 per cent per annum, as did real value added per worker. Thus, profit's share in value added was more or less constant. The increase in labour productivity was facilitated by a shift from low to high value-added types of production, characterized by increasing degrees of innovative competition.

- Other countries have entered manufacturing trade successfully, but have not achieved the step up to higher levels of technological competition that Korea has managed. They have kept up with international technological change. Exports have grown but less rapidly and less sustainedly than in Korea. Hong Kong is a case in point. There has been a much less spectacular growth in productivity and also in real wages. Wage pressure on profits' share has been a problem more frequently. There have been periods when real wages in Hong Kong have fallen, probably in response to a slowdown in productivity growth.

- In yet other countries – the large majority of developing countries in all probability – where entry into manufacturing trade has been in sectors or subsectors with a low degree of innovative competition, competitiveness is based on low real wages and relatively low rates of productivity growth are required. Many countries have shifted into a pattern of this kind after adjusting out of the import-substitution policy. Chile seems a particularly clear example. Entry on these terms is evidently much less demanding in terms of technological capability than in the preceding cases, but the economic and social outcomes are less favourable.

This differentiated pattern of entry is not stable. In a world of innovative competition, matters do not stand still for long. There is a tendency for areas of production which were hitherto calm backwaters of steady technology and fairly predictable price competition to be caught up in new rounds of innovative competition. When that happens, success depends on whether existing producers possess the technological capabilities needed to imitate process and product innovations. If they do not, they may be forced out of international markets, or they may hang on by cutting costs through real wage reductions. This pattern seems to be present in a number of low-wage sectors in developing countries. Successful industrialization depends increasingly not only on efficient production at today's technology and relative price patterns but also a capacity to keep up with an often unpredictable pattern of technological change. The success with which countries do this affects importantly the welfare implications of export-oriented industrialization. High rates of technological change permit increases in real wages without adverse implications for profitability and the incentive to invest. Lower rates often imply that the only way to succeed internationally is by forcing the real wage down, and turning the functional distribution of income against labour. The technological factor appears therefore to play a crucial role in determining the developmental outcomes of industrialization, and this is a direct consequence of the way competition works within a more open world economy with high rates of technological change across most sectors.

A very crude summary might be: industrialize without technological change and international competition forces the country into lower wage production and puts continual downwards pressure on the real wage, to maintain the share of profits in value added; industrialization with high rates of technological change at least opens up the option for increasing the real wage without necessarily reducing the share of profits. It seems there is what might be called a productivity knife edge: on one side there is a virtuous world of growth with socially acceptable distribution; on the other growth and competitiveness on which growth depends involves a vicious circle of declining real wages and increasing maldistribution of income between profits and wages.

## 3. Implications for Aid Policy

The increasing role of technological factors in trade and development has implications for countries of the North. These fall into two groups: first, there are two general implications which relate broadly to the position of Northern countries in the Bretton Woods system and in international trade questions; second, there are some more particular points about aid policy proper.

The first general implication relates to the influence of the Bretton Woods institutions on development policy and especially on industrialization policy. The main concern of both the World Bank and the

International Monetary Fund as expressed in structural adjustment policies has been to establish lines of production which are efficient at international prices. This is a natural corrective policy in the light of the extreme inefficiencies and misallocation of resources that were often found in import-substituting regimes. It is, however, a policy which puts the whole weight of resource allocation on static efficiency – on efficiency at today's factor prices and factor productivities. It leaves hardly any room for dynamic factors and especially for technological factors. It is perfectly possible, for example, to set up an export-free zone to make garments which will indeed be efficient in today's terms. But, if no account is taken of the international pattern of technological change as generic control technologies are applied to automate garments production elsewhere in the world, the only way to maintain today's efficiency in the face of tomorrow's higher factor productivities could well be simply to force wages down or to push down the prices of locally supplied inputs. The example is, of course, hypothetical, but it makes the point that by simply leaving out important dynamic factors because of the preoccupation with static efficiency, the World Bank is missing a very important aspect of industrial policy. The Northern countries, with the possible exception of Japan, have been extraordinarily slow and reluctant to open up debate on these issues. And yet it is clear and probably broadly agreed that success in industrial development has in nearly all cases been associated with the exploitation of these dynamic economies.

The second general implication for the North relates to industry and trade policy. A notable aspect of the impact of technology on trade is the rate at which it appears to change the pattern of trade and international production. At the innovative end of the system, new technologies open up new lines of production in Northern countries, whilst somewhat older ones are taken up by countries whose present capabilities are more geared to imitation than to innovation. The important point is that the flexibility of the Northern economies to move to new areas of production and to leave older ones to relatively new entrant developing countries becomes an important determinant of the trade opportunities of developing countries. It is also the case that protectionist pressures in the North are often associated precisely with inflexibilities associated with 'getting stuck' in old lines of production and therefore requiring the keeping out of low-cost imitative producers from the South. There are some obvious lessons here about the interrelation between trade and domestic production, but the past record of European economies suggests we have not learnt them very well.

As far as aid policy proper is concerned, the argument of this chapter leads to two main conclusions: one general, the other particular, though they are also related. The general one is simply that aid policy directed towards the accumulation of technological capabilities in the developing countries is likely to have a particularly high return. This implies considerably more than just training policies; it also relates to the development of the so-called national systems of innovation and technological change, which Freeman, Nelson and others have described. Further, it raises difficult questions about how to encourage and induce technological learning processes within firms; these are plainly central, but it is not really possible to answer them with confidence at present.

The particular implication for aid policy concerns those countries which, by virtue of history or policy or both, are on the wrong side of the productivity knife edge. These are usually amongst the poorest countries. As far as they are concerned, the processes of generic technological change which undermine existing production systems in what were once the technologically stagnant traditional industries are a serious threat. It is a threat that they can only meet by defensive measures, which have negative implications for welfare. These countries are an obvious target group. They demonstrate that the advent of the generic new technologies is both an opportunity, if countries are able to incorporate them, and a very considerable threat if they are not.

The question is of course: what would an aid policy directed to technological matters focus upon? It is a hard question to answer, because there is only a limited amount of agreement about the factors that lead some countries to situate themselves on the virtuous side of the productivity divide, whilst others fall on the wrong side. For one thing there is a major and still unresolved issue around the desirable role of the state (though since this is primarily an ideological debate, it is unlikely to be resolved anyway).

Broadly speaking, a policy in support of technological change in the most vulnerable economies would have to focus on the following main issues:

- Assistance to firms to make use of more innovative plant and equipment. What is in question for the low-productivity economies is obviously not incorporation of leading-edge innovations, but the attainment of some greater degree of dynamism in terms of productivity so as to meet international competition more successfully.
- Assistance in the development of an infrastructure of applied research institutes linked to production, and dependent on the production

sector for an important part of their institutional incomes. This is a dangerous area since there are many R&D systems in the developing countries which are essentially redundant in that they have had virtually no effects on production. However, there are also some successes: the Industrial Research institutes in Taiwan dealing with relatively small-scale producers are a case in point. Such R&D institutes are more likely to be useful if they are subjected to the test of the market; few of them have been in the past.
- Assistance in the development of engineering and management firms which can assist in building up more progressive approaches to production technology. The encouragement of new forms of production organization – even in countries at low levels of industrialization – may be a particularly hopeful experiment.

This is a very incomplete list of possible action. One problem is that a renewed aid policy in technology is bound to face the unsolved problems of the past; for example, it is hard to avoid proposing some system of R&D institutions, but it is also impossible to ignore the fact that many such institutes in developing countries have simply failed to deliver any significant technological advantages to local producers. There are, however, lessons to be learnt from the past. An important one, which is rather obvious in the new open-economy world, is that technology is about advance and survival in a competitive system. It does not follow at all that we can rely on markets to solve the problems we are discussing; but it certainly is the case that we must be concerned about competitive success.

## Notes

1. G. Dosi, K. Pavitt and L. Soete, *The Economics of Technological Change and International Trade* (Hemel Hempstead: Harvester Wheatsheaf, 1989).
2. The Dosi et al. approach to defining a spectrum of sectors assumes implicitly that the price competition end of their spectrum is associated with a lack of technological change. That is not the case in the way we have developed the idea here, where it is argued that price competition is essentially the norm in the supplier-dominated sectors – that is, where there is technological change but based on innovations obtained from suppliers.

*Part V*

Conclusions

# 16

## The Major Issues under Debate

## M.R. Bhagavan

We will begin this concluding section with a discussion of the issues that have arisen in the case of biotechnology, for it exemplifies many of the significant questions that have been raised in the debate over new generic technologies. We will then move on to present a number of issues that relate not only to information and communication technology and new materials technology *per se* but also to the environmental, economic and social contexts in which new generic technologies operate.

### Biotechnology: The Case *par excellence* of Controversial Issues

Compared to the very high expectations originally entertained in the industrialized countries concerning biotechnology, the actual achievements have been rather modest. A re-evaluation is taking place on what is feasible. While the high optimism of the last decade has abated and given way to more sober assessments in the industrialized countries, this does not seem to be true of certain developing countries, which continue to have unrealistic expectations and ambitions about what can be achieved through biotechnology in the short and medium term, given their limited resources. For instance, in pharmaceuticals it takes, on average, more than ten years to develop a new drug at a cost of over US$100 million. A more realistic attitude is thus called for, to take account of a country's actual and potential resources for the foreseeable future. Policy research, combined with the art of technology assessment, will help in assessing what is feasible, taking into account the role of infrastructure and market opportunities.

In several developing countries an awareness of the enormous potential of biotechnology seems to have taken root and led to the giving of very high priority to its indigenous development. These countries are Argentina, Brazil, China, Cuba, India, Mexico and Thailand. They are combining the promotion of indigenous R&D with the acquisition of technology from the USA, Japan and Western Europe. Substantial resources have been committed to creating national R&D programmes and specialist R&D institutions. These efforts, which were begun in the early 1980s, are now beginning to bear fruit. Some have achieved impressive results by international standards.

Nevertheless, an analysis of the experience of these 'pace setting' developing countries reveals a number of shortcomings in planning and performance, which should have a salutary effect on other developing countries preparing to follow suit. The R&D programmes formulated originally were much too broad and ambitious. Some of them were more in tune with prevailing international fashions than with local needs. The resources available were not commensurate with the number and variety of projects launched. The resulting lack of focus in national biotechnological programmes and lack of concentration of resources meant that some projects were doomed to failure. Sufficiently attractive incentives were not put in place in time to counteract the 'brain drain' of some of the best scientific and engineering talent to the industrialized countries.

## Creating and Strengthening National Capacity in Developing Countries

In the majority of developing countries very little has been done so far to create local capacity and capability in biotechnology. Lack of adequate resources cannot be blamed entirely for this. For one thing, substantial areas of biotechnology are relatively inexpensive to promote, being knowledge-intensive rather than equipment-intensive. For another, the importance of modern biology, biochemistry and other associated disciplines that together make up modern biotechnology was clearly recognized by the 1960s by the international scientific community. This awareness was available to developing-country universities, which could have advanced these disciplines in their teaching and research programmes two or three decades ago if they had wanted to.

Developing countries that want to use and produce biotechnology have to discard the tendency of promoting an odd project here and there. Instead, they need to take a 'systems approach' in order to build a base that can effectively import, disseminate and absorb existing 'off-the-shelf' technologies from industrialized countries.

Developing countries cannot hope to enjoy the potentially great benefits of the biotechnological revolution if they do not commit themselves to creating a local scientific base that can, one, absorb imported technology; and, two, conduct research appropriate to local resources and problems. The issue is how to go about doing this in countries that are waking up to this fact rather late in the day, such as for instance in sub-Saharan Africa and Central America. In this context, developing-country governments and the donor community need to address the following major questions:

- What are the optimal ways in which local resources and donor funding can be used for creating and strengthening national R&D capacity? What should be the division of labour between universities and specialist institutions outside the university system in this endeavour? What roles can regional and international research institutions play?
- Is it desirable and practicable for donors to coordinate their support? If so, how should the coordination be organized and what forms should it take?
- A great deal of biotechnology in the industrialized countries has been researched and developed with the help of public resources. Much of it is non-proprietary and is available in the public domain. What international mechanisms need to be set up to ensure that 'public domain technology' is available to developing countries freely and on concessional terms? How is it to be made possible for developing countries to obtain specialist information and advice on available technologies (both public and private) and on the means of acquiring them on the best possible terms?
- What lessons can the 'late starters' learn from the mistakes of the pace-setting 'early starters' mentioned above in identifying priorities and focusing on them, in concentrating resources and skills on a few carefully chosen high-priority areas, on incentives to scientific and engineering personnel, and so on?

## *Economic, social and trade issues*

The application of modern biotechnology in agriculture, industry and medicine will have profound repercussions on the lives of people all over the world, whether or not they are aware of the potential that biotechnology has for promoting or destroying their welfare.

Looking on the bright side first, the technology could lead to dramatic rises in *productivity* in agriculture and industry, which are furthermore environmentally benign and economically sustainable. It could lead to the provision of health services, medical diagnosis and vaccines against infectious and parasitic diseases, at far less expense than is the case now. But whether these technical and economic efficiencies can be translated into a sustainable improvement in the material and social conditions of life for the majority of the people in developing countries depends on putting in place appropriate economic and social policies inside and outside these countries. Very few national governments and international agencies have shown signs of urgency in confronting the question of the differential impact of biotechnology on the populations of developing countries – that is, which sections in society will benefit and which will suffer, which will gain and which will lose.

The dark side of the picture is currently more in evidence than the bright side, however. For instance, if the export crops of developing countries were to be displaced by biotechnologically engineered industrial substitutes in industrialized countries, it would spell catastrophe for those developing countries that have not prepared themselves economically, socially and politically to meet this contingency. A case in point is the ruin brought on the peasantry in the Philippines and the Caribbean by the steep drops in their exports of sugar, and in Madagascar of vanilla.

The lack of solidarity between developing countries is another deeply worrying feature. Two examples will illustrate this: (1) China developed a very promising rice hybrid technology. Instead of making this freely available to other developing countries, it sold the rights to a private company in the United States, which now exercises monopoly control over this technology. (2) India was once the world's leading producer and exporter of cardamom. But Guatemala overtook it at one stage. India is now poised to recapture the market through developing its own tissue culture for cardamom. The prevailing tragic competition between poor countries for each other's markets is bound to be exacerbated by the applications of biotechnology. How to redress this situation is an issue that needs to be confronted by social scientists and policy-makers in developing and industrialized countries.

Biotechnology opens up new avenues for the extension of control by agri-business transnationals over developing-country agriculture and agri-processing industry. Horticulture in Thailand and oil palms in Malaysia are examples of this.

The productivity of agriculture in industrialized countries is very high compared to that in developing countries. This productivity gap is likely to become even wider with the application of biotechnology to agriculture in industrialized countries. It raises the prospect of developing countries with growing populations becoming net importers of technology-intensive food from North America, Western Europe and East Asia.

The economic and social implications of the impact of industrialized-country agriculture and agri-business on developing countries in the biotechnological era need to be urgently investigated.

## The growing privatization of public research

Universities in industrialized countries have hitherto played the leading role in basic research in biotechnology. This work is funded in part by public money; however, the links between universities and the private sector are becoming stronger. University-based researchers in industrialized countries are setting up their own small companies, usually in the industrial parks adjoining universities, often with the assistance of large companies. They too have joined the race to acquire patents. The United States provides some dramatic examples of this tendency: roughly half the number of all biotechnology companies in the USA have links with universities, and one-quarter of university research is funded by private firms. Over 800 US scientists, comprising approximately one-third of the biomedical-sciences membership of the US National Academy of Sciences, have 'dual affiliation' — that is, they are linked to both universities and private companies.

Developing countries have expressed fears that the growing privatisation of public research may lead to restrictions on the free and open dissemination of research results. They feel that market considerations, rather than broader social and economic goals, tend to gain the upper hand in setting R&D agendas in the universities. In such a climate, they claim, research problems which are primarily of relevance to developing countries evoke little interest. However, the scientific community in industrialized countries feels that these dangers have been exaggerated, and that this issue needs to be 'demythologized'. It is claimed that research findings from universities and other publicly funded research in industrialized countries are openly available.

## The role of the private sector and the transnational corporations

About two decades ago, a fairly large number of small firms in the USA and Western Europe were in the forefront of biotechnology. Typically, a few scientists would pool their personal and borrowed capital to start R&D and consultancy work. Those that succeeded in their venture and expanded their activities went 'public' and offered shares. Transnational corporations (TNCs) were quick to perceive the importance of the work being done by the small firms. They took up the shares being offered as a 'risk minimizing' strategy of acquiring a stake in the R&D, without however upsetting the dynamism that comes from 'smallness' and the personal commitment of owner-scientists. However, this insight did not inhibit the TNCs from entirely taking over highly successful firms when it better suited their purposes. The cumulative result of this process is that today between 20 and 30 TNCs dominate the biotechnology scene in the fields of agriculture, pharmaceuticals, chemicals, bioenergy, food-processing and beverages. The huge costs involved in product development have meant that biotechnology could not remain a 'cottage industry', belying the initial trends where small companies played the major role.

The number of products of practical importance that biotechnology has led to is quite limited. One of the main reasons for this is the fact that while invention (the 'R' in 'R&D') is relatively inexpensive, product development (the 'D' in 'R&D' ) is much more expensive, requiring long-term capital of substantial magnitude. Further, the number of highly qualified scientists required at the product development stage tends to be several factors higher than at the invention stage. (The development of insulin illustrates this.) These cost and personnel factors also explain why TNCs dominate the field. Another reason is the non-survivability of genetically transformed organisms in competition with naturally occurring organisms.

The strategy employed by the TNCs has several dimensions. They want to assure themselves of maximum access to research results through linking-up with small firms and universities. They are lobbying governments in industrialized and developing countries for easier access to testing and marketing of their biotechnology products, and for co-operation in ensuring compliance with intellectual property rights, while at the same time dragging their feet on equitable conditions for technology transfer. They are unwilling to assume responsibility for environmental and health hazards. They do not want to be drawn into the question of accountability on critical issues like biosafety and bioethics.

Given the predominance of the private sector, in particular the TNCs, in biotechnology, one has to address the question of the arrangements that would permit developing countries to benefit from the technology owned and controlled by the TNCs.

Not all biotechnology need be prohibitively expensive. Much can be achieved by modest means, depending on the field of application and the type of problem tackled. Nitrogen fixation is an example.

## Conservation of biodiversity

One of the lessons learnt from the 'Green Revolution' is that a few high-yielding varieties of cereal crops bred 'in the laboratory' can rapidly displace hundreds of naturally occurring but lower-yielding indigenous varieties from farmers' fields, and thus cause them to disappear from the local biospheres as well. It is feared that a similar, though of several orders of magnitude larger, man-made loss in biodiversity in the plant and animal kingdoms is likely to result from the 'Bio-Revolution' brought about by modern biotechnology. (Other factors that are also contributing to the alarming acceleration in the global erosion of biodiversity and genetic resources are deforestation, soil erosion, climate change and ozone depletion.)

Establishment forces in developing countries and private-sector interests in industrialized countries are, in general, in favour of pressing ahead with the adoption of the techniques of the 'Bio-Revolution', citing the anticipated dramatic increases in productivity as being vital. But another body of opinion in both developing and industrialized countries is insisting that diffusion of biotechnology be made contingent on effective measures to conserve biodiversity.

The one solid response so far towards conserving plant genetic resources has been the creation of national and international gene banks. But gene banks, while recognized as being necessary, are seen as not being sufficient to halt and reverse the global trends in genetic erosion. The Keystone International Dialogue Series on Plant Genetic Resources, for instance, has identified important gaps in the present system of conservation and proposes further action along the following lines:

- *ex situ* conservation, including collecting, storage and regeneration, documentation and information systems, germplasm evaluation and enhancement, and exchange;
- on-farm community conservation and utilization;
- *in situ* conservation;
- monitoring and early warning of genetic erosion in specific locations;
- development of techniques for sustainable advances in agricultural productivity;
- research, training, and public education.

'In all these areas,' says the Keystone Dialogue further, 'there is a necessity to enhance the linkages between the formal and the informal sectors at the community, national, regional and global levels.'

One approach to the conservation of biodiversity, which goes beyond crop genes to include animals, insects, microorganisms, and so on, is to make certain demarcated natural habitats out of bounds for human exploitation by turning them into national parks or 'exclusion zones'. The United Nations Environment Programme and the International Union for the Conservation of Nature and Natural Resources are advocates of such a solution.

## Access to genetic resources versus access to biotechnology

A large number of developing countries are extremely well-endowed in genetic resources, whether it be in the realm of plants, animals, insects or microorganisms. These vast untapped reserves have tremendous potential for improving domesticated crops and animals; for biological control over plant diseases, weeds and insects; and for the production of new medicines and industrial bio-catalysts. To give but two examples: one-quarter of the genes that go into improved wheat varieties in the USA were obtained free from Mexico, and one-quarter of the world's pharmaceutical products are derived from tropical forests. It is therefore not surprising that industrialized countries are keen on maintaining free and easy access to developing countries' genetic resources, while developing countries are equally keen on obtaining fair compensation for their genetic resources.

The present controversy about plant genetic resources is but a forerunner of what is to come with respect to other genetic resources. Plant breeders and seed companies in industrialized countries need germplasm from developing countries to be able to improve their crop varieties. They argue for continued free access to developing countries' plant genetic resources, on the grounds that they constitute mankind's 'common heritage', while at the same time claiming plant breeders' property rights on their improved varieties. Developing countries have countered this by claiming farmers' rights over the genes that they have developed over centuries and insisting that the improved varieties based on these genes should be available to them freely. Debates, discussions and consultations on breeders' rights, farmers' rights and other related matters have been held in many international fora.

The issue has not yet been resolved, but some constructive steps have been taken. For instance, the FAO's Intergovernmental Commission on Plant Genetic Resources has proposed an International Undertaking which defines breeders' and farmers' rights and proposes the institution of an International Fund. The International Undertaking recognizes the rights of the biotechnology innovators, while simultaneously proposing that the innovators and the users of the biotechnology pay into the International Fund, which will be used for paying compensation to the developing countries for the use of their germplasm on which the technology is based. While many developing countries have joined the Intergovernmental Commission and are backing the International Undertaking, the major plant-breeding countries in the West have neither joined the Commission as full members (opting instead to remain as observers for the present) nor adhered to the Undertaking.

In general, developing countries want to create global institutional arrangements that will link access to biological resources to the equitable transfer of biotechnology. Industrialized countries are resisting such a linkage. Such were the positions taken, for instance, at an international meeting on biotechnology organized by UNEP in Nairobi in 1990. A global convention on the conservation and sustainable use of biodiversity was passed at the 1992 United Nations Conference on Environment and Development (UNCED) and has been ratified by many countries.

## Ownership and intellectual property rights (IPR)

In both the public and private sectors of some industrialized countries, there are powerful forces who are advocating the conferring of ownership and intellectual property rights with respect to biotechnological innovations. One of the arguments being advanced is that legal protection through patents and the like is necessary and legitimate for recouping the R&D investments in innovation. It is also claimed that patent rights promote innovation by safeguarding the economic returns of innovators. The USA has gone furthest along this road, by granting patent rights on thousands of innovations involving living organisms. It wants universal adoption of biotechnology patent laws and pushed very hard for it in the Uruguay Round of GATT negotiations. The EC, on the other hand, is not prepared to back the US position. It is limiting itself to protecting plant breeders' rights. Even among Western European countries there are considerable differences in existing laws and practices in the area of patents in general, with harmonization proceeding only slowly.

Developing countries are, in general, opposed to extending ownership and intellectual property rights to the realm of biotechnology. They maintain that, in so far as the innovations are based on the biological raw material of developing countries, exclusive legal and economic rights are inadmissible. Further, they fear that patents will extend to biotechnology the currently inequitable conditions operating in the transfer of other technologies.

The Keystone Dialogue Group has expressed 'strong concern about the imposition of IPR for plant genetic materials through GATT or bilateral trade negotiations'. It feels 'that every country has the right to decide whether and to what extent' to adopt IPR. It recommends that 'developing countries choosing to implement Plant Breeders' Rights should retain provisions allowing Farmer Plantback of protected varieties. This is especially important in developing countries where farmers cannot afford to buy seed every year or are not consistently reached by seed distribution infrastructure and must therefore rely on seed saved from the previous season.'

Non-establishment organizations in industrialized countries also reject the call for introducing proprietary rights into the realm of living organisms, both because of ethical considerations and because of the inequities they will bring in their wake.

## Biosafety and regulation

With few exceptions, developing countries have yet to formulate and introduce regulations pertaining to biosafety within their borders. This weakness is compounded by their inadequacy in a range of capabilities (for example, scientific, technological, juridical, organizational, administrative) required to enforce the regulations and to monitor whether they are being adhered to. Meanwhile, the pressure on developing-country governments to introduce regulations is mounting from different quarters: from private firms in industrialized countries which want to test and market their products and processes; from local and international non-governmental organizations concerned with environmental protection; from international research institutions located in developing countries which are keen to test their products and processes, sometimes in collaboration with domestic and/or foreign firms; and so on. Developing-country governments are obliged, to a certain degree, to take account of the expectations and the agenda of these lobbies, while at the same time not diluting their own interests too much, in any system of regulations they intend setting up. This is a formidable task.

Somewhat different considerations apply to donors. There is growing public opinion in many donor countries that biosafety regulations in developing countries should be as stringent as those in Western Europe and North America. It is argued that the testing and marketing of products and processes, as well as the transfer of technology, should be strictly contingent upon internationally agreed biosafety regulations being adopted. Such a conditionality would also have the merit of reducing the possibility that one industrialized country would 'steal a march' over another in unfairly penetrating a developing country market by being less demanding on biosafety.

Biosafety activists in industrialized as well as developing countries are greatly concerned about the high risk involved in releasing genetically engineered organisms into the environment. It would be very difficult to detect and to stop in time the onset of catastrophes. Systematic assessment of risks should be obligatory before approval is granted for tests and production facilities for biotechnology.

People concerned with biosafety point out that in the short and medium term most developing countries will not have acquired the capability to enforce biosafety regulations. They argue, therefore, that the onus is on industrialized-country governments to ensure that biotechnology exports to developing countries fully comply with the biosafety standards set within the industrialized countries. This standpoint, however, elicits some criticism by some developing-country policymakers and policy analysts. They fear that if the industrialized countries' stringent norms prevail, the diffusion of biotechnology to developing countries would be greatly hampered. They are therefore calling for a more lenient approach.

## Bio-ethics

Genetic engineering raises the fundamental issue of whether and how far human beings should intervene in 'the genetic order that nature has evolved', in order to transmute living organisms for their own purposes, in particular in animals and humans. While different religions, cultures, socioeconomic ideologies and nations will respond to this question in their own specific ways, the universal character of the science of genetic engineering, and the ease with which it can be transferred and acquired, makes it imperative to establish certain global conventions and regulations.

The donors, *as a community*, would have to arrive at a common understanding of the bio-ethical issues involved and agree on a common response. In deciding on which R&D they will support, they will have to address the ethical questions that arise, for example, in the following areas:

- growth hormones;
- contraception vaccines;
- *in vitro* fertilization of animal and human eggs, and associated experiments, including implanting the fertilized egg in the womb of the original or surrogate parent;
- transgenic animals;
- gene therapy – that is, treatment of diseases and genetic disorders by introducing compensatory genes into the patient's body cells;
- making of artificial organs (organoids) for implantation into a patient's body, using the patient's own cells which have been genetically modified outside the body to make the required product;
- the Human Genome Organization (HUGO) project: the ownership and patenting of the human gene sequencings discovered; and the use of gene sequencing to set up genetic profiles of individuals indicating proclivity to certain physical diseases and disorders.
- possibility of misusing apparently non-military R&D for the development of biological weapons.

The issues of biosafety and bioethics are as relevant and important to developing countries as to industrialized countries. The shortcomings in safety practices in the laboratories of some developing countries are a cause for great concern. The growing use of genetic engineering in certain medical fields raises bioethical questions. They need to be addressed with the same urgency in the developing countries as they are being in the industrialized world.

Some of the fear surrounding biosafety and bioethics is due to both the inadequacy of the information available and the credibility of the sources of information. It is also a question of power: information emanating from sources with certain vested interests tends to be treated with scepticism.

## Sustainability and the 'Cleaner' Technology Trajectory

The models of economic growth and accumulation of wealth that have dominated the industrialized world so far have taken it for granted that material and fossil fuel resources are, for all practical purposes, inexhaustible. Further, these models are underpinned by the conviction that the physical and the living environments have almost infinite capacity for 'absorbing' the wastes generated by the current practices of production, distribution and consumption in the North. These models and convictions have been embraced by the elites in the developing world and implanted in the South. But it is now abundantly clear that these roads to economic growth and accumulation have to be abandoned, if the planet is not to be overwhelmed by irreversible catastrophes. Future human activity must tread environmentally sustainable paths, which are also inseparable from economic and social sustainability on a global basis.

Environmental sustainability will require the development and deployment of 'cleaner' technologies. Here the new generic technologies can cut both ways. Both IT and NMT have in various ways contributed to higher energy efficiency and greater energy savings, with attendant reduction in pollution. Through the use of appropriate microorganisms, BT has made it possible to remedy and rehabilitate badly polluted soils and water bodies, without in turn producing hazardous waste. With its ability to deliver greater strength and durability of stuctures for less material input, NMT can slow down resource depletion. On the other hand, the production technologies that generate NMT and the hardware of IT can themselves be significant polluters. There exists the risk that BT can introduce dangerous and uncontrollable microorganisms, both naturally occuring and genetically modified, into the physical environment and the food chain.

This duality uncovers the urgent need to arrive at strategies for shifting generic technologies on to the 'cleaner' technology trajectory and to make them stay there. IT, BT and NMT have to a significant degree become 'solutions in search of problems' – that is, in search of ever newer areas of application and newer markets. In sharp contrast, the issue of 'cleaner' technologies is one of 'problems in search of solutions'.

Environmental problems are both very diverse and often highly location specific. But this need not mean that 'cleaner' technologies would share the same degree of diversity and non-generality. There are already 'clusters' of cleaner technologies, sharing common traits, such as pollution monitoring equipment and control systems, energy-efficient hardware and energy saving processes, with IT and generic new management methods integral to all these. However, on present trends, 'cleaner' technologies are unlikely to become entirely generic, developing explosively out of a few key scientific discoveries and technological innovations. They are more likely to retain a certain degree of nongenericness, not least because of the variability and fluidity of the social and economic parameters that cause and shape environmental problems.

Even the cleanest and most recyclable of technologies leaves behind some irreducible waste, some of it toxic. Therefore the strategy for sustainable development should not only aim at the cleanest possible technology but also at reducing non-essential luxury consumption. The latter is a part of the growing consciousness of the need to alter 'consumerist' lifestyles. The promotion of environmental rehabilitation, on the one hand, and of less consumerism, on the other, can only be achieved through cooperative strategies involving governments, companies, producers and the public at large.

## Free Trade versus Intellectual Property Rights

In close collaboration with their private sectors, the governments of most of the economically strong countries have formulated and implemented technology policies that protect and promote their own industries. This is an old tradition that continues apace. Meanwhile, the technology content of foreign trade has grown rapidly, blurring the boundaries between trade and technology issues. In parallel, under the pressure of the leading powers of the North, the 'free trade' concept has been modified to make room for the protection of intellectual property rights, which covers, among other things, the knowledge-content of new technologies. In the face of this reality, what policies should developing countries adopt to reap some benefit from the new technological revolution?

### *Investment patterns, technological behaviour and industrial reorganization*

As pointed out in the Introduction to this volume, countries need to have reached a certain threshold level in economic, industrial, scientific and technological infrastructure before they can effectively absorb imported new technology and use it to their own advantage. In other words, the technological challenge cannot be addressed without simultaneously confronting a host of issues in other sectors of the economy and society. Nevertheless, we can identify some technological policies that need to be in place. There ought to be more public and private investment in science and technology areas in higher education, R&D and management. Given the long gestation periods involved, investment must have a consistent character and be sustained over the long term. Apart from generating sufficient numbers of people with the necessary expertise and skills, the investment must be geared to ensuring consistently high quality in training and output.

The deployment of new technologies should in the first instance be geared towards the domestic and near-abroad markets, these being easier to penetrate than world markets exposed to fierce competition. Since the use of technology is almost always at the enterprise level, governments and corporate sector institutions need to go beyond the macro-policy level to promote the availability and acceptability of the technology at the company level. Similar approaches ought to prevail in scientific, technological and research cooperation between the North and the South that bilateral and multilateral donor agencies support.

New material technologies have added to the variety of possible transitions to less energy- and resource-intensive production sytems. They are also part of a transition from mass production to customized batch production, where IT plays a key role. NMT is a prime example of a combination of different types of innovations: radical and incremental, technical and organizational. Its development has induced substantial changes in investment patterns, technological behaviour, industrial organization, institutional structures and competitive strengths. NMT has made it possible for the more advanced developing countries to develop niche areas and break into world markets. Such strategies are not immune from the pressure of severe short-term problems, which can however be overcome, given long-term and firm policy commitment by the public and private sectors.

Over the last few decades, industrial corporations based in the North have been moving away from the national character of their manufacturing and marketing networks to transnational ones. Industries like automobiles, electronics, pharmaceuticals and heavy electrical equipment, to name but a few, are becoming increasingly 'borderless'. The same process is now underway in technological R&D, and in the organization and control of technological resources. Parts of the strong South (for example East Asia) are displaying the same trends.

## 'Borderless' R&D

To the degree that countries of the South get drawn into the net of the globalized economy, they cannot avoid responding to the promise and the threat of 'borderless' R&D. Too deep an integration may critically weaken their fledgling national research efforts, while closing the door unselectively would mean forgoing access to key technological advantages. The South has to craft suitable policies and strategies for finding the appropriate entry points into the transnational R&D networks.

Systems of innovation in industrialized countries are no longer predominantly national in character. Precisely like production and marketing, R&D is also being internationalized, with industries relying on networks of researchers and laboratories. One can see a marked difference in the approaches adopted by national governments and transnational corporations (TNCs). In collaboration with the corporate sector, the governments of the leading economies of the OECD region have tried to foster breakthroughs in high-tech R&D either within the national context or in regional groupings, such as the European Union. These attempts have been only modestly successful. TNCs have taken a different route. On the one hand, they are translocating parts of their R&D effort from their home base to other countries, including a few of the technologically more advanced developing countries. On the other hand, they are creating internationally linked project teams, involving universities and industries, to tackle specific problems. In high-tech, the TNC approach has delivered more impressive results than the national one.

The 'technological landscape' in the advanced industrialized countries has changed considerably over the last decade. With decreasing expenditure on the military at the national level, R&D alliances are being forged among the military across national boundaries to use the reduced resources more effectively. Military R&D is becoming less dominant in achieving advances in high-tech, with civilian R&D now taking the lead in many areas.

Through takeovers and mergers at a global level, TNCs are acquiring control over civilian R&D establishments in countries other than their home bases, for their own corporate interest. In contrast, governments are insistent that their contribution to nationally based R&D should be linked to the advance of the national economy and the general welfare of the country.

How these changes, and the contradictions between them, will impact on developing countries' technological situation needs to be examined. For instance, global competition in products and services based on new generic technologies calls for economies of scale achievable through huge production volumes, which in turn creates pressures on TNCs to invest heavily on continuous innovation at the technological forefront. To reduce the costs of R&D, as well as to guard their shares of national markets, TNCs will want to pool the financial, scientific, technological and managerial resources across national boundaries. This would certainly run up against developing countries' wish to promote a nationally oriented technological development.

Transnational corporations and banks account for most of the current investment and trade in the world, the bulk of it confined to the OECD region. The small fraction that finds its way into the developing world is concentrated in a few rapidly growing economies in Asia and Latin America. The ideology behind this globalizing economy is still 'business as usual' for maximizing growth and profits, irrespective of almost everything else. However, at the rhetorical level, the TNCs have slowly begun to concede the non-viability of this dogma, as they are confronted by the catastrophic environmental, economic and social stresses that 'business as usual' has brought about in both the industrialized and the developing worlds.

Deliberate, strong and sustained intervention is required to move the TNCs beyond rhetoric to action. New generic technologies can be harnessed to this task, for their innate characteristics enable them to be shifted on to the trajectory of sustainable technology. Given the present huge asymmetries in the world in economic and political power, market forces will not by themselves bring about such a shift. This calls for innovative and effective worldwide coalitions between a range of actors and institutions, both public and private, from both the state and civil society.

## Price Competition and Innovative Competition

In a world where the impact of rapid technological change can no longer be halted at national frontiers, countries have to resort to either price competition or innovative competition, or a mix of the two, to hold their own in both domestic and foreign markets. Industrial branches which hitherto felt safe in the calm waters of technological and price stability can suddenly find themselves facing the stormy winds of change. One way of staying afloat would be to acquire the capability to assimilate the new technology and thus raise productivity. If linked to higher real wages for an increasingly skilled workforce, this strategy could ensure long-term growth in market share and profitability. If, on the other hand, companies were to try and maintain price competition, principally through cutting real wages rather than technological innovation, they may stave off the evil day for a while. But they are soon likely to find themselves caught in a vicious downward spiral of decreasing real wages, attendant loss of motivation and morale by the skilled workforce, and lower productivity and quality – a sure recipe for terminal decline.

## Static versus Dynamic Efficiency

Within the framework of the structural adjustment policies that the World Bank and the IMF have imposed on developing countries, the emphasis is on resource allocation and production which are efficient at current international prices and open markets. This is a static strategy, which takes little account of what happens when generic technology-driven efficiency elsewhere successfully undercuts current price efficiency in a given country, leading to crises in specific industrial branches in that country. This has happened time and again, not only in developing countries but in industrialized ones as well. There is a strange lack of urgency in the Bretton Woods institutions and in most governments of the leading economies of the North in addressing the challenge of how to devise policies and strategies that benefit both the North and the South which take proper account of dynamic technology-driven efficiencies.

One essential element of such an approach is how the North should balance its entry into new products and processes made possible by generic technologies by its facilitating exit from older lines of production in favour of developing countries. This calls for radical rethinking of and action on current trade policies of the North, which are loaded against the South, in particular through protectionist measures.

The aid policies of individual governments in the North are, among other things, aimed at promoting economic growth characterized by increasing efficiency, productivity and competitiveness, in combination with more equitable income distribution. As argued above, for such an outcome to materialize, developing countries need assistance to absorb effectively and use the global technological changes brought about by new generic technologies. Here, as elsewhere, differential aid policies are called for, depending upon whether it is the weak or the strong South that is targeted.

## The Indispensability of the Entrepreneurial Class

No technology can establish itself for wider use unless there are entrepreneurs willing to take on the risk involved in promoting it. If the technology is mature from a commercial point of view, and if there is an obvious unfulfilled demand for it, entrepreneurs are usually willing to take the jump. But a range of generic technologies that ought, in principle, to be within the reach of small- and medium-scale enterprises, have not yet reached this stage. Their commerical maturity has to be tested locally, the resulting adaptations to suit local conditions have to be made, and local expertise has to be generated through training, before they become sufficiently attractive to local investors in developing countries. Official development assistance, which has so far shown little interest in facilitating this vital process, ought to be rethinking its priorities and strategies in this era of new generic technologies.

## Capacity Strengthening in the 'Excluded' South

In the 'weak and excluded' South, the primary task is to rehabilitate, maintain and strengthen existing scientific and technological capacity, and to build new capacity. These efforts have to be directed at various levels in society to be able to strike root and grow of their own accord in the future; this is unlikely to happen if capacity strengthening is restricted to the top of the educational pyramid and to so-called 'centres of excellence'.

Accumulation of technological capabilities is likely to yield high returns. This argues in favour of developing the so-called national systems of innovation and technological change. However, one should keep in mind the fact that national systems of innovation are being slowly eroded by a global system of innovation, in which the interests of the large transnational corporate sectors take precedence over national interest. But the TNC sector is not monolithic. Its members compete severely with each other in the global arena. In such a climate, they are likely to be open to proposals from both the strong and the weak South on addressing technological problems whose solution would benefit both partners. Handling such negotiations requires sophisticated local capacity in technology assessment and policy analysis.

Both in the North and the South, and between the North and the South, networking between individual professionals as well as institutions has emerged as a resource-efficient and relatively rapid way of strengthening capacity and solving problems. The networking process has greatly expanded and accelerated due to the advances in information technology. To the 'excluded' South, which is hamstrung by lack of access to capital for investment in S&T infrastructure, the linking of institutions of higher learning and research in neighbouring countries with one another and with advanced institutions in the wider world should prove particularly attractive, while providing breathing space to carry on with the task of building local infrastructure.

A necessary, but by no means sufficient, condition for a developing country to set forth on the road to the new technological era is to acquire and strengthen its national capacity to manage the choice, import and absorption of highly modern technologies. This capacity to manage significant technological change is made up of a number of components:

- good-quality higher education and specialist training in selected areas of science, engineering, economics and management;
- a national community of leading practitioners from the natural, engineering and social sciences;
- a broadly constituted independent policy community drawn from the private, public and the non-governmental sectors.

The capacity that has been sketched above is, of course, an 'ideal' picture, far from the actually existing situation in most of the South. Even the capacity that has been built up in the strong South is full of gaps in terms of the above definition. Nevertheless, we need to be conscious of the indispensability of creating this kind of capacity, however slowly and imperfectly, if developing countries are to be the beneficiaries rather than the victims of technological advance.

The governments of the South and the donor community need to address the following questions:

- What are the optimal ways in which local resources and donor funding can be used for creating and strengthening the kind of capacity in the 'excluded' South that would form the base for technological advance? What should be the division of labour between universities and specialist institutions outside the university system in this endeavour? What roles can regional and international research institutions play in this?
- Is it desirable and practicable for donors to coordinate their support?
- A good deal of the advanced technology in the North has been researched and developed with the help of public resources. Substantial parts of it are non-proprietary and ought to be available in the public domain. What international mechanisms need to be set up to ensure that 'public domain technology' is available to the 'weak' South either freely or on concessional terms? How is it to be made possible for the 'excluded' South to obtain specialist information and advice on available technologies (both public and private) and on the means of acquiring them on the best possible terms?
- What lessons can the 'late starters' in the South learn from the mistakes of the pioneering efforts of the 'early starters' in the South?

## Institutional Reform and Institutional Innovation

Several approaches have been tried in the developing world in creating and empowering institutions to help orchestrate industrial and technological advance. If one looks at the handful of countries that have clearly outdistanced the others in the developing world in terms of industry and technology, one finds two paradigms.

In one, the state intervened in the economy at almost all levels, striving to exercise detailed control even on minor activities. China, India and Mexico are prime examples. It is now acknowledged that this paradigm, while successful in creating a strong and integrated base in modern industry and infrastructure, was not reformed in time in the direction of a less statist and more free-enterprise approach, resulting in colossal waste of resources (natural, human and financial) and crippling underperformance.

In the other paradigm, although the state intervened strongly, it did so very selectively, and mainly at the macro-policy level, to steer the development process in the direction of high productivity and good quality, aimed at both the domestic mass market and export markets. It instituted a system of rewards and penalties to boost good performance and penalize underperformance. South Korea, Taiwan and Singapore travelled along this 'East Asian Route', achieving spectacular economic growth and relatively more equitable income distribution.

Both paradigms totally disregarded the impact on the environment, allowing extensive environmental damage to take place. In thinking about future institutions that would be appropriate for meeting the new technology challenge in an environmentally sustainable way, developing countries may want to appraise critically both paradigms to extract what could still prove useful in the present rapidly changing circumstances. They may also want to look at the institutional and policy models currently under debate in the Netherlands and Scandinavia, where efforts have begun to steer economy and society in the direction of environmental sustainability. A key feature of both the East Asian and the Northern European models is the active cooperation of the public and private sectors in setting a mutually acceptable policy framework.

The institutional reform and innovation that would have to take place should be geared to the following tasks:

- division of responsibilities between, and coordination of, publicly and privately funded R&D;
- incentives for, and mobilization of, domestic and foreign investment in new technologies;

- advancement of skills and professions appropriate to new technologies through formal education and training, as well as in-house training for employees at enterprise level;
- incorporating and enforcing international standards for products, processes and services:
- design, testing and quality control services.

## Policy Approaches to Development Assistance

Traditionally, development aid agencies have dedicated the bulk of their support to projects in agriculture, health, and primary and secondary education. Such support, while essential, will not by itself initiate the process of technological advance.

In the past and at present, some assistance has been made available for project aid (including technical assistance) in industry and infrastructure, from both bilateral and multilateral aid agencies. This support gives the impression of being somewhat haphazard and rather lacking in consistency of direction. This probably stems from having to reconcile the wishes of the recipient governments with the priorities of the donors, including, in the case of some bilateral assistance, the rider that procurement of equipment and technical and managerial know-how must take place through donor-country firms. Further, the volumes of industrial and infrastructural aid will never be more than a small fraction of the huge magnitudes of capital and skills needed for contemporary forms of industrialization.

Under these circumstances, aid can at best only catalyse the 'right kind of development' in infrastructure, industry and technology. We use the term 'catalyse' advisedly, because the large magnitudes of capital and skills that are required for industrialisation have to be mobilized by the developing countries themselves. What forms should this catalytic support take and in what areas should it be placed? In our opinion, the following will be of strategic significance.

The first in order of importance would be to initiate and strengthen the processes of institutional reform and institutional innovation that would be appropriate to the tasks ahead.

National capacity has to be in place to absorb, use, maintain and manage the environmentally benign advanced technologies that will have to be imported from the OECD region. This calls for an improvement in the relevance and quality of vocational education, higher education and specialist training in science, engineering, economics and management.

In addition to the strengthening of S&T infrastructure, networking should be encouraged as a means of more efficiently using existing resources.

An analysis of the experience of the 'strong' South tells us that the ability to absorb imported technology depends critically on the indigenous capacity to 'unpackage' the technology so as to be able to understand the know-how and the know-why, and thus to be able to use the technology efficiently. This is usually done at the firm level.

Direct assistance to individual firms would generate very little in the way of multiplier effects, as seen from the overall national point of view, besides being very expensive in terms of unit costs. A more resource-efficient form of assistance, which would have long-standing and widespread effect, would be the funding of specialist technology institutes that would train key technical personnel sponsored by firms in the techniques of absorbing major types of advanced technologies.

Such technology institutes should also be funded (i) for providing training in technology assessment, choice and management, environmental impact assessment and management, and cost-effectiveness management, and (ii) for setting up databases that would help mount international searches for the relevant state-of-the art technology, and to identify technology suppliers who would be globally competitive in price, quality and servicing.

Independent indigenous technology policy analysis is crucial for generating country-specific priorities, strategies and scenarios for technological advance. Of equal importance is social-science research into the economic, social and trade impact, actual and impending, on the lives of people in the South brought about by the advent of the new technology. In the 'weak' South, technology-oriented policy analysis and social-science research have been severely neglected. Donors should redress this situation.

# Notes on the Contributors

**Richard Adams** has been Director of New Consumer, the UK corporate social responsibility research charity, since 1988. In 1972 he set up an agricultural imports company working with horticultural groups in developing countries; this led to the founding of the public company Traidcraft, which to date has marketed over £30m of craft, food and clothing products from the Third World on fair-trade terms. He has written about the concepts of alternative trade in *Who Profits?* (Oxford: Lion, 1989), and co-authored *Changing Corporate Values* (London: Kogan Page, 1990) and *Shopping for a Better World* (London: Kogan Page, 1991). He is a director of the Fairtrade Foundation; of the UK Social Investment Forum, a non-profit group promoting ethical investment; and of the Third World investment society, Shared Interest. He is an honorary Fellow of Dundee University's Centre for Social and Environmental Accounting.

**Keith A. Bezanson** has a Ph.D. from Stanford University, for his work on cognitive learning in primary and secondary schools in Ghana. He was appointed President of the International Development Research Centre (IDRC), Ottawa, Canada, for a five-year term starting in 1991. In spring 1997, Dr Bezanson became the Director of the Institute of Development Studies at the University of Sussex, Brighton. Prior to joining IDRC, he was the Administrative Manager in the Inter-American Development Bank, Washington DC (1988–91) and Canadian Ambassador to Peru and Bolivia (1985–88). Between 1973 and 1985, he held a number of senior positions in the Canadian International Development Agency (CIDA), working in CIDA's East African, Latin American and multilateral divisions, among others. He has written and published numerous papers, and received several awards and fellowships.

**M.R. Bhagavan** took his doctorate in nuclear physics from the University of Munich, and taught physics in the Universities of Manchester and London. He has also researched and published in the fields of technological, industrial and energy policies in the context of developing countries, teaching these subjects for several years in Eastern and Southern African universities. In addition, he has worked as a visiting professor at several research institutions in Western Europe and India. At present, he is a Senior Research Advisor in Science and Technology in the Department for Research Cooperation (SAREC) of the Swedish International Development Cooperation Agency (Sida). He also holds an associate professorship in industrial economics and management at the Royal Institute of Technology in Stockholm. He has worked as a consultant on technology and energy issues to the EEC, ILO, UNCTAD, UNIDO and UNU-WIDER.

**José Eduardo Cassiolato** is the Research Director at the Institute of Economics and Professor of Public Policies and Industrial Economics at the Federal University of Rio de Janeiro. He is also a Research Fellow at the Economics Institute of the University of Campinas, Brazil. He was Secretary for Planning at the Brazilian Ministry of Science and Technology from 1985 to 1988. He has held visiting fellowships at the Massachusetts Institute of Technology (1977), and at the Institute of Development Studies and the Science Policy Research Unit of the University of Sussex (1989–92). His recent work includes the co-editing of the book *Hi-Tech for Industrial Development* (London: Routledge, 1992).

**Charles Cooper** is Director of the United Nations University Institute for New Technologies (UNU/INTECH) in Maastricht. For the past 25 years, he has conducted research, training and consulting in technology, economics and industrial development in developing countries. As Director of UNU/INTECH, he is responsible for planning and the technical overseeing of all research programmes. He served for 10 years as Professor of Development Economics at the Institute for Social Studies in the Hague, and before that was for 12 years a Joint Fellow of the Institute of Development Studies and the Science Policy Research Unit at the University of Sussex. His research has concentrated on technology transfer and the environmental and socioeconomic impacts of technology and macroeconomic policy for development.

**Carl-Göran Hedén** is Professor Emeritus at the Karolinska Institute, Stockholm, where he held the Professorship in Bacteriology for several decades, as well as a Research Professorship in Biotechnology. He is a Fellow of the Swedish Royal Academy of Sciences and the Royal Academy of Engineering Sciences. He has served in leading positions on many international organizations, including the International Association of Microbiological Societies, the UNESCO Panel on Applied Microbiology, the International Council of Scientific Unions, the International Federation of Institutes of Advanced Study, the International Development Research Centre, and the Club of Rome. He is the author of numerous publications in bacteriological chemistry and physiology, biotechnology and science policy.

**Calestous Juma** received his doctorate in science and technology policy studies from the University of Sussex. He is the Executive Secretary of the Convention on Biological Diversity in Montreal. He was the founding Executive Director of the African Centre for Technology Studies (ACTS), Nairobi. Dr Juma won the 1991 Pew Scholars Award in Conservation and Environment, and the 1993 United Nations Environment Programme (UNEP) Global 500 Award. He serves, or has served, on the boards of directors of the World Resources Institute, the United Nations University Institute for New Technologies, and the Centre for International Environmental Law. His research interests include systems and evolutionary theories of social change with emphasis on the role of technological innovation. He has written widely on issues of science, technology and environment. Among his many published works are *The Gene Hunters: Biotechnology and the Scramble for Seeds* (London: Zed Books and Princeton, N.J.: Princeton University Press, 1989), and *Coming to Life: Biotechnology in African Economic Recovery* (London: Zed Books and Nairobi: ACTS Press, 1995), co-edited with John Mugabe.

**Stephen Karekezi** is the Director of the African Energy Policy Research Network (AFREPREN) as well as the Executive Secretary of the Foundation for Woodstove Dissemination (FWD), Nairobi. Prior to his current appointment, he was a Senior Energy and Natural Resources Advisor to USAID's Regional Office for Eastern and Southern Africa in Nairobi. He is an engineer with postgraduate qualifications in management and economics. He has written extensively on energy policy, renewable energy technology, and energy efficiency, as well as on a host of issues linking energy to sustainable development, in particular in the context of Africa. In 1990, he received the Development Association Award in Stockholm in recognition of his work on the development and dissemination of energy-efficient household energy technologies. In 1995, he was appointed a member of the Scientific and Technical Advisory Panel (STAP) of the Global Environment Facility (GEF) Programme administered by UNDP, UNEP and the World Bank.

**Helena M.M. Lastres** is a Senior Researcher at the Brazilian National Research Council (CNPq), and a Lecturer in the Postgraduate Programme in Information Science conducted at the Federal University of Rio de Janeiro jointly with CNPq. She was Head of the Centre for Studies and Planning in New Materials at the National Institute of Technology, and Executive Secretary of the National Commission for New Materials in the Brazilian Ministry of Science and Technology. She has worked as a consultant on innovation and advanced materials to several Brazilian and international institutions. Among her recent publications are: *Advanced Materials Revolution and the Japanese System of Innovation* (London: Macmillan, 1994); and *Novos Materiais – Desafio e Oportunidade* (INT, 1992).

# Contributors

**Hugo F. Lopez** is an Associate Professor in the Materials Department of the University of Wisconsin-Milwaukee (UWM), where he has spent the last eight years. His field of expertise involves solidification of metallic materials and hydrogen embrittlement and degradation of engineering materials. Before joining UWM, he was a Research Professor at the Saltillo Institute of Technology in Mexico, where he worked for two years. While in Mexico he was made a Member of the National System of Mexican Researchers. He has an in-depth understanding of Mexican official policy as it relates to R&D in materials engineering.

**P.K.B. Menon** is a Director in the Department of Science and Technology (DST), Government of India, New Delhi. He is a biologist by training, with over 15 years' research experience. Prior to joining DST, he worked for many years as a research scientist in the Indian Council of Medical Research, New Delhi, and in the Commonwealth Institute of Biological Control, Indian Station, Bangalore. His expertise extends to issues concerned with science communication, technology transfer, innovation and entrepreneurship development. He is also currently serving as Member Secretary to the National Science and Technology Entrepreneurship Board, Government of India, New Delhi.

**Ian Miles** is Associate Director of PREST (the Programme of Policy Research in Engineering, Science and Technology) at the Victoria University of Manchester. His original training was in social psychology, but in a research career of over 20 years he has examined a broad range of topics, including social forecasting and social indicators, the management of technology, the 'service economy', and changing ways of life. Among his many publications are the following co-authored books: *The Shape of Things to Consume* (Aldershot: Avebury, forthcoming); *Development, Technology and Flexibility: Brazil Faces the Industrial Divide* (London: Routledge, 1992); *Telematics in Transition: The Emergence of New Interactive Services* (London: Longman, 1989); *Information Horizons: The Long-term Social Implications of New Information Technology* (Aldershot: Edward Elgar, 1988); *Worlds Apart: Technology and North–South Relations in the Global Economy* (Brighton: Wheatsheaf, 1985).

**John Mugabe** holds a doctorate in technology and environmental policy from the University of Amsterdam. He is the Executive Director of the African Centre for Technology Studies (ACTS), Nairobi, and was the founder director of the Biopolicy Institute of ACTS in Maastricht. He has authored a number of papers, and has co-edited with Calestous Juma *Coming to Life: Biotechnology in African Economic Recovery* (London: Zed Books and Nairobi: ACTS Press, 1995).

**Donald V. Roberts** received his training in civil engineering at Stanford University, followed by postgraduate training in geotechnical engineering at Imperial College, University of London. Between 1951 and 1987 he was a senior partner in the US engineering firm Dames & Moore, holding several senior operational and managerial positions, including the managing of some 500 environmental projects in about 20 countries. During 1987–94, he was vice president of the US consulting engineering firm CH2M Hill, and since 1994 has been active as an independent consulting engineer. He holds membership of several US and international professional societies in the environmental engineering area, and is currently the president of the World Engineering Partnership for Sustainable Development.

**Pradeep Kumar Rohatgi** has been a Professor in the Materials Department of the University of Wisconsin-Milwaukee (UWM) since 1985. He is the Founder and Director of the Research Laboratories for Solidification, Composites, and Tribology at UWM. He has received a number of awards including the Samuel C. Weaver Professorship, and the Ford/Briggs and Stratton Professorship. Before going to work in the USA, he was a Director of Research at the Council of Scientific and Industrial Research in India (1977–85) and Professor in Materials Science at the Indian Institute of Science, Bangalore (1972–77). He has considerable consulting experience with manufacturing organizations in various countries, and with international development agencies including UNIDO, UNESCO, the UN Office of Science and Technology and the World Bank. He has written several reports on technology policy, materials and education in the context of developing countries.

**Jon Sigurdson** was formerly Professor of Research Policy at Lund University in Sweden, and is now Professor of Science and Technology at the European Institute of Japanese Studies at the Stockholm School of Economics. He was Director of the Research Policy Institute at Lund University during 1978–92. Trained in economics and electronics, his long research career in policy science has spanned the following specializations: technology management at national and company level; the emerging character of globalizing companies; global structures for R&D and their interaction with national systems of innovation; structural changes in the electronics and information technologies industries. He has held posts in the Swedish Ministries of Foreign Affairs, Finance and Industry, and undertaken consulting assignments in Sweden, South Korea, India, China and Mexico, with extensive stays in China, Hong Kong and Japan.

**Jian Song** is a State Councillor (Senior Cabinet Minister) in the Chinese government, holding several portfolios, among which are science and technology, and environment. He is Chairman of the State Science and Technology Commission, and of the State Environmental Protection Committee. He is a Member of the Chinese Academy of Sciences and a Foreign Member of the Russian Academy of Sciences. Trained as an electronics engineer, with two doctoral degrees from Moscow, Professor Song specialized in cybernetics and systems control, making important contributions to systems-control development in the Chinese missile programme. He has held several professorships in Chinese technical universities, published numerous scientific articles on systems control and received many Chinese and foreign awards in recognition of his distinguished work.

**T.J. Wessels** is a graduate of the Agricultural University of Wageningen, where he specialized in tropical agronomy and agricultural economics. He served with FAO as a regional agricultural planning specialist in the northeast of Brazil, in the Farm Management and Production Economics Service at the FAO headquarters in Rome, and in Peru as an agricultural economist. He then joined the Latin America Division of the Netherlands' Technical Cooperation Programme, and later became the coordinator for international agricultural research in the Research and Technology division. In that capacity he was also the Netherlands' delegate to the CGIAR. Since 1989, he has been involved in developing Dutch policy for support to biotechnology applications in developing countries. In 1992 he was appointed Head of the Special Programme in Biotechnology at the Directorate General for International Cooperation of the Netherlands' Ministry of Foreign Affairs.

**Anders Wijkman** graduated in political science and economics from the University of Stockholm in 1967. He was a Member of the Swedish Parliament from 1970 to 1978, where he focused his attention on foreign affairs and foreign aid, in addition to energy and environment. As Secretary General of the Swedish Red Cross from 1979 to 1988 he became increasingly involved in relief and development activities in a number of regions in the South. In parallel with this responsibility, he also guided the disaster relief, prevention and preparedness work of the Red Cross Headquarters in Geneva as the President of the International Red Cross Disaster Relief Commission from 1981 to 1989. On becoming Secretary General of the Swedish Society for Nature Conservation in 1989, he devoted his energies to environmental conservation and protection activities on a national scale in Sweden. Between 1992 and 1995 he was Director General of SAREC, where, among other things, he promoted assistance to research on environmental sustainability in developing countries. He moved in 1995 to New York to join the United Nations Development Programme (UNDP) as Assistant Administrator and Director of the Bureau for Policy and Programme Support. Mr Wijkman is a Member of the Royal Swedish Academy of Sciences and the Club of Rome. He is the author of several books on issues related to disasters, development and environment.

# Index

acetylation, 148
acid rain, 36, 180
advanced materials (AM), 1, 25, 29–36, 97, 99, 106, 164; customized applications of, 68; demand for, 11; diffusion of, 68; high rate of growth of, 70; improved qualities of, 74; in Brazil, 85, 86 (policy concerning, 83–7; research into, 87); influence on materials markets, 68; prices of, 74; relation to information technology, 77; revolution of, 79, 87, 140 (effects on developing countries, 68–92); theoretically predictable nature of, 73, 76, 88
aerospace, as borderless industry, 265
Africa, 199, 204, 210, 228; declining support for research, 138; R&D expenditure in, 56, 197; rethinking of development policies, 132; sub-Saharan, 13, 14, 299
*AgBiotech* journal, 263
Agenda 21, 127, 247
agricultural waste, 148; as construction material, 147; use of, 149, 167
agricultural wastelands, 261
agriculture, 8, 17, 244, 245, 255, 274, 302; and biotechnology, 116, 118, 120; capital-intensive, 117; cold-water, 227; commercial, 251; controlled-environment, 224–7; demand for products of, 104; industrialization of, 252; monocultural, 222; research in, 275; small-scale, 251, 252, 254, 256, 257, 258; urban, 227
agroforestry, 261
aid: bilateral, tied, 58; focused on capital-intensive equipment, 202; of advanced countries, 58 *see also* donor agencies
aid policy, 291; differential, 314
aircraft, materials advances in, 75
airline tickets, processing of, 209
alcohol, produced for fuel, 182
alloys: amorphous, 74; high-melting-point, 168; high-performance, 149; hydrogen-absorbing, 74; shape memory, 74, 168; superplastic, 73
aluminium, 70, 77, 148, 149, 156, 164, 184; direct reduction of, 150, 167; production of, 106
aluminium-graphite, 149
aluminium-silicon carbide, 149
Amazon minerals province, 81
antibiotics, 34
Appropriate Technology International, 228
aquaculture, 227
Arab League, 241
Arabization: of networking, 241; of systems, 237
aramid fibre, 110
Argentina, 4, 247, 298
ARISNET system, 241
Armauer Hansen Research Institute (AHRI) (Ethiopia), 122
arms sales, 206
ARSONET system, 238
artificial intelligence, 243
artificial organs, making of, 308
Asia, R&D expenditure in, 55
aspartame, 209
atomic bomb, discovery of, 281
Atomic Energy Commission (India), 159
Australia, 79, 99, 262
automation, 2, 3, 53, 97, 151; in Sweden, 48; of Brazilian banking system, 51; of Indonesian banking system, 50; systemic advantages of, 52
automobile industry, 70, 265
automobiles, 1, 183; maintenance of, 183; materials advances in, 75

bagasse, 180; for power production, 181, 182
bamboo: *in vitro* flowering of, 100; as construction material, 147
bananas, 118

# Index

Bangladesh, 242, 245; garment industry in, 210
BARC company (India), 108, 112
BellaNet system, 248
beneficiation technologies, for coal, 176
Benin, 198
beryllium, 84
beverage container industry, 70
Bhaba Atomic Research Centre (BARC) (India), 107
Bharat Immuno Biologicals Co (BIBCOL) (India), 102
bilateral assistance, 16
bio-catalytic methods, 8
bio-centres, 226
bio-energy, 7, 8
'Bio-Revolution', 303
bio-villages, 226
biochemistry, 130
biodiversity, 18, 38, 115, 137, 226, 247, 254; and biotechnology, 123–6; conservation of, 125; erosion of, 303; pricing of goods, 128 *see also* Convention on Biological Diversity
bioengineering, 224
bioethics, 20, 302, 307–8
biofertilizers, 254
Biofocus Foundation, 226
Bioindustry Association (UK), 228
biological weapons, 308
biomass, 164; as energy source, 34, 147, 178, 217 (hurdles to deployment of, 181); briquetting of, 176; for energy, 151, 180
biomass integrated/gas turbines (BIG/GT), 176
biomass integrated/intercooled-steam injected turbines (BIG/ISTIG), 180, 181, 182
biopesticides, 254, 259
biosafety, 14, 18, 20, 120, 253, 254, 255, 258, 259, 302, 306–7; activism, 307
biosensor technology, 105
BIOTASK *see* Taskforce on Biotechnology
biotechnology (BT), 1, 7–8, 10, 12, 25, 29–36, 37, 97, 99, 202, 227, 264, 266, 268, 274, 297–308, 309; access to, 126, 127; and agriculture, 13–14, 116, 118, 120; and biodiversity, 123–6; and legislation, 254; applied to varieties of crops, 118; as means of combating poverty, 253, 254; company links with universities, 301; costs of, 303; dualistic impact of, 251; entry barriers for, 129; global developments in, 116–19; in developing countries, 251–63 (impact of, 300); in India, 100; in sub-Saharan Africa, 115–39; international cooperation in, 115, 126–31; legislation on, 258; medical, 116; national programmes in, 298; new, 34–5; not adapted to small producer, 251; partnerships in, 131; pervasiveness of, 34; policy-making in, 131–3; products of, demand for, 104; R&D in, 119–23 (in Cameroon, 123); re-evaluation of, 297; research, in East Africa, 119–23; support for research, 115; systems approach to, 298; threat to African cash crops, 117; training in, 130, 228
Biotechnology Advisory Committee: India, 261; Kenya, 120
*Biotechnology and Development Monitor*, 263
*Biotechnology Monitor*, 227
bismuth, production of, 107
blended technologies, 16
blood cell production, 35
blood products, making of, 34
Bolivia, 79, 245
borderless industries, 311
bottom-up approach, 18, 255, 257, 260
Brazil, 4, 5, 12, 69, 79, 178, 180, 182, 199, 202, 245, 298; advanced materials policies in, 83–7; optical fibres in, 83, 88; strategic materials in, 84
breeding techniques, traditional, 116
Bretton Woods system, 57, 291, 314
buffalo cart, development of, 204
building design, energy-efficient, 178
Burkina Faso, 243, 245
Burundi, 119
'business as usual' approach, 228, 313
Business Council for Sustainable Development, 198, 202

cacao, effects of biotechnology, 117
cadastral mapping, 245, 247
Cameroon, 14, 121, 243; biotechnology R&D in, 123
Canada, 16, 17, 114, 145, 233–50; R&D expenditure in, 55
Canada Centre for Remote Sensing, 247
cancer, 35
capacity building, 16, 115, 234, 255, 298, 315–16
Capacity Building in Electronic Communications for Development in Africa (CABECA), 238, 239
capital markets, 60
Carajás project, Brazil, failure of, 81
carbon fibre, 109, 110
cardamom seeds, tissue culture for, 300
carrots, 119
cars, battery-powered, 75
cartography, 245
cassava, 118, 258, 260

Cassava Biotechnology Network (CBN), 18, 263
castor, 261
cauliflowers, 119
celery, 119
cell-based therapeutics, 104
cement, 142, 155, 156, 168, 184; reduced energy costs of, 167; shortages of, 147; high-tech, 147
Cement Research Institute (India), 160
Central Electrical and Electronic Research Institute (CEERI) (India), 106
Central Electro Chemical Research Institute (CECRI) (India), 106
Central Glass and Ceramic Research Institute (India), 159
Central Machine Tools Institute (India), 112
Centre for Artificial Intelligence and Robotics (India), 112
Centre for Iron and Steel (India), 159
Centre on Integrated Rural Development for Asia and the Pacific (CIRDAP), 242
Centro Internacional de Agricultura Tropical (CIAT), 256, 263
ceramics, 68, 70, 88, 150, 156, 167; advanced, 38, 73, 74, 76, 85, 143, 149; for electronics, 143; (production figures for, 72; production in Japan, 72; sales of, 71); structural, 149
chemistry, applied, 158
Chile, 4, 79, 242, 243, 245, 290
China, 4, 12, 56, 155, 156, 161, 176, 177, 199, 203, 245, 277, 281, 290, 298, 300, 317; divorce between research and production, 96; high technology programmes in, 95–8
chlorofluorocarbons (CFCs), 38; substitutes for, 39, 110
chocolate, new production methods of, 117
chromium, 8, 149
civilian products, as driving force of progress, 276
clays, 157
climate change, 303
clothing, materials for, 150
co-generation technologies, 176, 180, 181
$CO_2$ emissions, 39
coal, 157, 176, 180; for power production, 181
cobalt, 8, 149
cocoa, 118; butter, produced in laboratory, 117
coconut, 118
coffee, 118, 122
Cold War, 205, 206, 280, 282
Colombia, 17, 245, 255, 258, 260, 263
Comité Campesino Regional (Colombia), 260

commodities, falling prices of, 43, 78
Commonwealth Secretariat, 248
communications, 237; capacity for, 53; development of systems, 148; improvement of infrastructures, 128; protocols, problems of, 237
comparative advantage theory, 288
competition: innovative, 288, 290, 313; price, 287, 288, 289, 313; standard definitions of, 287
competitive advantage, national patterns of, 289–9
complementarity between developed and developing countries, 44, 58
Composite Product Centre (COMPROC) (India), 109
composites, 68, 70, 73, 75, 84, 85, 88, 143, 149, 150, 167, 168; production of, 108–9
composting, 102
computer-aided design (CAD), 48, 111, 151, 169, 210
computer-aided manufacturing (CAM), 210
computer-integrated manufacturing (CIM), 97
computers, 233; manufacture of, 111; material composition of, 70; personal, 287 (demand for, 112)
condensing/extraction steam turbines (CEST), 181
conferencing systems, 237
Congo, 79
conservation, 207, 303; of biodiversity, 125, 304, 305; of natural resources, 270; of resources, 161
conservation groups, 123
conserver society, 199
Consultative Group on International Agricultural Research (CGIAR), 18, 125, 237, 254, 262, 263; Statement on Intellectual Property, 262
consumer society, 199, 214
consumerism, 196, 310; price of, 196
consumption, non-essential, reduction of, 310
contraception vaccines, 308
Convention on Biological Diversity, 14, 115, 124, 125, 126, 129, 130
cookstoves, wood-fuelled, 177
cooperation, in technology, 128
copper, 70, 148, 208; dependence on export of, 152; new uses for, 152; reduced demand for, 209
cordless systems, 111
Costa Rica, 207, 242, 243, 245
Côte d'Ivoire, 245
cotton, 119, 208; reduced demand for, 209

# Index

Council of Scientific and Industrial Research (CSIR) (India), 100, 159
cowpeas, 259
crisis management of economics, 54
crystals, artificial, 97
Cuba, 206, 298
cucumbers, 119
cultural lags, and innovation, 29
cytokine, market, 104

data processing, 209
databases: access to, 135; establishment of, 235, 319
de-skilling, 51
debt, foreign, 43, 54, 82
decentralization, 6, 7; of energy production, 185; of power generation, 15; of processing facilities, 32
decolonization, 1, 4
Defence Metallurgical Research Laboratory (DMRL) (India), 107, 108, 159
Defence Research Development Organization (DRDO) (India), 112
Defence Service Centre (DSC), 108
defence spending, declining, 275
deforestation, 303
demand, definition of, 199
demand management, in energy, 14
democratic development, 270
Denmark, 179, 180
Department of Biotechnology (DBT) (India), 100
Department of Science and Technology (DST) (India), 100
desulphurization of crude oil, 100
detailed intervention in matter, 31, 73
develease, concept of, 226
developing countries, as full partners, 234
development, cooperation in, 222
development aid, 115, 315; policy, 285–94, 318–19
diagnostic tools, 8, 102, 103, 116, 119, 129; for livestock, 259
dialysis, 74
DIEL company (India), 111
diesel generators, 176
digitalization of information, 7, 32
Directorate General of Development Cooperation (DGIS) (Netherlands), 125
disaster relief, 270
disease-free plants, 8, 118, 258
distributed systems, 111
division of labour, 44, 78, 81, 279
DNA: diagnostic probes, 102; discovery of molecular structure of, 7; fingerprinting, 100; understanding of operation of, 34
donor agencies, 9, 13, 202, 262, 299, 300, 307, 311, 316, 318; interest in information technology, 248
drought tolerance of plants, 261
drugs: delivery systems, 74; development time of, 297
durability, 9
dyes, chemical, 209

e-mail, 237
East–West relations, 235
ecology, industrial, 215
Economic Community of West Africa (ECOWAS), 184
ecosystems: man-made, 216; natural, 214
education, 6, 140, 164, 185, 221, 229, 310, 318; general, 53; graduate, 156; higher, 5, 133, 316; in developing countries, 46; postgraduate, 160; spending on, 54, 59; university, 153, 155, 156, 160
efficiency, 199, 209; dynamic, 314; static, 292, 314
effluent management, 102
Egypt, 243, 245
863 Programme (China), 96–7
electric motor, 175
electricity, 1, 196
electron beam micro fabrication equipment, 106
Electronic Atlas, 247
electronics industry, 265; building of, 54; sales in USA, 45
empowerment, 234
Endod plant, 207
energy: alternative sources of, 122; cheaper, 75; dependence on, 174; fossil fuel, 39; fusion, 150, 167; generated from materials, 150; global consumption of, 175; in Africa, 174–91; investment in, 174, 175; management of, 38; renewable, 39; savings, 44, 183; solar, 147, 150, 151, 164, 176, 217; waste, recycling of, 15 *see also* biomass *and* nuclear energy
energy analysis, 40
energy control systems, 33
energy efficiency, 9, 14, 76, 217, 218
energy saving, 68, 73, 108, 150, 151, 167, 178, 184; technologies, breakthrough in, 184
energy technologies, 97; and technological change, 175–9; capital-intensive, risks of, 185; management-intensive, 185; modular, 185
energy-intensiveness, 148
engineering, 214–20, 294
engineers, 4, 5, 9, 15, 52
engines, efficiency of, 76
entrepreneurial activity, 157

entrepreneurial class, 9; assistance through, 16; indispensability of, 314–15
entrepreneurship, 228, 229; and absorption of technologies, 224; biotechnology-based, decision-making in, 225
environment, restoration of, 216, 219
environmental change, 216
environmental degradation, 36, 147, 153, 154, 222, 313, 317
environmental impact studies (EIS), 154, 219, 220
environmental issues, 37, 98, 153, 196, 200, 201, 202, 204, 205, 210, 270, 300, 302, 309
environmental management, 17, 40, 116, 255
environmental monitoring, 220
environmental planning, 219
environmental protection, 244
environmental regulations, compliance with, 73
environmental rehabilitation, 310
environmental studies, 217, 218
epitaxial processing, 106
ESANET systems, 238
ethanol, as fuel, 182
Ethiopia, 14, 119, 121, 122
Ethiopian Science and Technology Commission (ESTC), 121
Eureka, 265
Europe, 43, 265; production moving to, 210; R&D expenditure in, 55, 282; single market, implications for Africa, 135; Western, 10, 298
European Centre for Nuclear Research (CERN), 281
European Community (EC), 57, 253, 254, 255
European Union, 19, 271, 305, 312; R&D Framework Programme, 265, 276, 280
expert systems, 243–4
export crops of developing countries, displaced, 300
export processing zones (EPZ), 207, 208, 292

factories, organized around power sources, 28–9
farmer plantback of protected varieties, 306
fermentation processes, 102, 103, 130, 198; effluents from, 182
fertilizers, artificial, 148
fibre-resin production systems, 109
fibres: available from trees, 148; extraction of, 149; pyrolysis of, 149
firm-specificity of technology, 49, 50, 59, 60
fish, falling catches of, 221

fishing, 245
FJ Industrial company, 208
flax, 119
flexibility, 200; of production, 51, 82; of products, 73; of systems, 32
flexible manufacturing systems (FMS), 51
Flosolver system, 112
fluidized-bed combustion, 176
fluorescent lamps, compact, 177
Food and Agricultural Organization, 221; Intergovernmental Commission on Plant Genetic Resources, 304–5
food, technology-intensive, 301
food productivity, 221
food supplies, concern for, 221
foreign direct investment (FDI), 80, 210
forest products, 157, 160
Forest Research Institute (India), 159
forestry, 7, 8, 245
free trade, 310–22
freon, 227
fuel efficiency of vehicles, 183
fullerenes, 35, 168
functional markets, 195–213
funding options for research, in Africa, 137–8

garment industry, 208
gas turbines, advanced, 176
gene banks, creation of, 303
gene therapy, 308
General Agreement on Tariffs and Trade (GATT), 57, 253, 255, 306; Uruguay round, 57, 197, 206, 305
generic technologies, 234, 264–84, 285, 286, 289, 292, 293; definition of, 266; new, 5–6, 313, 315 (in India, 99–114); scenarios for, 273–5; support for, 269, 270, 278; transfer to developing countries, 221
genetic base, narrowing of, 252
genetic diversity, 254
genetic engineering, 7, 34, 38, 148, 167, 224, 259, 266, 307; in medical fields, 308; legislation on, 255–6
genetic erosion, 303
genetic resources: access to, 130, 304–5; fair compensation for, 304
genetically modified organisms, 279; dangers of releasing, 307, 309; non-survivability of, 302; risks of, 252
Geographic Information Systems (GIS), 236, 242, 244, 246
geomatics, 17, 236, 244–7; R&D in, 246
geothermal energy, 176, 217
Germany, 227, 281, 289
Gesellschaft für Biotechnologische Forschung (GBF), 227, 228

# Index

Ghana, 243
Ghana Investment Centre, 203
glass-fibre reinforced plastics (GFRP) industry, 109
Global Environment Facility programme (Mauritania), 180, 182
global positioning systems (GPS), 17, 236, 245, 247
globalization, 43, 44, 58, 99, 275, 313; of financial markets, 282
GLOBESAR project, 246
governments, role of, 44, 48, 300 (in regulating technology, 202; in support of R&D, 95 (reduced, 276); in support of technology infrastructure, 59; in technological change policy, 58 green economics, 199
Green Revolution, 123, 303
greenhouse, zero-energy, 227
greenhouse gases, 36, 180
gross national product (GNP), growth of, 211
groundnuts, 261
growth, maximization of, 313
growth hormones, 308
Guatemala, 300
Guinea, 79
Guinness Nigeria plc, 198

habitats, global list of, 124
hardware, 53, 234, 246, 247; decreasing size of, 32; design of, 47; falling costs of, 6, 33; industry in India, 113
HCL-HP Ltd (India), 111
health care, 17, 102, 104; materials for, 150; primary, 259; traditional healers, 259–60
health services, 244, 300
HealthNet system, 238, 239
heartland technologies, 9, 26, 27, 28, 29, 31, 33, 36, 38, 39, 40
Heavy Industries Corporation of Malaysia (HICOM), 200
hepatitis: detection of, 102; vaccine for, 97
Hershey company, 117
Hewlett Packard company, 208
high fructose corn syrup (HFCS), 117, 209
high temperature superconducting materials (HTSM), 107
high-yielding seed varieties, 303
Hindustan Aluminium, 159
Hindustan Cables Ltd, 107
HMT company (India), 112
Hong Kong, 290
housing: materials for, 147–8, 160, 167
human genome, sequencing of, 116
Human Genome Organization (HUGO), 308
human health care, 116, 255

human resources, 46, 52–4, 85, 185, 239
human rights, respect for, 270
human–machine interaction, 33
hydro energy, 178, 182
hydrogen car fuel, 74
hydrological cycle, 214
HYL steel reduction process, 142
hypertexts, 33

ICIM company (India), 111
IIT-Delhi company, 109
immuno-diagnostics, 100, 104
import substitution, 204, 292
*In Search of Excellence*, 201
*in vitro* fertilization, 308
incentives for investment, 317
India, 4, 5, 12, 13, 17, 140, 176, 177, 178, 199, 202, 245, 255, 258, 261, 270, 275, 277, 298, 300, 317; advantages in science and technology, 106; as research superpower, 99; biotechnology in, 100; new materials technology in, 147–52; power sector investment in, 174
India Rare Earths Ltd, 107
Indian Council of Agriculture Research (ICAR), 100
Indian Council of Medical Research (ICMR), 100
Indian Petrochemicals Corporation Ltd, 109
Indian Space Research Organization, 109
Indonesia, 4, 243; automation of banking system, 50
industrial ecologies, 15
industrial policy, in developing countries, 197
Industrial Research institutes, Taiwan, 294
industrialization, 197, 289, 291; complexity of process, 82; export-oriented, 291; relation to technological change, 291
industrialized countries, interests of, 278–9
inflation, 54, 82
informatics, 17, 70, 236, 240–44
information: access to, 128, 135; dissemination of, 137; provision of, 135
'information age', 31
information and communications technology (ICT), *see* information technology
information highway, 112
information processing, 31, 33; decreasing cost of, 32; speed of, 32
information technology (IT), 1, 5, 6–7, 10, 11, 12, 16, 17, 25, 29–36, 37, 72, 77, 97, 99, 233, 247, 266, 309, 311; access to, 237; as key ingredient of organizational change, 247; as transformative technology, 247; diffusion of, 45, 46, 235 (hampered in developing world,

46); donor agencies' interest in, 248;
falling prices, 45; falling profit rates, 45;
importance for development of
advanced materials, 84; in developing
countries, 43–67; in India, 111–14;
integration into systems, 14; intensity of
technical change, 46–9; investment in,
247; new, 31–4 (key features of, 32);
problems of spread to developing
countries, 17; slowing of growth of, 45;
special programmes in R&D, 236;
support to developing countries,
233–50; widening gap with developing
countries, 17
infrastructures, development of, 148
innovation, 3, 49, 58, 195, 201, 236, 252,
256, 292, 306, 309; all-pervasiveness of,
27; competition of, 29; energy-saving,
39; evolutionary view of, 25; fostering
of, 157; global systems of, 315;
incremental, 26; intrinsic part of
competition, 287; key factors of, 27;
process-, 287, 288, 289, 291; product-,
287, 288; radical, 26; role of
information technology in, 48;
swarming of, 27; unsuccessful, 28; *see
also* national system of innovation
innovative learning, 50
Institute for Scientific Information (US), 99
Institute of Agricultural Research (IAR)
(Ethiopia), 122
institutional arrangements for research,
122–3
institutional development, 83, 133, 134
institutional reform, 317–18
insulin, 302
integrated manufacturing processes, 146
integrated resources planning (IRP), 183
integration–exclusion pattern of world
trade, 10, 43
intellectual property rights, 14, 18, 20, 123,
124, 126, 127, 128, 197, 206–7, 252,
253, 254, 255, 256, 258, 305–6, 310–11;
and living organisms, 306; protection
of, 15, 120, 277, 279
intelligent systems, 111
interactivity, 33
interconnectivity in IT, 111
interferon, 97
interleukin, 97
Intermediary Biotechnology Service, 18, 262
Intermediate Technology organization
(UK), 204
internal combustion engine, 175
International Agricultural Research Centres
(IARCs), 118
International Biotechnology Program, 228
International Centre for Genetic

Engineering and Biotechnology
(ICGEB), 125
International Centre for Insect Physiology
and Ecology (ICIPE), 123
International Centre for Research on
Agroforestry, 123
international collaboration, in
biotechnology, 262–3
international conventions, 136
International Council of Scientific Unions
(ICSU), 125, 281
International Development Research Centre
(IDRC), 16, 17, 233–50; Information
and Communications Technologies, 236;
Information Policy Research Program,
235; Information Sciences and Systems
Division, 235; Software Development
and Applications, 236; Telematics
programme, 237
International Laboratory for Animal
Disease (ILRAD), 123
International Livestock Centre for Africa
(ILCA), 122
International Monetary Fund (IMF), 20, 57,
292, 314
international organizations, importance of,
59
international relations, redefinition of, 132
International Service for National
Agricultural Research (ISNAR), 262
International Society for Plant Molecular
Biology, 256
Internet, 237, 238
inventiveness, 53
investment, 310, 311, 313, 315; direct, 197,
205; in R&D, 279; in technology, 60;
needed by R&D, 278 *see also* foreign
direct investment
iron, 164; direct reduction of, 150, 167
IRRIDIUM system, 239
irrigation systems, 216
Italy, 99, 281

Japan, 10, 52, 71, 228, 265, 277, 289, 290,
292, 298; foreign direct investment in,
80
Japan Fine Ceramics Association (JFCA),
72
joint ventures, 202, 203; in China, 203
Jordan, 245
just-in-time management, 40

Kenya, 14, 17, 119, 122, 126, 132, 177, 182,
183, 238, 245, 255, 258, 259, 261;
national conference on biotechnology,
120
Kenya Agricultural Research Institute
(KARI), 122

Kenya Forestry Research Institute (KEFRI), 122
Kenya Industrial Property Office (KIPO), 132
Kenya Medical Research Institute (KEMRI), 122
Kenya Motor Industry Association (KMIA), 183
Kenya Trypanosomiasis Research Institute (KETRI), 123
Kenyan Agricultural Biotechnology Platform (KABP), 259
Kenyan National Tuberculosis Programme, 260
Kevlar fabrics, 109
Key Science and Technology Achievements Dissemination Programme (China), 96
Keyhole Point, Hawaii, 227
Keystone International Dialogue Series on Plant Genetic Resources, 303, 304, 306
knowledge production, problem-oriented mode of, 26, 269
knowledge, 236; corporatization of, 15, 206–7; definition of, 2; generation of, 267 (organization of, 266); generic, role of, 265–70; means to empowerment, 234; new, generation of, 48, 266, 267; organization-embodied, 50; people-embodied, 50, 129; processing of, 33; shift in production of, 266; tacit, importance of, 47; 'traditional', owners of, 207; transfer of, 49, 266
knowledge content of technology, 8
knowledge production, disciplinary mode of, 269–70
knowledge-based systems, 243, 244

labour: as resource rather than cost, 53; demand for, new paradigm of, 53; non-skilled, 80
labour costs, low, 49, 151, 210
land reclamation, 39
lanthanium, production of, 107
lasers, 97; use of, 169
Latin America, R&D expenditure in, 55, 56
Latin American Demographic Centre (CELADE), 242, 243
lead, 142
leading sectors, 3
learning by innovation, 52
legislation, on biotechnology, 258, 306
leishmaniasis, 122, 260
leprosy, 122
less developed countries (LDCs), 52, 77, 79, 87; articulation with advanced countries, 82; energy requirements of, 174

liberalization, 10, 19, 57, 124, 203
libraries: provision of, 135; strengthening of facilities, 135
licensing authorities in materials, delays of, 158
Light Combat Aircraft (LCA) project, 109
lighting systems, advanced, 178
*Limits to Growth*, 199
literacy, need for, 53
livestock husbandry, 120, 122
Locally Grounded Transnational Research Sites (LGTRS), 281
low-earth-orbit (LEO) satellites, 237, 239

MAb technology, 102, 105
Madagascar, fall in vanilla exports, 300
Madhya Pradesh State Electronics Development Corporation, 107
magnets, 107–8, 110; neodymium-iron-boron, 107; samarium cobalt, 107
maintenance work, 5, 15, 45, 48, 51, 129, 183, 202; high standard required, 184
maize, 119, 258, 260; insect tolerance in, 259
malaria, 260; detection kit, 102
Malawi, 182
Malaysia, 4, 199, 204, 243, 245, 300
Mali, 243, 245
management, 15, 33, 195, 310; and technology choice, 39; changes in, 7, 10; generic character of, 14; in African energy sector, 174–91; in new energy technologies, 179–85; Japanese techniques, 286; new methodologies, 40; of technical change, 47
marker-assisted breeding, 259, 260
materials processing, 151–2
materials science: development of institutions, 152–4; education in, 159; institutions of, 154–5; monitoring of, 146
Materials Science and Engineering, as discipline, 159
materials technologies: home-grown, 109–10; in Mexico, 142–7
materials: and energy generation, 150; better utilization of, 147; bio-processing of, 148–9; costs in relation to incomes, 141; declining growth of, 87; developing strength of, 76; for housing, 147–8, 167; global demand for, 11; imports of, to India, 162; institutions of (in developing countries, 155–6; in development of new materials, 157–8); manufactured, in possession of rich elites, 141; markets for, 169; policy concerning, 153, 161; problem areas of, 170; quality of, standards for, 158;

science and technology institutions (in India, 158–60); international collaboration in, 161–4); standards for, 161; strategic, 149; substitutes for, 251; trends in usage of, 68, 69–77
Mauritania, 180.
Mauritius, 180, 181, 182
medical diagnosis equipment, 75
membrane switches, production of, 208
Merck corporation, 207
mercury, 142
mergers and acquisitions, importance of, 276
metal extraction, by biotechnology, 34
metal-intensive goods, 78
metal-organic chemical vapour deposition (MOCVD) devices, 106
metal-saving industries, 70
metals, 68, 69; changes in industries, 81; declining consumption of, 11, 69, 70, 78, 79; displacement by advanced materials, 73; falling prices of, 78, 80; global consumption of, 165; imports of, to India, 162; low consumption of, indicative of mature economies, 78; microbiological extraction of, 148, 167; regional consumption of, 166; strategic, 8
metcars, 35
methanol, 182
Mettur Chemicals (India), 108
Mexico, 4, 5, 12, 13, 140, 298, 304, 317; materials technology in, 142–7; science and technology research in, 143
microbial metal extraction *see* metals
Microbial Resource Centre (MIRCEN) (Kenya), 122
microbial technology, 148
microbiology, 130; applied, 226
microelectronics, 31, 34, 38, 43, 51, 75, 84, 129, 196, 202, 264, 268, 276; diffusion of, 33
microenterprises, 204
microprocessors, 6, 143, 150
micropropagation, 226, 259
migration, to cities, 227
militarization, 205
military sector, conversion of, 205
military spending, 205; declining, 312
military technology, 196
military–industrial complex, 205
millet, 258
minerals, 157; falling prices of, 80; leaching of, *in situ*, 216; processing of, 216; source of export earnings for developing countries, 78
miniaturization of components, 111, 167
mining operations, 216

MINISIS system, 240–41
Mishra Dhatu Nigam Company (MIDHANI) (India), 108
modified chemical vapour deposition (MCVD) technology, 107
Modo Olivetti company, 111
molasses, conversion to alcohol, 100
molecular beam epitaxy, 168
molluscicides, 207
monitoring equipment, 38
monopolies, temporary, 20
Montreal Protocol on Substances that Deplete the Ozone Layer, 130
Moore's Law, 6, 31
Morocco, 245, 246
motor power, development of, 26, 27
motor transport, effectiveness of, 183
motor-pump systems, efficiency of, 183
motors, electric, 28; efficient, 178
multi-party politics, introduced in Africa, 132
multilateral procurement, 271
multimedia, 112
multinational corporations (MNC), 49, 50, 251, 277

nanostructure materials, 168
nanotechnology, 266
National Aeronautical Laboratory (NAL) (India), 109, 112
National Biotechnology Forum, model of, 257
National Biotechnology Programme (Zimbabwe), 120
National Chemical Laboratory (India), 159
National Commission in New Materials (Brazil), 83, 84
National Council for Science and Technology (NCST) (Kenya), 120
National Council of Research and Technology (CONACYT) (Mexico), 143
national industries, and technological advance, 278
National Informatics Centre (NIC) (India), 112
National Materials Advisory Boards, proposed formation of, 155, 160
National Metallurgical Laboratory (India), 159
National Physical Laboratory (India), 109, 159
National Planning Commission (India), 156
national sovereignty, and issue of biodiversity, 124
National Superconductivity Programme (NSP) (India), 107

national systems of innovation, 265, 279–82, 293, 312, 315
natural resources, management of, 17
Nepal, 245
Nestlé company, 117
Netherlands, 17, 18, 179, 227, 251–63, 317; biotechnology policy in, 18 (and development dimension, 253–4)
networking, 6, 7, 33, 133, 240; between professionals and researchers, 135, 136, 315, 319; in R&D, 112, 265, 315, 319; problems of, 239
networks, 17, 46, 47, 49, 209, 224, 238; computer, 278; externalities of, 47; industrial reliance on, 265; of innovating firms, 58; of satellites, 114
new materials technology (NMT), 5, 8–9, 10, 12, 13, 35–6, 202, 208–9, 264, 266, 268, 309, 311; characteristics of, 37; development in India, 147–52; in developing countries, 140–73; marketing of, 195; pervasiveness of, 36; profitability of, 12
new technologies: absorption process of, 50; diffusion of, 11, 50, 52 (in developing countries, 49–52; hampered, 45); directed to local market, 60; efficient use of, 60; in developing countries, effective, 59
new world order, 10, 11
NGONET system, 237
nickel, 8, 149
NICNET network, 112
Nigeria, 203, 245
nitrogen fixation of plants, 148, 303
non-governmental organizations (NGOs), 9, 15, 16, 133, 134, 135, 136, 204, 223, 237, 256, 257, 261, 271, 273; and aid-management, 223; aid provision through, 224; training packages for, 134
non-tariff trade barriers, 10, 43, 44, 128
North: development in, 58; flexibility of economies in, 292; trade policies of, 314
North American Free Trade Agreement (NAFTA), 13, 145
North Atlantic Treaty Organization (NATO), 205
North Korea, 206
North–North relations, 57
North–South relations, 11, 13, 44, 49, 57, 199, 222, 226, 246, 314, 315; in research cooperation, 311; widening disparity in, 54
Norway, 262
nuclear energy, 154, 177, 178, 217
Nuclear Fuel Complex (NFC) (India), 107
Nucleus of Study and Planning in New Materials (NMAT) (Brazil), 84, 85

off-line data facilities, provision of, 135
oil, 142, 176
oil palm production, 204, 300
oil spills, degradation of, 100
oligonucleotides, synthesis of, 100, 102
open economies, 197, 286, 289, 294; move to, 19
open systems, importance of, 45
optical fibres, 7, 75, 84, 148, 209; diffusion of, 33; in Brazil, 83, 88
optoelectronics, 74
organization, 286; changes in practices of, 7, 10; generic nature of, 14; in African energy sector, 174–91; in new energy technologies, 179–85; industrial, 82
Organization for Economic Cooperation and Development (OECD), 10; concentration of trade within, 44
orphan technologies, 254
ORSTOM research institutes, 228
OTEC electricity device, 227
'overchoice' of materials, 11, 77, 251
oxygen, 164
ozone layer, diminution of, 36, 303
ozone-related technologies, 130

Pakistan, 4, 275
Pan African Development Information System (PADIS), 238
papaya, 260
Papua New Guinea, 79
paradigms, 28; shifts of, in technology, 9, 10; transition across, 29
participation, 255, 257; of workforce, 53
partnerships, 16, 234, 247, 257; in biotechnology, 131; military, 276
parts consolidation, 152, 167
patent law, in developing countries, 256
patent rights, 305
patenting, 252, 277, 279, 301; in biotechnology, 305
patents: harmonization of practices, 305; numbers granted, 197
peace dividend, 205
Pertech Computers company (India), 111
Peru, 79, 243
petroleum, use of, 178
pharmaceutical sector, 278, 297
phenolics, 109
Philippines, 174, 243, 245, 300
photovoltaic energy, 168, 176, 177; costs of, 177
pigeon peas, 261
plant breeders, 304; rights of, 252, 306
plant disease, combating, 259
plant products, indigenous, 167
plant propagation, 118
plantain, 260

plastics, 70, 77, 148; global consumption of, 71
Pockets of Poverty study (Chile), 242
policy science, teaching of, in Africa, 132
policy-making, in biotechnology, 131–3
policy-oriented research, 14
pollution, 38, 76, 151, 164, 180, 220, 222; cleaning up of, 8; of water, 38, 216, 219
polyesters, 109
polymeric membranes, 74
polymers, 68, 70, 74, 75, 84, 85, 88, 102, 149, 156, 167, 168; conductive, 74; new, 73
poplar, 119
post-industrial capitalism, 196
post-modernity, 31
potatoes, 118, 119
poverty, 95, 205, 210, 293; alleviation of, 201; biotechnology as means of combating, 253, 254
powder metallurgical processing, 107
power sector load management, 176
private sector, role in biotechnology, 301–3
privatization, of technology, 20, 287
problem-identification, capacity for, 53
process technology, 218
productivity paradox of information technology sector, 45
product development, costs of, 302
production workers, role of, 52
productivity, 27, 29, 45, 51, 202, 204, 314; gap with developing countries, 300; in agriculture, 299, 300, 303; knife edge of, 291, 293; low, 293; of food, 221; of labour, 290
profit, desire for, 196
programmability, 32
Programme of Scientific and Technological Development in New Materials (Brazil), 85
propagation of plants, 119, 226; *in vitro*, 261
prosperity, expectations of, 196
protectionism, 57, 222, 292, 314
protein engineering, 100
Proton/Mitsubishi, Malaysia, 198, 200
public domain technology, access to, 299, 316
public health, 275
publishing houses, in Africa, 137
pyrethrum, 118

quality circles, 40
quartz, 85

r-DNA techniques, 102
radar, 281
Radar Remote Sensing, Costa Rica project, 246

RADARSAT satellite, 246
railway engines, development of, 27
railways, 1, 109; growth of, 210
rapeseed, 119
rare earth, 84, 107
'reaching the unreached', 223
Reagan administration, 179
recycling, 9, 16, 150, 154, 164, 214–20, 221, 310; of paper, 150; of waste energy, 15; of waste products, 15
REDATAM software package, 242–3
Reduced Instruction Set Computer (RISC), 111
rehabilitation of polluted land, 309
reliability of equipment, 32
remote sensing (RS), 17, 236, 244, 246, 247
research: access to, 301, 302; commercialization of results of, 97; diffusion of, 301; government sponsorship of, 156; in Mexico, 143, 144; liberalization of, 128; postgraduate, 156 (in materials, 159); practical, 235; regionalization of, 280
research and development (R&D), 3, 11, 13, 15, 19, 40, 49, 51, 82, 96, 106, 109, 116, 126, 153, 197, 222, 264, 276, 277, 301, 302, 310; and advanced materials, 68, 85; and transnational corporations, 206; 'borderless', 311–13; biotechnological, in Cameroon, 123; civilian, 312; corporate, 276, 277, 282; costs of, 8; creating of specialized institutions, 298; declining funding by state, 282 (in Africa, 138); EU Framework Programme for, 265; funding of (by government, 99, 275; from foreign sources, 282; in USA, 179; public and private, 317); generic, 265; global character of, 278; global figures for support of, 55, 56, 146; globalization of, 20, 265, 282; in biotechnology, 119–23; in geomatics, 246; in India, 99; in materials, 155, 159, 164; in Mexico, 143, 144 (spending on, 145); in Nigeria, 203; in South Korea, 203; in steel, 159; in 'strong' South, 4–5; in USA, 206; indigenous, 298; industry's contribution to, 100; internationalization of, 18; military, 275, 281, 312; networking of, 112; on Intel chip, 111; private sector commitment to, 57; recouping of investments, 305; redundant systems, 294; sourced in universities, 276; spending on, 197 (in Africa, 197; of private sector, 60); state-controlled, 281; system of institutions proposed, 294; transnational, 312; transnationals' stakes in, 302; research

institutes, 276, 299, 316; development of, 293
research intensity, 77; of new materials technology, 37, 68, 152; of technology, 2, 3, 8, 11, 31, 52, 88, 221, 298
resistance breeding, 259
Resources Development Foundation (USA), 228
resources: conservation of, 153, 161; extraction of, 214; local, 152, 153, 200 (development of, 129; use of, 156, 157, 158, 160); management of, 261; non-renewable, 215; renewable, 215
restriction endonucleases, 102
restriction fragment length polymorphism (RFLP), 119, 262
restructuring, 45; of sunset industries, 201
Reunion, 180
rice, 118, 119, 260; husk as source of fibres, 149; hybrid technology, 300
robotics, 33, 97
robotization, 151
Rolta India Ltd, 111
rubber, 208; reduced demand for, 209
Rural Investment Overseas Ltd, 228
Russia, 277
Rwanda, 243
rye, 119

salination of soil, 221
Sandö Training School (Sweden), 271
satellites, 7, 114, 245; low-earth-orbit (LEO), 237, 239; RADARSAT, 246; UOSAT-2, 239
schistosomiasis, 260
scholarships, provision of, 134
science: commitment to, 58; denationalization of, 280, 282; importance of, 58; internationalization of, 280, 281; nationalization of, 281; spending on, 46; transnational nature of, 279; trends in, 54–7
science and technology: China's three-tier development strategy, 12, 96
Science and Technology Dissemination Programme, 97–8
science and technology policy, 124; analysis of, 132; capacity-building for, 133; framework for, 121; in Africa, 14; studies of, 135; trends in, 134
science parks, 266; creation of, 12, 97, 266
science-relatedness of technology, 2, 3
Scientific and Industrial Research and Development Centre (SIRDIC) (Zimbabwe), 122
Scientific Committee on Genetic Experimentation (COGENE), 125
scientific travel, global, 282

scientists, 4, 5, 9, 26, 52, 302; and policy-making, 131; in China, 95
Secretariat for New Materials (Brazil), 86
security concerns of nation-states, 276
sediments, recovery of, 219
seed, price of, 252
seed companies, 304
Semi Conductor Complex Ltd (SCL) (India), 106
semiconductors, 84, 106–7
semiconductor industry, as polluter, 38
Senegal, 245
services, 10, 44, 141
sewage treatment, 217
shellac, reduced demand for, 209
Sieflex Ltd (India), 112
silicon, 85, 106, 164
silicon carbide, 149
silicon nitride, 149
Silitronics company (India), 108
silver, 142
Singapore, 155, 208, 239, 317; Local Industry Upgrading Programme, 208
single crystals, production of, 106, 108, 168
single crystal superalloys, 73
skilled labour, 52, 53
skills, creation of, 54, 58
'small is beautiful' technology, 204
snail infestation in Great Lakes area, 207
social dumping, 222; electronic, 209
social science research, importance of, 319
software, 2, 17, 32, 34, 48, 53, 208, 234, 236, 244, 245, 246, 247, 279; decreasing costs of, 7; design of, 47; for development, 240–44; industry in India, 113–14; maintenance of systems, 50; production as craft work, 33; suppliers, 50
soil erosion, 221, 259, 303
solar energy see energy, solar
sorghum, 258, 261
South: 'strong', 3, 4, 12, 13, 311, 314, 316, 319 (R&D in, 4–5); concept of, 3; 'weak and excluded', 5, 9, 13, 20, 314, 315–16, 319 (exclusion of, 3)
South Africa, 4
South Asia, 60
Southeast Asia, 56
South Korea, 155, 156, 161, 202, 203, 205, 206, 290, 317; R&D expenditure in, 56
South–North relations, 235
South–South relations, 222, 235
Southeast Asian Regional Center for Tropical Biology (BIOTROP), 243
soybean, 119
Space Research Organization (India), 159
space technology, 97
spatial information technologies, 244, 245

Special Programme in Biotechnology (Netherlands), 17; achievements of, 255–63
Special Secretariat for Informatics (SEI) (Brazil), 83, 86
specialist technology institutes, funding of, 319
specialized skills, need for, 51
species, loss of, 124
SQUIDS, development of, 107
Sri Aurobindo Institute for Rural Development (SAIRD) (India), 261
St Lucia, 243
standard-tech, 3, 4, 8
standardization: national institutes of, 160; of nomenclatures, 280; of scientific instruments, 280
standards, development of, 164
state: funding of laboratories, 12; intervention by, 317; role of, 12, 293; see also governments, role of
steam power, 26, 27, 29, 175, 196
steel, 70, 71, 142, 155, 156, 169, 184; advanced, 143; high yield point, 77
Steel Authority of India, 159
Stimulation Programme for Biotechnology and Development Cooperation (Netherlands), 253
strategic alliances, 276
structural adjustment, 59, 292
stylosanthus, 119
sub-Saharan Africa, see Africa
Sudan, 245
sugar: consumption of, in USA, 118; falling demand for, 300; substitutes for, 117, 209
sugar industry, alliance with power industry, 181
sugar-cane: and energy production, 181; costs of, 182; effects of biotechnology on production of, 117; transportation of, 182
Super Semi-conductors company (India), 108
superalloys, 76; nickel-based, 143
superconductors, 75, 151, 168, 266; development of, 107; high-temperature, 140; room-temperature, 167
Superior Footwear Company (Zimbabwe), 184
superplastic forming, 168
sustainability, 16, 214, 216, 300, 303, 308–10; definition of, 200
sustainable development, 215, 222; indigenous capacity for, 195–213
sustainable technology, 15
Swedefund International, 273
Sweden, 9, 16, 48; aid for technology development, 264–84; aid policy of, 222–4; management capability of, 275; objectives of development assistance, 270, 270; R&D in, 19; technological strength of, 274
Swedish Agency for International Economic and Technical Cooperation (BITS), 271, 272, 273
Swedish Agency for Research Cooperation with Developing Countries (SAREC), xii, 2, 271, 272, 273–20
Swedish International Development Authority (SIDA), 2, 125, 273
Swedish International Development Cooperation Agency (Sida), xii, 2, 271
Swedish International Enterprise Development Corporation (Swedecorp), 271, 272, 273
sweet potato, 258
Switzerland, 262
Synthetic Aperture Radar (SAR), 246
system integration, 47
systems design of new products, 51
systems integration in information technology, 111

Taiwan, 205, 317
Tanzania, 14, 121, 238, 270
tariff barriers, reductions of, 197
Taskforce on Biotechnology (BIOTASK), 262, 263
Tata Iron and Steel Company (India), 155, 159
Tata Unisys Ltd (India), 112
tax: exemptions for research foundations, 138; incentives for R&D, 279
technicians, in China, 95
techno-economic paradigm, 39, 59; changes in, 28, 72, 82
techno-globalism, 277
technological change, 313; acceleration of, 19, 285, 286; and energy technologies, 175, 175; supplier-dominated, 290
technological colonialism, 158
technological dependence, 4
technological forecasting, 234
technological revolution, 1, 9, 10, 29, 34, 36, 44, 54; characteristics of, 25–42; definition of, 29
technological trajectories, 9, 10, 28, 37, 38, 39, 40; cleaner, 308–10
technology: 'appropriate', 60, 200, 203–5; and defence, 205–6; as labour substitute, 209–10; clean, 36–40; complexity of; management of, 199; cooperation in, 128, 198; definitions of, 2; diffusion of, intrafirm, 47; emerging, 30; foreign, dependence on, 49, 58, 143;

global sourcing of, 278; import of, 160; importance of, 58 (in relation to development policy, 285–6); in international economy, 287–91; industrial-era, 30; link with market, 204; management of, 316; measurement of, 3; monitoring of global trends, 134–5; priority projects, 266; traditional, 30; transfer of, *see* transfer of technology; *see also* generic technology *and* new technology
technology assessment, 140, 157, 161, 297
technology forecasting, 140, 153, 154, 160, 161
Telebrás company, 83
telecommunications, 33, 70, 75, 84, 233, 234, 274, 278; mobile, 112
teleconferencing, 112
TELEDESIC LEO system, 239
telematics, 17, 236, 237–40
terminals, growing processing power of, 32
text communication systems, 239, 240
Textile Research Associations (India), 160
Thailand, 4, 178, 202, 243, 298, 300
thalium, production of, 107
tissue culture, 103, 117, 118, 129, 259, 261
titanium, 84, 108
Titanium Development Advisory Committee (TDAC) (India), 108
titanium tetrachloride, 108
tobacco, 119
Togo, 79
tomatoes, 119
Torch Programme, China, 96
trade: barriers *see* non-tariff trade barriers; fair, 57; impact of technology on, 292; international, changes in, 44, 286; managed, 44; patterns of, 10; policy, 292; technology content of, 310
trade balances, of developing countries, 77
trade not aid, 222
trade-related aspects of intellectual property rights (TRIPS), 206
trademarks, as international guarantee, 207
training, 5, 15, 133, 140, 203, 227, 293, 316, 318, 319; in bioscience, 130; in biotechnology, 227, 228; in context of institutional development, 134; in materials science and technology, 153; of information technology workforce, 48; of technology experts, 224; packages for NGOs, 134
Transborder Data Flows, 237
transdisciplinarity, 267
transfer of technology, 16, 49, 59, 98, 126, 127–31, 137, 144, 145, 154, 195–213, 243, 246, 307; in India, 202; legal provisions in, 129, 130; of biotechnology, 305
transgenic animals, 308
transnational companies (TNCs), 15, 19, 43, 195–213, 265, 311, 313, 315; agribusiness, 300; and research-led technology, 206; estimated numbers of, 210; role in biotechnology, 301–3
transportation, 148, 178
tuberculosis, 260; detection of, 102
tungsten, 8, 149
Turkey, 4

Uganda, 14, 121, 123, 199, 238
Unilever company, 204
Union of Soviet Socialist Republics (USSR), 199, 281
UNISPACE '82, 239
United Arab Emirates, 114
United Kingdom (UK), 71, 161, 228, 262
United Nations (UN), 125, 135, 136, 137, 164, 221, 254, 271, 281
UN Commission on Trade and Development (UNCTAD), 199, 253, 255
UN Conference on Environment and Development (UNCED), 127, 255, 305
UN Development Programme (UNDP), 158
UN Economic Commission for Africa (UNECA), 238
UN Economic Commission for Latin America and the Caribbean (UNECLA), 242
UN Educational, Scientific and Cultural Organization (UNESCO), 281
UN Environment Programme (UNEP), 304, 305
UN University Institute for New Technologies, 285
United States of America (USA), 10, 13, 43, 45, 71, 114, 116, 145, 159, 161, 180, 184, 262, 276, 277, 289, 298, 301, 304, 305; chocolate manufacture in, 117; dependence on imported materials, 163; patents granted in, 197; R&D expenditure in, 155, 179
universities, 133, 155, 156, 159, 208, 223, 256, 267, 268, 269, 270, 276, 281, 282, 298, 299, 301, 316; changing role of, 266; links with corporate world, 301; materials research in, 155; no longer controlling major knowledge, 267; no longer research institutions, 268
University of Yaoundé, 123; Biotechnology Centre (BTC), 121
US Agency for International Development (USAID), 125, 174, 228

US–Japanese agreements, 57
user-friendliness of systems, 33, 34
user-producer relations, 52, 59; in technology development, 46, 50

vaccine, 100; development of, 104; DPTP, 102; measles, 102; polio, 102; production of, 100
vanilla, 300
vehicles, energy efficiency of, 218
venereal disease, 260
video games, 33
Vietnam, 206, 242

wages, declining, 291, 292, 313; increases in, 291; low, 80; pressure on profits, 290
walnuts, 119
waste, 309; industrial, 216, 218; nuclear, storage of, 217; reduction of, 215; remediation of, 34, 38; toxic, 310
waste heat, use of, 184
waste management, 7
waste products, recycling of, 15
water: conservation of, 227; purification of, 74, 149, 167
weed identification, 243, 244

weevil, as pollination aid, 204
wheat, 118; genetic improvement of, 304
wind turbines, 176, 177, 217; advantages of, 180; siting of, 179; US budget for, 179
'windows of opportunity', 82
windows technology, advanced, 178
WIPRO Infotech company (India), 112
women, 223, 256; and biotechnology, 254; technical knowledge of, 204
wood, 157
woodfires, 178
World Bank, 20, 57, 78, 113, 174, 208, 247, 254, 262, 291, 292, 314

yams, 260
Youth for Action group (YFA) (India), 261
yttrium, production of, 107

Zaire, 79
Zambia, 79, 238
Zenith Computers company, 112
Zimbabwe, 14, 17, 119, 120, 182, 184, 238, 255, 258, 261
Zimbabwe Biotechnology Advisory Committee (ZIMBAC), 258
Zimbabwean Research Council, 256
zinc, 70, 142